CREATING A WINNING E-BUSINESS

Second Edition

CREATING A WINNING
E-BUSINESS

Second Edition

H. Albert Napier, Ph.D.
Ollie N. Rivers
Stuart W. Wagner
JB Napier

THOMSON
COURSE TECHNOLOGY

Australia • Canada • Mexico • Singapore • Spain • United Kingdom • United States

Creating a Winning E-Business, Second Edition

by H. Albert Napier, Ollie N. Rivers, Stuart W. Wagner, and JB Napier

Senior Acquisitions Editor:
Maureen Martin

Product Manager:
Beth Paquin

Development Editor:
Saher Alam

Marketing Manager:
Karen Seitz

Associate Product Manager:
Jennifer Smith

Editorial Assistant:
Allison Murphy

Production Editor:
Kelly Robinson

Manufacturing Coordinator:
Justin Palmeiro

Cover Designer:
Laura Rickenbach

Copy Editor:
Mary Kemper

Proofreader:
Brooke Booth

Indexer:
Rich Carlson

Compositor:
GEX Publishing Services

BRIEF CONTENTS

TABLE OF CONTENTS

PREFACE

Creating a Winning E-Business, Second Edition provides general business students, graduate students, continuing education students, executive education seminar participants, and entrepreneurs with practical ideas on planning and creating an e-business. We assume that readers have no previous e-business knowledge or experience. This book is designed to help you learn about the key business elements involved in planning and starting an e-business from the ground up.

When we began teaching both an executive education seminar and an MBA-level course in planning and starting an e-business, we found many textbooks that emphasized the technological aspects of electronic commerce, but we did not find a textbook that focused on the planning and startup phases of an e-business. *Creating a Winning E-Business, Second Edition* is our attempt to fill that void.

Organization and Coverage

This book takes a practical, case-based, hands-on approach to planning and starting an e-business. Numerous real-world e-business examples are used in each chapter to illustrate important concepts. Additionally, the evolution of a real-world e-business, Rackspace Managed Hosting, is featured throughout the book, from the e-business's conceptual stage through its Web page design and operations phases. Important topics covered in the book include:

- the global e-business economy and e-business models (Chapter 1)

- the e-business entrepreneurial process and how to exploit e-business advantages (Chapter 2)

- the elements of an e-business plan (Chapter 3)

- financing options for an e-business startup (Chapter 4)

- e-business startup challenges, such as identifying legal issues, selecting office space, hiring employees, identifying electronic payment methods, and selecting appropriate e-business technologies (Chapter 5)

- effective tools for branding and marketing an e-business and its products or services (Chapter 6)

- affiliate marketing programs as both a marketing tool and a source of revenue (Chapter 7)

- Web site accessibility, usability, and design (Chapter 8)

- Web site development and performance measurement tools, including markup languages and Web analytics (Chapter 9)

- E-business risk management and security (Chapter 10)

Features

Creating a Winning E-Business is unique in its field because it includes the following features:

- **Opening and Closing Case:** A real-world e-business case opens and closes each chapter and provides a unifying theme for the chapter. At the beginning of the chapter, the case establishes background elements and introduces relevant issues. At the end of the chapter, the case concludes with a discussion of whether those relevant issues were resolved and how they were resolved.

- **E-Case in Progress:** A real-world e-business, Rackspace Managed Hosting, is used throughout the text to illustrate the processes involved with starting a new e-business.

- **E-Cases and E-Pioneers:** Other real-world e-business examples are used throughout the text to illustrate key concepts and to show the impact of decisions made by real-life entrepreneurs. A table of these e-cases appears on the inside front cover of the book.

- **Numerous illustrations:** The text of the chapter is well supported with many conceptual figures and screenshots of e-business Web sites.

- **Quotes on Success:** Quotes are presented from real-world entrepreneurs and others involved in starting, building, operating, and securing e-businesses.

- **Tips**: Each chapter has multiple tip boxes that contain useful information about individual topics.

- **Chapter Summary:** Each chapter concludes with a summary that concisely recaps the most important concepts in the chapter.
  - **Checklist:** Following the Chapter Summary is a checklist of the major concepts discussed in the chapter. Students can use this checklist to establish reference points from the chapter concepts to their own e-business—whether it is a real-world or classroom-based e-business.
  - **Key Terms:** Following the Checklist is a list of key terms used in the chapter. These key terms are bolded in the chapter text and defined in the Glossary at the end of the book.
  - **Review Questions and Exercises:** Every chapter concludes with meaningful review materials that include both objective questions and hands-on exercises. Many exercises guide students toward investigating chapter topics and creating hardcopy materials (in the form of reports or tables) from their findings, and then using these materials to engage in discussions with their fellow classmates. One of the exercises in Chapter 9, for example, asks students to use links from this text's student online companion to visit e-business Web sites that focus on providing Web analytics products. Students are asked to take a product tour at each site, summarize what they have learned about each Web analytics vendor and its products, and then use their summaries to direct a discussion with a group of classmates on the Web analytics tools available to a startup e-business.

- **Case Projects:** The chapters contain two to three case projects, each of which is based on a fictional but realistic startup e-business scenario. Students are required to apply concepts discussed in the chapter to the scenario to create a case project solution. These projects can be completed individually or in groups.
- **Team Project:** Each chapter contains a specially designed team project that allows two to three students to work toward the project solution and then make a formal presentation of that solution to fellow classmates. Team members must work together to name and describe a startup e-business, and complete the project by implementing key concepts from the chapter. This format allows multiple teams to work on the same project and arrive at different solutions. An important aspect of the Team Project is that students are required to prepare a 5-10 slide presentation illustrating their project solution, and then use these presentation materials to formally present this project solution to the class. This allows students to both understand the practical application of key concepts and to experience and practice important presentation skills.
- **For Further Study:** Every chapter contains a list of references to online magazine articles and reports, print magazine articles, newspaper articles, journal papers, and books that students can read to learn more about the topics discussed in the chapter.
- **End Notes:** Every chapter contains an extensive list of references tied directly to chapter concepts.

- **Glossary:** A glossary containing key terms and their definitions appears at the end of the text.

- **Appendix, Microsoft FrontPage Tutorial:** A short appendix is included that introduces how a Microsoft FrontPage wizard can be used to create a Web site. Students may be directed to use this appendix to create a simple e-business Web site to complement various case or team project solutions.

- **Student Online Companion:** Connects concepts from the book to the Web at **www.course.com/mis/napier2e**.

Teaching Tools

When this book is used in an academic setting, instructors may obtain the following teaching tools from Thomson Course Technology:

Instructor's Manual. The Instructor's Manual has been carefully prepared and tested to ensure its accuracy and dependability. The Instructor's Manual is available through the Course Technology Faculty Online Companion on the World Wide Web. (Call your customer service representative for the exact URL and to obtain your username and password.)

ExamView®. This textbook is accompanied by ExamView, a powerful testing software package that allows instructors to create and administer printed, computer (LAN-based), and Internet exams. ExamView includes hundreds of questions that correspond to the topics covered in this text, enabling students to generate detailed study guides that include page references for further review. The computer-based and Internet-testing components allow students to take exams at their computers, and they also save the instructor time by grading each exam automatically. In addition, Thomson Course Technology is proud to present online test banks in WebCT and Blackboard, as well.

Classroom Presentations. Microsoft PowerPoint presentations are available for each chapter of this book to assist instructors in classroom lectures. The Classroom Presentations are included on the Instructor's CD-ROM.

ACKNOWLEDMENTS

Creating a quality text is a collaborative effort between author and publisher. We work as a team to provide the highest quality book possible. The authors want to acknowledge the work of the seasoned professionals at Thomson Course Technology. We thank Maureen Martin, Senior Acquisitions Editor; Beth Paquin, Product Manager; Kelly Robinson, Production Editor; Karen Seitz, Marketing Manager; and Allison Murphy, Editorial Assistant, for their tireless work and dedication to the project. We also thank Saher Alam, our wonderful Development Editor, for her insightful suggestions and unflagging support.

We want to thank the following reviewers for their very helpful comments and suggestions at various stages of the book's development: Andrew Chen, Arizona State University; Linda Durkin, Delaware County Community College; Jo-Anne Romano, Briarcliffe College; Robert Griffin, Art Institute; Dan Connolly, University of Denver.

- H. Albert Napier
- Ollie N. Rivers
- Stuart W. Wagner
- JB Napier

ABOUT THE AUTHORS

H. Albert Napier is the Director of the Center on the Management of Information Technology and is a professor at the Jones Graduate School of Management at Rice University, where he teaches graduate and executive development courses related to information technology, entrepreneurship, and e-business. Dr. Napier also makes numerous management development program presentations on e-business and related topics. Additionally, he was a principal of Napier & Judd, Inc., a company engaged in computer training and consulting for 20 years. Dr. Napier is on the board of directors of three e-business companies. He holds a Ph.D. in Business Administration, an M.B.A., and a B.A. in Mathematics and Economics, all from The University of Texas at Austin. He is the author of more than 20 articles related to management information systems and the application of computer-based decision processes in business, and he is the co-author of over 60 textbooks.

Ollie N. Rivers has more than 20 years of business experience in financial and administrative management and more than 10 years of experience as a corporate trainer. She is a co-author of two e-business textbooks, a contributing author on more than 15 software applications and Internet textbooks, and has developed numerous continuing education seminars for accounting professionals. Ms. Rivers holds an M.B.A. and a B.S. in Accounting and Management, both from Houston Baptist University.

Stuart W. Wagner manages an IT department for a large private company in Houston, Texas. He has also managed Web marketing and online strategies at Hewlett-Packard, as well as taught a graduate e-business course and executive seminars at the Jones Graduate School of Management at Rice University. Mr. Wagner has an M.B.A. from Rice University, which he attended while starting his first e-business. He has started five different e-businesses.

JB Napier has over seven years of experience in the high technology and consulting industry. He has led early stage software companies in product launches, business development, and partner integration activities. Prior to working in operational roles in the software industry, Mr. Napier provided financial, operation, and strategic consulting services to Fortune 1000 companies for a global consulting firm. Mr. Napier holds a B.S. in Economics from Trinity University and an M.B.A. from the McCombs School of Business at the University of Texas at Austin.

CHAPTER 1

UNDERSTANDING
E-BUSINESS

LEARNING OBJECTIVES

In this chapter, you will learn to:

- Discuss e-business basics
- Describe the Internet and World Wide Web
- Discuss the role of e-business in the global economy
- List e-business advantages and disadvantages
- Explain e-business value chains and value activities
- Identify e-business models

SCRAMBLED EGGS. . .

From its inception in 1984 to the early 1990s, Spokane, Washington-based Egghead Software enjoyed great success as the first nationwide chain of stores that sold only software. The chain grew to comprise more than 200 stores, employing thousands of workers and earning millions of dollars in revenues. Egghead's success attracted competitors, including mass merchandisers such as Wal-Mart and CompUSA, who began selling software at discounted prices. As software prices fell, Egghead was unable to compete successfully and began suffering operating losses.

Needing a change in direction, in the mid-1990s the Egghead board of directors turned to George Orban, a longtime Egghead investor with a track record for turning around failing businesses. Orban, who joined Egghead as chairman and CEO, believed that the only way to save Egghead was to simplify the business by closing stores, cutting costs, and better identifying the company's target market. He began Egghead's turnaround with a massive cost-cutting program that involved closing more than 70 stores and a distribution center, selling real estate holdings, and laying off hundreds of employees. The merchandise at some of the remaining stores was switched to a more profitable product mix that included computer systems and accessories. Hoping to attract new customers and increase sales, Egghead also began selling software and computer systems through an online store.[1] Would these changes be enough to save Egghead?

E-BUSINESS BASICS

Commerce, the exchange of valuable goods or services, has been conducted for thousands of years. Traditionally, commerce involved bringing traders, buyers, and sellers together in a physical marketplace to exchange information, products, services, and payments. Today, many business transactions occur across a telecommunications network where buyers, sellers, and others involved in the business transaction (such as the employees who process transactions) rarely see or know each other and may be anywhere in the world. This process of buying and selling of products and services across a telecommunications network is often called **electronic commerce** or **e-commerce**.

Many people use the term "e-commerce" in a broader sense: to encompass not only the buying and selling of goods, but also the delivery of information, the providing of customer service before and after a sale, the collaboration with business partners, and the effort to enhance productivity within organizations. Others refer to this broader spectrum of business activities that can be conducted over the Internet as **e-business**. Most people today use the terms "e-commerce" (in its broadest sense) and "e-business" interchangeably. In this book, we use the term "e-business" to indicate the widest spectrum of business activities that use Internet and Web technologies.

TIP

One of the first companies to use the term e-business was IBM, which launched an e-business marketing campaign directed at selling services to companies that needed to connect their current electronic systems to the Web.[2]

The initial development of e-business transactions began more than thirty years ago when banks began transferring money to each other by using **electronic funds transfer (EFT)**, and when large companies began sharing transaction information with their suppliers and customers via **electronic data interchange (EDI)**. Using EDI, companies electronically exchange information that used to be traditionally submitted on paper forms, such as invoices, purchase orders, quotes, and bills of lading. This exchange occurs both with suppliers and customers (often called **trading partners**). These transmissions generally occur over private telecommunications networks called **value-added networks**, or **VANs**. Because of the expense of setting up and maintaining these private networks and the costs associated with creating a standard interface between companies, implementing EDI has usually been beyond the financial reach of small and medium-sized companies. Today, companies of all sizes use a less expensive network alternative to VANs for the exchange of information, products, services, and payments—the Internet. Global access to the Internet and the Web has changed the way people and businesses around the world communicate.

Almost a billion people worldwide use the Internet to shop for products and services, listen to music, view artwork, conduct research, get stock quotes, keep up-to-date with current events, chat with each other, upload and download electronic files, send e-mail, and much more.[3]

What Is the Internet?

To understand the Internet, you must first become familiar with computer networks. A computer **network** is a group of two or more computers linked by cables, telephone lines, or other wired or wireless media (Figure 1-1). A network of linked computers usually includes special computers called **servers** that give users access to shared resources such as electronic files, programs, printers, and connections to other networks, such as the Internet. The **Internet** is a worldwide public network that connects private networks (Figure 1-2).

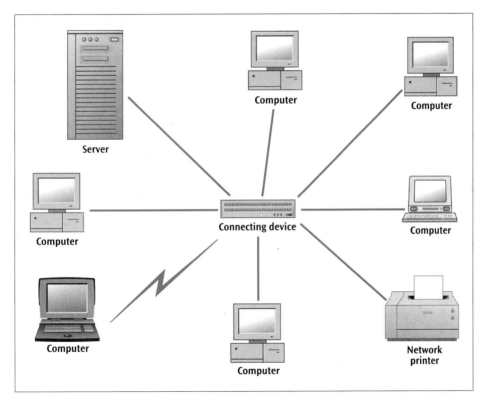

FIGURE 1-1   Computer Network

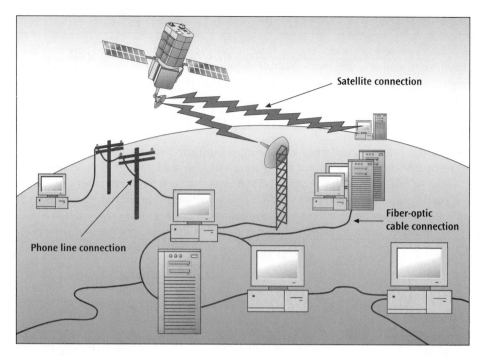

FIGURE 1-2   The Internet

A Brief History of the Internet

The Internet originated in the late 1960s, when scientists sponsored by the U.S. Department of Defense developed a network called the Advanced Research Projects Agency Network or ARPANET. Quickly realizing the usefulness of such a network, researchers at colleges and universities soon began using ARPANET to share data. In the 1980s, the military portion of ARPANET became a separate network called the MILNET. The National Science Foundation (NSF) began overseeing the remaining nonmilitary portions, called the NSFnet, which connected hundreds of colleges, universities, and research centers. During this time, commercial activity over the NSFnet was prohibited by law. Restrictions on commercial activity over the NSFnet were lifted in 1991, and in the spring of 1995, the NSFnet was replaced by high-speed telecommunications backbones, or main networks, operated by commercial network providers, such as AT&T and MCI—thereby creating the Internet as we know it today.[4]

To connect to the Internet, individuals and businesses generally use some type of physical communications media, such as a network cable, phone line, or, as is increasingly true these days, wireless media. In addition to some form of media, individuals and small-to-medium sized businesses seeking access to the Internet usually need the services of an **Internet service provider (ISP)**—an e-business that provides access to the Internet for a fee. Examples of ISPs include America Online, Netscape Network, EarthLink, NetZero, and Road Runner. Large businesses, colleges, universities, and government institutions may have a computer network that is connected directly to the Internet. Figure 1-3 illustrates an example of an ISP.

FIGURE 1-3   EarthLink

Table 1-1 describes some of the many types of communication that are available to businesses and individuals using the Internet.

TABLE 1-1   Internet Communications

Service	Description
E-mail	Electronic messages sent by one computer user and received by another
Instant messaging	Online messages exchanged instantaneously between two or more parties simultaneously connected to the Internet
Newsgroups and Web-based forums	Electronic "bulletin boards" or discussion groups where people with common interests (such as hobbies or professional associations) post messages to which participants around the world can respond
Mailing lists	E-mail on a specific topic that is periodically mailed to a list of interested parties
Internet Relay Chat (IRC)	Online conversations in which two or more participants key in messages and receive responses on their screen within a few seconds
FTP	Service based on the File Transfer Protocol (FTP) that enables users to upload or downloaded electronic files from a server on the Internet

TABLE 1-1 Internet Communications (continued)

Service	Description
VoIP telephony	Telephone calls transmitted over networks using Internet technologies
Peer-to-peer file sharing	Service that allows music and other files on one user's computer to be listed and shared with other Internet users
World Wide Web pages	Multimedia documents stored on Web servers so that their content can be accessed and downloaded
Weblogs or blogs	Web-based diaries kept by participants who write their thoughts about specific topics and make them available for visitors to read on an ongoing basis
RSS	Really Simple Syndication or Rich Site Summary; a communication standard that syndicates Web-based content such as news headlines

A Brief History of the World Wide Web

In 1989, a software consultant named Tim Berners-Lee was working at CERN (the European Laboratory for Particle Physics) in Switzerland, where he was trying to find ways to improve information sharing and document handling between the lab's research scientists. Building on the concept of **hypertext**, where text on one page links to text on another page, Berners-Lee developed the first computer programs that could be used to store, access, and view hypertext-linked pages. Berners-Lee called his system of linked documents the "WorldWideWeb," which we now call the **World Wide Web** or simply the **Web**.[5]

The Web is a subset of the Internet consisting of computers called **Web servers** on which documents that are linked together by hypertext links, or **hyperlinks**, are stored. A hyperlink can be text or a picture that is associated with the location (path and filename) of another document. The hyperlinked documents, called **Web pages**, can contain text, graphics, video, and audio, as well as hyperlinks to other Web pages. A group of related Web pages is called a **Web site**. The program used to access and view Web pages stored on a Web server is called a **Web browser**. The most popular Web browser at this writing is Internet Explorer®. Other popular Web browsers include Mozilla Firefox™, Netscape®, and Opera®. Figure 1-4 illustrates the Mozilla Web page.

FIGURE 1-4   Mozilla

Many companies are increasingly using a company intranet for internal business trans-actions and the exchange of information between employees. An **intranet** uses Internet and Web technologies to allow employees within a company to view and use internal Web sites that are not accessible to anyone outside the company. Intranet Web sites provide an interface to internal systems such as accounting, customer relationship management, and other company operating systems. Transactions between two or more companies can be conducted over the Internet or over a private network called an extranet. An **extranet** con-sists of two or more company intranets connected via the Internet. Using an extranet, the participating companies can view the other company's data and complete business transactions.

QUOTES ON SUCCESS

"I suspect it [founding Amazon.com] wouldn't have made very much difference [to the Web]. The big forces at work on something like this are much larger than any individuals."

Jeff Bezos, founder of Amazon.com

Internet and Web Demographics

Determining how many individuals are online around the world and which Internet services they are using can be difficult. Research and marketing groups publish various estimates on a regular basis. However, because of the dynamic growth of the Internet, these esti-mates become outdated quickly. Additionally, there are differences among the various esti-mates due to differences in how Internet access and Web content are defined, which survey and calculation methods are used, and how the data are gathered. Furthermore, the sheer speed of change in Internet access and the rapid evolution of online content make estimating current growth and predicting future growth difficult at best.

However, there is one thing all growth estimates and predictions have in common: They indicate that the remarkable growth of Internet access from year to year shows no sign of abating. For example, World Internet Stats reports that worldwide Internet access increased from approximately 360.9 million people in 2000 to approximately 888.7 million people in 2005—an increase of more than 146 percent in five years![6]

Along with increased Internet access, the amount of information available on the Web is growing very rapidly. In fact, IDC Research expects that in the near future, Internet users worldwide will exchange information equivalent in volume to the entire Library of Congress more than 64,000 times every day.[7]

TIP

Internet users are eagerly adopting wireless Internet access as the number of wireless devices and Internet applications grows. In the U.S., the number of wireless Internet subscribers is expected to explode to more than 84 million, with businesses making up almost 60 percent of all subscribers.[8]

Given that global Internet access and the growth of Web content has changed our everyday lives, it should not be surprising that the Internet and the Web have also transformed the way buyers, sellers, employees, and business intermediaries interact with each other to conduct business transactions.

E-BUSINESS AND THE GLOBAL ECONOMY

The widespread electronic linking of individuals and businesses around the world has created an economic environment in which time and space are no longer limiting factors; the business value of information is more important than before and information itself is more accessible; traditional business intermediaries are being replaced by new business intermediaries; and buyers are growing more powerful. In the past, some large companies were able to conduct their business transactions electronically using EDI and private networks, but the high costs associated with EDI prevented most businesses from using the technology. The Internet has leveled the playing field by making it easier and cheaper for companies of all sizes to transact business and exchange information electronically.

As many of the business limitations of space and time disappear with the emergence of the Internet, businesses that once had geographically limited customer and competitor bases are finding that the whole world is now both customer and competitor. In addition, millions of companies that previously engaged in business transactions only during traditional hours now conduct those transactions online 24 hours a day, 7 days a week.

E-CASE

Competing for Customers

The Tattered Cover Book Store, a popular independent bookstore with three stores in the Denver, Colorado area, has been doing business successfully for over 30 years. The Tattered Cover Book Store has traditionally competed with other local Denver bookstores, but these days, it must also compete with online bookstores such as Amazon.com and Barnes&Noble.com. To contend with this online competition, the Tattered Cover Bookstore added its own online bookstore (Figure 1-5), where customers can buy books and gift certificates. The online store also provides special customer services around the clock, including a schedule of upcoming special events and personalized search requests for hard-to-find books.[9]

FIGURE 1-5   The Tattered Cover Book Store

While the Internet and the Web are providing online opportunities for sellers, it is buyers who are dramatically gaining new economic power. Internet and Web access has changed many buyers' expectations about how quickly transactions can be processed and how convenient it should be to access information about competing products and services. With Internet and Web access, buyers no longer have to travel to various physical locations to compare prices and features of the products or services they need—competing businesses that offer unique services, lower costs, or products with the best features are just a mouse click away!

In addition to accessing the Web sites of various sellers directly, buyers can take advantage of online shopping sites such as Shopzilla shopping search and NexTag. Buyers use these online shopping sites to quickly locate and compare the prices and availability of products offered by multiple online stores. Generally, these shopping sites gather and publish buyer ratings on the products (which may include items such as digital cameras, laptop computers, and flat screen LCD projection TVs) as well as the online stores that sell them. Figure 1-6 illustrates an online shopping site.

FIGURE 1-6 Shopzilla

Because information is easier to customize than hard goods, many companies are finding that the information portion of their products or services is becoming a larger part of the total value they offer customers. For example, office product suppliers, such as Staples, have discovered that they can increase online sales by creating custom options for buyers, such as customized product catalogs that list only frequently purchased items or those items and prices negotiated by contract.

The ready access to consumer information via the Internet and the Web has empowered buyers and dramatically changed how many business transactions are conducted—even relatively substantial ones such as the buying and selling of cars and trucks. An automotive industry information provider that pioneered the offering of automotive information online is Edmunds. Edmunds, founded in 1966, publishes a number of automobile and truck reviews and pricing guides. In the mid-1990s, Edmunds introduced its Web site, Edmunds.com, which is a valuable online source for information about new and used vehicles, including pricing, dealer cost and holdbacks, reliability, buying advice, and product reviews. The Edmunds.com Web site provides a marketplace for buying and selling used vehicles, as well as information about financing, insurance, and other areas of interest for someone involved in buying or selling a vehicle. Competing with Edmunds.com are other automotive industry information and sales e-businesses such as eBay Motors,

Automotive.com, Autobytel.com, and CarsDirect.com. In addition to consulting these information and sales Web sites, buyers can visit sites such as CARFAX and AutoCheck to purchase a complete history on a given used car before making the purchase. Figures 1-7 and 1-8 show automotive information Web sites.

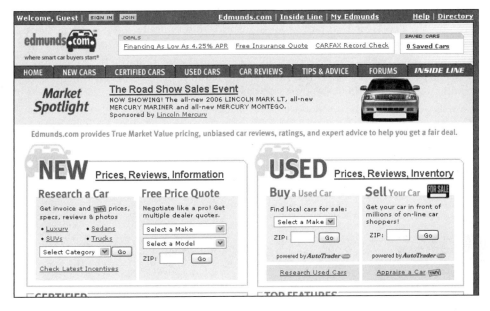

FIGURE 1-7 Edmunds.com

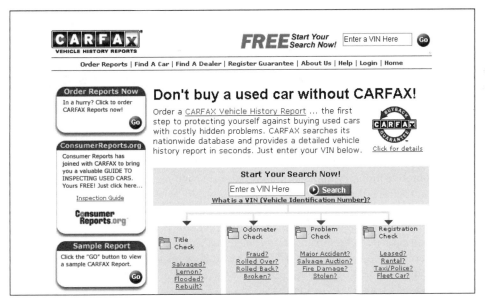

FIGURE 1-8 CARFAX

In a global economy where buyers located anywhere in the world are linked directly to sellers who are also located anywhere in the world, traditional business intermediaries, such as distributors and agents, are being threatened—and new kinds of intermediaries are emerging. For example, brick-and-mortar travel agents are being replaced by travel industry e-businesses, such as Travelocity, and traditional brokerage firms are losing business to online trading services, such as E*TRADE Financial.

Global access to the Internet and the Web has led to the development of a new kind of e-business intermediary, one that organizes information on the basis of customer needs. The Internet Truckstop is an example of this new type of middleman—an online service that gathers information from truckers, trucking companies, brokers, shippers, freight forwarders, and others in the trucking industry and then makes that information available to subscribers for a fee. Figures 1-9 and 1-10 illustrate examples of new e-business intermediaries.

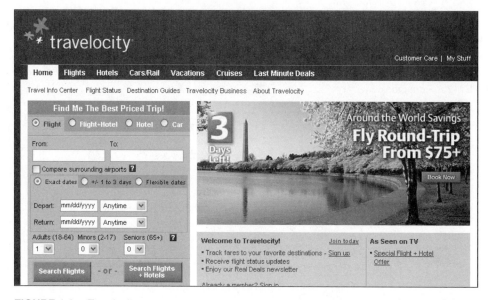

FIGURE 1-9 Travelocity

Access to the Internet and the Web has created a business environment in which time and distance have less meaning, information has greater value, traditional intermediaries are being replaced or eliminated completely, and buyers hold more power than ever before. While buyers are enjoying greater access to markets, sellers are also finding tremendous advantages in doing e-business.

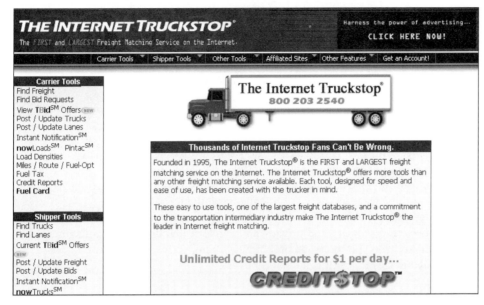

FIGURE 1-10 The Internet Truckstop

E-Business Advantages and Disadvantages

Like buyers, sellers also benefit tremendously from the global e-business-based economy. Sellers can increase sales and operations from local to worldwide markets, improve internal efficiency and productivity, enhance customer service, and increase communication with both suppliers and customers. Table 1-2 illustrates some e-business advantages for sellers and buyers.

TABLE 1-2 E-Business Advantages

Advantages for Sellers	Advantages for Buyers
Increased sales opportunities	Wider product availability
Decreased costs	Customized and personalized information and buying options
24 hours a day, 7 days a week sales	24 hours a day, 7 days a week shopping
Access to narrow market segments	Easy comparison shopping
Access to global markets	Access to global markets
Increased speed and accuracy of information delivery	Quick delivery of digital products and information
Data collection and customer preference tracking	Access to rich media describing products and services

QUOTES ON SUCCESS

"The information highway will . . . carry us into a new world of low-friction, low-overhead capitalism, in which market information will be plentiful and transaction costs low. It will be a shopper's heaven."

Bill Gates, co-founder of Microsoft Corporation

Unfortunately, however, the global e-business economy also poses some disadvantages for both sellers and buyers. Businesses may find it difficult and expensive to keep up with rapidly changing technologies, and exploiting the global marketplace means businesses must confront diverse language and cultural issues, conduct transactions in different currencies, and navigate unknown political environments.

The global e-business economy poses disadvantages for consumers in the form of concerns about transaction security and privacy and vendor reliability. Many consumers still prefer to touch and feel products before buying them. Table 1-3 illustrates some e-business disadvantages for both sellers and buyers.

TABLE 1-3 E-Business Disadvantages

Disadvantages for Sellers	Disadvantages for Buyers
Growing competition from other e-businesses	Difficulty differentiating among so many online sellers
Rapidly changing technologies	Unpredictable transaction security and privacy
Greater telecommunications capacity or bandwidth demands	Dealing with unfamiliar, possibly untrustworthy, sellers
Difficulty of integrating existing business systems with e-business transactions	Inability to touch and feel products before buying them
Problems inherent in maintaining e-business systems	Unfamiliar buying processes and concerns about vendor reliability
Global market issues: diverse languages, unknown political environments, and currency conversions	Issues with state sales tax charges and logistical difficulties of product returns

E-Business Value Chains

The widespread access to the Internet and the Web by suppliers and customers has encouraged many companies to reevaluate their value chains. In his book, *Competitive Advantage: Creating and Sustaining Superior Performance*, Michael E. Porter of the Harvard Business School first introduced the concept of a value chain. A company's **value chain** consists of all the primary and support activities, called **value activities**, performed to create and distribute its goods and services.[11] Primary activities include all the activities necessary to produce, sell, and support the company's products. Support activities include purchasing, human resources, technology, and other functions necessary to support the primary activities. At each link in an e-business's value chain, Internet and Web technologies improve communications and transaction speed. Figure 1-11 illustrates a generic value chain.

FIGURE 1-11   Generic value chain

Value chains are also used to represent the value activities of any transaction starting with a product or service and ending with a customer. Internet and Web access is redefining the relationships among manufacturers, suppliers, distributors, and customers. Increasingly, a company's value chains can be seen as value networks of the multiple relationships the company depends on to produce and sell its products and services (Figure 1-12).

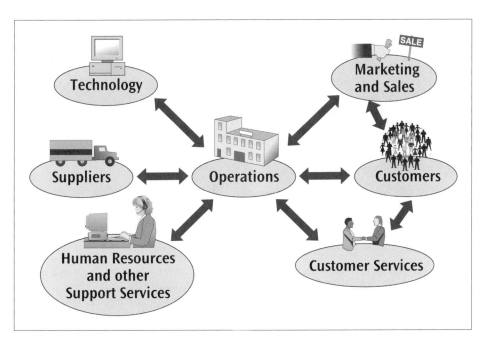

FIGURE 1-12   Value network

The process of rethinking and redefining value chains enables companies to develop new ways of conducting business using the Internet and the Web.

E-BUSINESS MODELS

A company's business model is the way in which the company conducts business in order to generate revenue. Widespread access to the Internet and the Web is driving companies to adapt old business models and create new ones. Although there are many different ways to categorize e-business models, they can be broadly categorized as business-to-consumer (B2C), business-to-business (B2B), business-to-government (B2G), consumer-to-consumer (C2C), and consumer-to-business (C2B). Table 1-4 describes the general features of these e-business models and provides examples of e-businesses that follow them.

TABLE 1-4 E-Business Models and Examples

Model	Description	Examples
B2C	Business-to-consumer: business sells products or services directly to consumers	Amazon.com Tattered Cover Book Store eDiets.com
B2B	Business-to-business: business sells products or services to other businesses or brings multiple buyers and sellers together in a central marketplace	AirParts.com Jayde.com Rackspace Managed Hosting
B2G	Business-to-government: business sells to local, state, and federal agencies or creates a marketplace to bring government agency buyers and sellers together	B2GMarket ScanPlanet.com SupplyCore
C2C	Consumer-to-consumer: consumers sell or trade directly with other consumers	eBay swapvillage.com
C2B	Consumer-to-business: consumers submit bids for products or services that competing businesses accept or decline	priceline.com

Within these broad categories, there are a number of variations in the way the models are implemented; in fact, many e-businesses follow some combination of these e-business models. Take, for example, Interstate Batteries, a Dallas, Texas-based company that has been marketing and distributing batteries of all kinds through a system of wholesale distributors since 1952. Interstate Batteries' original Web site focused primarily on its traditional B2B marketing and sales. But in 2004, the company made major modifications to its Web site—including the addition of new search and navigation tools—to help individual consumers more easily find and purchase batteries at the site. These Web site modifications, together with an advertising program targeting individual consumers, added substantial online B2C sales to the mix.[12, 13]

Other e-businesses that combine B2B and B2C e-business models are Dell, Office Depot, and Staples. To appreciate the competitive advantage of these variations, you must first become more familiar with how each of these e-business model works.

Business-to-Consumer (B2C)

Consumers are increasingly going online to shop for and purchase products, arrange financing, prepare shipment and delivery of digital products such as software, and get service after the sale. Business-to-consumer, or **B2C**, e-business includes retail sales, often called **e-retail**, of goods and services, as well as online purchases of items such as airline tickets, entertainment venue tickets, hotel rooms, and shares of stock.

Businesses that conduct their transactions from a physical location are sometimes known as **brick-and-mortar** enterprises. Many traditional brick-and-mortar retailers—from nationwide companies such as Sears, Best Buy, Barnes & Noble, and the Gap, to regional or local stores such as The Sunglass City—are now **e-retailers** who maintain online stores at which their customers can also view merchandise and make purchases. Companies such as these, which combine brick-and-mortar business facilities with e-business operations, are sometimes called **brick-and-click** companies.

Some B2C e-businesses provide high-value content for a subscription fee. Examples of e-businesses following a **subscription model** such as this include the Wall Street Journal Online (for financial news and articles), Consumer Reports (for product reviews and evaluations), and eDiets.com (for nutritional counseling).

The B2C e-business category also includes **virtual malls**, which are e-business Web sites that host a number of online merchants. Virtual malls, which may also offer transaction handling services and marketing options, typically charge online merchants setup, listing, or transaction fees. Most virtual malls allow consumers to search for a specific product, compare the product features and prices offered by various stores, and even check out each store's consumer satisfaction ratings. Two examples of virtual malls are MSN Shopping and Yahoo! Shopping.

Pure-play e-retailers, merchants that offer traditional or Web-specific products or services only over the Internet, are sometimes called **virtual merchants**, and they provide another variation on the B2C model. Amazon.com—a company that sells books, electronics, toys, music, and more—is one of the most successful original pure-play e-retailers. Other successful pure-play e-retailers include eBags, which specializes in bags and luggage of all types, and Hometown Favorites, an e-business that offers hard-to-find foods.

Some businesses supplement a successful traditional mail-order business with an online shopping site, or move completely to an online store. These businesses are sometimes called **catalog merchants**. Examples of catalog merchants include Avon.com (cosmetics and fragrances), CHEF'S (cookware and kitchen accessories), Omaha Steaks (premium steaks, meats, and other gourmet food), and Harry and David (gourmet food gifts). Figures 1-13 through 1-16 depict various B2C Web sites.

FIGURE 1-13   eDiets.com

FIGURE 1-14   Yahoo! Shopping

FIGURE 1-15 eBags

FIGURE 1-16 Harry and David

While B2C may be the most familiar form of e-business, worldwide transactions among businesses account for the majority of e-business transactions.[14]

Business-to-Business (B2B)

Like B2C e-business models, business-to-business, or **B2B**, e-business models take a variety of forms. There are basic B2B Internet storefronts, such as Staples and Office Depot, which provide business customers with purchasing, order fulfillment, and other customized services. Some B2B e-businesses offer Internet and Web products such as Web site hosting and Web page design, networking hardware and software, or e-business consulting services.

Another B2B model is an online trading community that acts as a central source of information for a vertical market. A **vertical market** is a specific industry in which similar products or services are developed and sold using similar methods. Examples of broad vertical markets include insurance, real estate, banking, heavy manufacturing, and transportation. The information available at online trading community Web sites includes buyer's guides, supplier and product directories, industry news and articles, schedules for industry trade shows and events, and classified ads. MediSpeciality.com (healthcare industry), Hotel Resource (hospitality industry), and Elance (IT industry) are examples of B2B e-businesses that support vertical markets. Figure 1-17 shows a B2B Web site.

FIGURE 1-17 Elance

In addition to supporting vertical markets, Elance also serves as a B2B exchange. **B2B exchanges** are e-businesses that bring multiple buyers and sellers together in a virtual centralized marketplace, sometimes called a **marketspace**. B2B exchanges may aggregate information from multiple sellers, allow participants to post buy or sell opportunities on an electronic bulletin board, provide auction services that enable multiple buyers or sellers to enter competitive bids on contracts, or provide access to expert information for a specific field.

B2B exchanges that deal in products or services include XSAg.com (agricultural industry), freightquote.com (shipping industry), Covisint (automotive industry), VIPAR Heavy Duty (truck parts), and Dairy.com (dairy industry). One example of a B2B expert information exchange is the ATLA Exchange Expert operated by the Association of Trial Lawyers of America (ATLA). ATLA members can login and access the ATLA Expert Exchange database to locate legal experts who can help them prepare for a case. Legal experts wishing to become part of the database pay a subscription fee. Figures 1-18 and 1-19 are examples of B2B exchange Web sites.

FIGURE 1-18 Dairy.com

FIGURE 1-19   atla.org

E-PIONEERS

Business.com

Jake Winebaum and Sky Dayton, entrepreneurs formerly associated with Disney Online and EarthLink, respectively, set a record in the late 1990s by purchasing the Internet domain name "business.com" with a common stock transaction valued, at the time, at $7.5 million. Their plan was to build a B2B exchange that aggregated business information from various industries. Winebaum and Dayton wanted the Business.com Web site (Figure 1-20) to be the first place a businessperson looked to find all types of business information on the Web. Launched in 2000, Business.com initially provided news, statistics, company profiles, financial data, and product/service directories for 57 industries. Since then, Business.com has evolved into a business-oriented search tool, with more than 60,000 industry, product, and service directory subcategories.[15, 16]

continued

FIGURE 1-20   Business.com

Another subcategory within in the B2B model is a B2B auction, where products and services are exchanged through online bidding. B2B auctions include both online **forward auctions**, where many buyers bid on products or services offered by a single seller, and online **reverse auctions**, in which a single buyer offers to purchase products and services from multiple competing sellers. One B2B auction site that offers both forward and reverse auction services for the retail, construction, travel, and manufacturing industries is HedgeHog. Figure 1-21 illustrates a B2B auction Web site.

Business-to-Government (B2G)

A variation on the B2B model is the business-to-government, or **B2G**, model. These e-businesses create a marketspace for sellers wanting do business with government agencies. B2G e-businesses provide information on government contracting and bring suppliers and government agencies together. E-businesses, such as Bidmain and B2GMarkets, follow the B2G e-business model. Figure 1-22 illustrates a B2G Web site.

Now that you've become familiar with how B2C, B2B, and B2G e-businesses interact with individual consumers and other businesses, you can examine another type of business activity fostered by the growth of the Internet and Web access, one in which consumers interact directly with other consumers to buy, sell, and trade items, personal services, and information.

FIGURE 1-21 HedgeHog

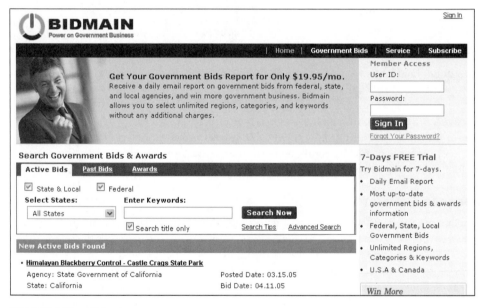

FIGURE 1-22 Bidmain

Consumer-to-Consumer (C2C)

In the consumer-to-consumer, or **C2C**, e-business model, consumers sell products, personal services, and expertise directly to other consumers through a number of methods: by placing online classified ads, by participating in forward and reverse auctions, or by making trades. Examples of e-businesses that involve consumers selling directly to consumers are American Boat Listing, an online boat listing servive; eBay, which offers both fixed price items and auctions; TraderOnline.com, which hosts classified ads; and AllExperts.com, an expert information exchange. Figures 1-23 through 1-26 show C2C Web sites.

FIGURE 1-23 American Boat Listing

In addition to enabling consumers to interact with each other directly, the growth of Internet access and Web content has shifted marketplace power from sellers to buyers. This shift has transformed the relationships between these two groups in dramatic ways, and has led to the creation of the final business model listed in Table 1-4: the consumer-to-business model.

FIGURE 1-24   eBay

FIGURE 1-25   TraderOnline.com

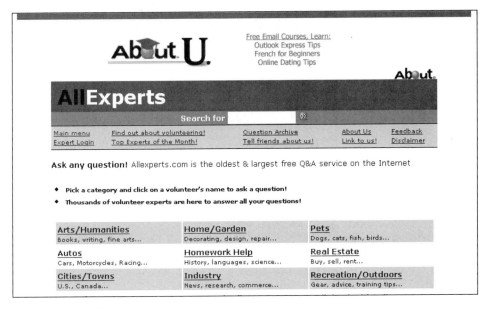

FIGURE 1-26 AllExperts

Consumer-to-Business (C2B)

Like the B2B reverse auction model, the consumer-to-business, or **C2B**, e-business model uses reverse auctions to enable consumers to name their own price for a specific good or service; once the bid is offered and accepted, it is often binding. An e-business following the C2B model collects an individual consumer's bid for a product or service, such as an airline ticket, rental car, or hotel room, and then offers the bid to multiple competing sellers who either accept or decline the consumer's bid. The most well-known e-business following the C2B e-business model is priceline.com (Figure 1-27).

In addition to broad categories such as B2B and B2C, there are various subcategories of e-business models. Many of these subcategories were created by organizations such as government agencies, non-profit institutions, and social or religious groups that decided to reduce their operating expenses and improve customer service by adapting e-business models to their specific needs. National Public Radio is an example of a non-profit institution following an e-business model.

In Chapter 2, we introduce Rackspace Managed Hosting, an e-business located in San Antonio, TX. In subsequent chapters, we will follow Rackspace's progress from idea to launch to operation.

FIGURE 1-27 priceline.com

. . . SCRAMBLED EGGS

Despite the cost-cutting changes implemented by CEO George Orban, the Egghead stores continued to lose money. A notable exception to this trend was the online store, whose sales were growing dramatically. The message was clear: Egghead had to change its business even more dramatically or go under. The decision was made to close the remaining brick-and-mortar stores, lay off store employees, and begin doing business online only—that is, as the pure-play e-retailer now known as Egghead.com.[17] As part of this initiative, Egghead.com began courting small- and medium-sized businesses as its customers by offering new software licensing options and volume pricing discounts. It also improved customer support by adding an automatic service that sent order status updates via e-mail. Finally, Egghead.com joined in new strategic reseller partnerships with Compaq Computer Corporation, Sony, and others.[18]

By changing its business model and rethinking its value chains, Egghead Software reincarnated itself as Egghead.com, and the company's online sales continued to grow. In 1999, Egghead.com merged with the online auction company OnSale to form a new company that kept the Egghead.com name. But a combination of factors forced the new Egghead.com to file for bankruptcy in two short years. One such factor was that competitors continued to undercut Egghead.com by selling the same or similar products at lower prices. Also, a general downturn in the overall economy further reduced sales, and in trying to stay afloat, the company began hemorrhaging cash.

continued

Egghead.com cut costs again by laying off employees and moving many of its jobs from California to Washington State. Unfortunately, however, all this was too little too late. The company suspended operations, and its assets—including its URL www.egghead.com—were ultimately purchased from the bankruptcy court by Amazon.com.[19, 20] As you can see in Figure 1-28, Egghead.com is now an Amazon.com storefront.

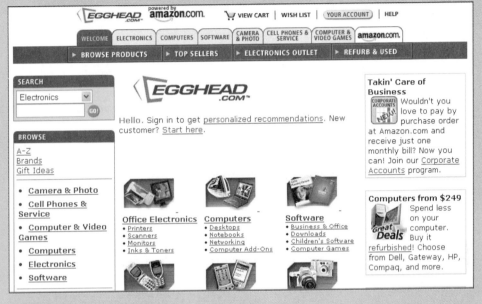

FIGURE 1-28 Egghead.com

Chapter Summary

- In traditional commerce, buyers and sellers come together in a physical marketplace to exchange information, products, services, and payments.

- The terms "e-commerce" and "e-business" both refer to the process of conducting a broad spectrum of business activities over a telecommunications network.

- The earliest forms of electronic commerce were electronic data interchange (EDI) and electronic funds transfer (EFT).

- The Internet is a worldwide public network connecting thousands of private networks and individuals.

- The World Wide Web is a subset of the Internet and consists of servers that store hyper-linked documents called Web pages.

- Traditional business limitations of time, space, and geographical location are being over-come by the flexibility of doing business over the Internet.

- Access to consumer information and competing sellers through the Internet and the Web has led to a shift in market power from seller to buyer.

- A company's value chain comprises all the primary and support activities necessary to create and distribute the company's goods and services.

- Because of the growth of the Internet and the Web and the way they have redefined busi-ness relationships, companies are rethinking their traditional value chains and consider-ing new ways of doing business.

- A company's business model is the way in which it conducts business in order to gener-ate revenue.

- Global access to the Internet and the Web is driving businesses to adapt old business models and create new ones.

- E-business models can be broadly categorized as business-to-consumer (B2C), business-to-business (B2B), business-to-government (B2G), consumer-to-consumer (C2C), and consumer-to-business (C2B).

Checklist

Thinking about Doing Business Online?

- ❏ Consider the current costs involved in providing information and services to your customers that they could then get for themselves by accessing a Web site.

- ❏ Determine what additional information or transaction services you could provide to your existing customer base through a Web site.

- ❏ Think about how you can use the information you have about your customers to make it easier for them to continue to do business with you if you were to be online.

- ❏ Identify how valuable your customers might find the expertise of your employees or your other customers, and look for ways to deliver this expertise online.

- ❏ Consider the competitive disadvantages if your competitors were to offer online services and information before you do.

- ❏ Discover ways to improve your business by rethinking your value chains.
- ❏ Identify the capabilities your company needs to take over the functions that are currently provided by other businesses in your value chains.
- ❏ Consider new ways to generate revenue by enhancing sales or attracting new customers to your Web site.
- ❏ Develop ideas for repackaging current information to attract new customers or create new business opportunities on your Web site.
- ❏ Determine whether online competitors can significantly harm your business by providing some of the value you currently offer customers in the traditional way.

Key Terms

B2B

B2B exchanges

B2C

B2G

brick-and-click

brick-and-mortar

C2B

C2C

catalog merchants

e-business

e-commerce

electronic commerce

electronic data exchange (EDI)

electronic funds transfer (EFT)

e-retail

e-retailers

extranet

forward auctions

hyperlinks

hypertext

Internet

Internet service provider (ISP)

intranet

marketspace

network

pure-play e-retailers

reverse auctions

servers

subscription model

trading partners

value activities

value chain

value-added networks (VANs)

vertical market

virtual malls

virtual merchants

Web browser

Web pages

Web servers

Web site

World Wide Web (Web)

Review Questions

True/False Questions

1. The World Wide Web is a subset of the Internet. True or False?

2. Amazon.com is an example of a B2B exchange. True or False?

3. It is easy to keep track of how many people are using the Internet and how many Web sites exist on the World Wide Web. True or False?

4. The traditional factors that can limit business transactions—time, space, and geographical location—are less limiting for e-business transactions. True or False?

5. One disadvantage to online buyers is the lack of trust when dealing with unfamiliar sellers. True or False?

Multiple Choice Questions

1. A traditional business environment in a physical location is called a:

   a. brick-and-click marketplace.

   b. brick-and-mortar marketplace.

   c. brick-and-mouse marketplace.

   d. brick-and-Web marketplace.

2. The Internet is:

   a. a public worldwide network of private networks.

   b. the same as the World Wide Web.

   c. a brick-and-mortar marketplace.

   d. none of the above.

3. Which of the following e-businesses follows the B2B exchange model?

   a. Autobytel.com

   b. Amazon.com

   c. eDiets.com

   d. freightquote.com

4. An e-business that offers a marketspace in which consumers submit bids for products and services to multiple competing sellers is following the:

   a. B2B e-business model.

   b. B2G e-business model.

   c. C2B e-business model.

   d. C2C e-business model.

5. Which of the following is not an advantage of doing business online?

   a. increased sales opportunities

   b. rapidly changing technologies

   c. 24/7 operations

   d. access to global markets

Exercises

1. Using online search tools or other relevant resources, research the origins and history of the Internet and the World Wide Web. Then write a one- or two-page paper describing at least five major events in this history, and how these events led to the growth of global e-business.

2. Using online search tools and other relevant resources, locate information about two people whose contributions have had a significant impact on the growth of the Internet and the World Wide Web as a business medium. Write a one- or two-page paper describing each person and his or her contribution.

3. Using online search tools and other relevant resources, research the rise and fall of Pets.com. Then write a one- or two-page paper explaining what went right and what went wrong at Pets.com.

4. In addition to the primary e-business models discussed in detail in the chapter, additional models, such as P2P, M2M, G2C, and B2E, continue to be defined. Using online search tools and other relevant resources, research these and other e-business models not discussed in this chapter. Then write a one-page paper describing at least three of the models.

5. Using online search tools and other relevant resources, find a representative example of each of the e-business models discussed in detail in this chapter. Do not use examples already discussed in the chapter. Then write a one- or two-page paper describing each example and how the example fits the e-business model.

Case Projects

1. While cleaning out your grandmother's storage shed, you find several old items, including carnival glass plates, china dolls, baseball cards, and soft drink signs. You think some of the items may be valuable as collectibles. First, you want to determine the value of the items, and then you want to find a place to sell them. Locate at least three C2C Web sites that might be helpful in determining the items' values and in selling the items. Write a brief summary of each site and how it could be useful to you.

2. You are the executive assistant to the president of a brick-and-mortar company that sells equipment for extreme sports. The president is considering adding an online store and asks you to prepare a report on existing e-business sites for similar stores. Locate at least three extreme sports e-business Web sites and write a brief summary of each site, including an explanation of its e-business model. Make a recommendation to the president on the type of e-business model he should consider.

3. You maintain a file of Internet statistical data for your supervisor, the online sales manager. She asks you to prepare a report for the next sales meeting. She would like current estimates of the number of people who are online in the United States and worldwide; estimates for global and U.S. B2C and B2B online sales; and other relevant data about the state of the e-business economy. Using Web search tools and other relevant resources, gather useful data estimates. Then write a brief report containing the data estimates and their sources for your supervisor.

Team Project

You and two classmates are eager to start your own e-business. Meet with your classmates, and using brainstorming as well as other applicable techniques and resources, decide on an e-business idea, including the e-business model your e-business will follow. Then, using presentation software, create a 5–10 slide presentation that describes your e-business and its e-business model. Include in your presentation an analysis of the advantages and disadvantages you expect to face by doing business online. Present your e-business idea to your classmates and your instructor.

For Further Study

Here are some resources that might help you in further investigating the topics covered in this chapter.

Student Online Companion

Check out the *Creating a Winning E-Business, Second Edition* student online companion Web site for links to the sites discussed in this chapter and to other useful Web sites.

Articles and Books

Amor, Daniel. *The E-Business (R)evolution: Living and Working in an Interconnected World*. Upper Saddle River, NJ: Prentice Hall PTR. 2000.

Sculley, Arthur B. and Woods, W. William A. *B2B Exchanges*. USA: ISI Publications. 1999.

Segaller, Stephen. *Nerds 2.0.1: A Brief History of the Internet*. New York: TV Books, L.L.C. 1999.

U.S. Department of Commerce. "Digital Economy 2003." www.esa.doc.gov/2003.cfm. June 2003.

End Notes

[1] Guglielmo, Connie. "He's Unscrambling Egghead." *PC Week*, 14(8); A1(2). February 24, 1997.

[2] Lohr, S. "IBM Plays Up Its On-Demand Computer Service." *Naples Daily News, New York Times News Service*. answers.google.com/answers/threadview?id=253727. October 31, 2002.

[3] *IDC Research*. "Internet Commerce Market Model." www.idc.com. 2005.

[4] *National Science Foundation*. "The Internet: Changing the Way We Communicate." www.nsf.gov/about/history/nsf0050/internet/internet.htm. 2005.

[5] Berners-Lee, Tim. *Weaving the Web*. New York: HarperCollins. 1999.

[6] *Internet World Stats*. "Internet Usage Statistics—The Big Picture." www.internetworldstats.com/stats.htm. February 3, 2005.

[7] *IDC Research*. "Worldwide Traffic to Rise." www.nua.ie/surveys/index.cgi?f=VS&art_id=905358733&rel=true. March 3, 2003.

[8] Pastore, M. "Businesses Will Lead Wireless Net Adoption." *ClickZ*. www.clickz.com/stats/sectors/wireless/article.php/906771. October 18, 2001.

[9] *The Tattered Cover Book Store Info Desk. www.tatteredcover.com.*

[10] *American International Automobile Dealers Association (AIADA)*. "J. D. Power Reports: More Internet Users Visiting Manufacturer Web Sites First." www.aiada.org/article.asp?id=22751. September 10, 2004.

[11] Porter, Michael E. *Competitive Advantage: Creating and Sustaining Superior Performance.* New York: Free Press. 1985.

[12] *Interstate Batteries.* "About Us." www.ibsa.com/www_2001/content/about_us/default_aboutus.asp?js=1. 2005.

[13] *Internet Retailer.* "New Site Search Takes Interstate Batteries Into B2C and Triples Web Sales." www.internetretailer.com/dailynews.asp?id=13168. October 19, 2004.

[14] *BuddeComm.* "Global-Business Users—B2B Market Statistics." www.budde.com.au/Reports/Contents/Global-Business-Users-B2B-Market-Statistics-1925.html. January 9, 2005.

[15] *Business.com.* "About Us." www.business.com/info/aboutus.asp. 2005.

[16] *BusinessWeek.* "The Be-All and End-All of B2B Sites?" Issue 3684; p 56. June 5, 2000.

[17] Guglielmo, Connie. "He's Unscrambling Egghead." *PC Week*, 14(8); A1(2). February 24, 1997.

[18] Cox, B. "Egghead.com Scrambling But Not Fried." *E-Commerce News.* www.ecommerce-guide.com/news/news/article.php/10375_920641. November 9, 2001.

[19] Wolverton, T. "Egghead to File for Bankruptcy." *CNET News.com.* http://news.com.com/2100-1017-271685.html?legacy=cnet. August 15, 2001.

[20] Liu, B. "Egghead.com Becomes Amazon.com Property." *E-Commerce News.* http://news.earthweb.com/ec-news/article.php/932871. December 3, 2001.

CHAPTER 2

DEFINING YOUR E-BUSINESS IDEA

LEARNING OBJECTIVES

In this chapter, you will learn to:

- Identify entrepreneurial abilities
- Describe the entrepreneurial process
- Understand the factors affecting e-business success
- Identify ways to exploit e-business advantages

IDEAS! IDEAS! . . .

Kelby Hagar, a fifth-generation Texan, grew up in Hereford, Texas, a small town in the Texas panhandle. After graduating from Angelo State University with a degree in prelaw and accounting, Hagar married and moved east to study law at Harvard in the early 1990s. Looking for a snack late one night, Hagar opened the refrigerator and found it was practically empty! But who had time to shop for groceries? The Internet-savvy Hagar, who was accustomed to going online to shop for books and make travel arrangements, began wondering why he couldn't order his groceries online and have them delivered to his doorstep like his books. After graduating from Harvard, Hagar and his wife moved to Dallas, where he began working for a major law firm. But the time-saving idea of ordering groceries online and having them delivered just wouldn't go away.[1, 2] Could Hagar turn his e-business idea into reality?

Do you have an e-business idea? Are you excited about taking your e-business idea to market? One of the first things you should do is consider the entrepreneurial abilities needed to start any business, including an e-business.

The Entrepreneur

An **entrepreneur** is someone who assumes the risks associated with starting and running his or her own business. **Entrepreneurial abilities** include:

- leadership traits
- a high-energy personality
- self-confidence
- organizational skills
- the ability to act quickly and decisively

To successfully start, operate, and grow any business—including an e-business—an entrepreneur must be able to lead others. Leadership traits common to most successful entrepreneurs include intelligence, determination, integrity, listening skills, and the ability to relate to others in a positive way. As a leader, an entrepreneur must be able to communicate the mission and goals of his or her business to investors, suppliers, customers, employees, and interested others.[3]

QUOTES ON SUCCESS

"I'm convinced that about half of what separates successful entrepreneurs from the non-successful ones is pure perseverance."

Steve Jobs, founder of Apple Computer, Inc.

Starting any new business requires considerable energy and the ability to focus that energy on accomplishing objectives. Because starting and running a business is more time-intensive than working for someone else, following an appropriate stress management plan (including exercise and a healthy diet) is a must for an entrepreneur.

Successful entrepreneurs must believe in their business ideas and have the self-confidence to accomplish their goals. Self-confidence is enhanced when the business idea is closely related to an area of the entrepreneurs' interests or expertise. Along with self-confidence, entrepreneurs need the ability to organize business activities successfully, as this is essential to get things done on time, locating business-critical information quickly, and staying on schedule.

TIP

Entrepreneurship is also taking place within large corporations, such as Sun Microsystems and Boeing, where project leaders are encouraged to use their entrepreneurial skills to turn new product ideas—especially the kind that strive to take advantage of the Internet and Web—into profits. When employees exhibit entrepreneurship behaviors within a large corporation, this phenomenon is sometimes called "intrapreneurship."

In addition, successful entrepreneurs are generally independent, goal-oriented, creative, and competitive. Many successful entrepreneurs display their competitive natures early in life—that is, in their approach to school, hobbies, or sports. Finally, entrepreneurs must be ready to make short-term sacrifices, such as spending limited time with family and friends, in return for long-term success.[4]

TIP

Some people consider starting their own business so that they can set their own working hours. While entrepreneurs often do set their own work schedules, they are seldom easy ones. To achieve a business's startup goals, an entrepreneur may need to work 10 to 12 hours a day, 6 or 7 days a week!

E-PIONEERS

Failure—A Successful Idea

A few years ago, Jason Zasky, a survivor of three failed entrepreneurial adventures, was walking with his cousin in Manhattan, when his cousin stopped and said: "I have a great idea for a magazine—failure." Zasky, who also thought the idea was great, began working "serious 18-hour days" with his partner, Kathleen Ervin, in a small building in Scarsdale, New York. Together the two of them launched an online magazine named *Failure* in July, 2000.

According to Zasky, *Failure Magazine*'s target market consists of adventurous, risk-taking individuals between 25 and 55 years old, for whom the word "failure" carries little stigma. *Failure Magazine* is designed to be "thought-provoking rather than provocative," and includes articles and features on topics such as the world's greatest golfer never to succeed on the pro tour and an interview with the player who made one of the greatest baseball errors in history.

Because failure is a universal experience, Zasky believes that people are interested in failure-related stories. When asked if it were possible to create a successful e-business by publicizing the failures of others, Zasky responded that he and his partner believe that "the greatest failure would be in not trying to make *Failure Magazine* a success." Even with a history of other failed entrepreneurial attempts, Zasky was not afraid to use his entrepreneurial abilities and hard work to try a new e-business idea![5, 6, 7, 8]

QUOTES ON SUCCESS

"I think we write about winners because those are the people out there taking risks, making things happen, even if they fail."

Jason Zasky, founder of Failure Magazine

After considering entrepreneurial abilities, the next step is to understand the entrepreneurial process.

The Entrepreneurial Process

The **entrepreneurial process**, which can be divided into several steps or stages, is a methodology in which you begin by determining whether or not you are an entrepreneur and then deciding whether to start your own business or purchase an existing business. If you choose to start your own business, you must define the business idea, create a business plan, and perhaps, secure financing before you can begin operating the business. Figure 2-1 illustrates the five stages of the entrepreneurial process.[9]

- *Stage 1*: Decide whether or not you are an entrepreneur. It is critical that you assess your entrepreneurial abilities and evaluate the time and effort that will be required. Consider seriously the effect of this commitment on you and your family before going to the next stage.
- *Stage 2*: Decide whether to buy or start a business. If you determine that you are an entrepreneur, the next step in the entrepreneurial process is to decide whether to start a new business or purchase an existing business.
- *Stage 3*: If you decide to start a new business, you must first define the business idea or concept, including the products or services to be sold. Then you must create a business plan that includes your assessment of the business environment in which you will operate, an identification of the business need, estimates of expected profitability, and a description of the legal form the business will take. Part of the business planning process includes defining stakeholders, such as customers, suppliers, distributors, and possible investors. After developing the business plan, you must then secure financing and resources from various sources: you (a business owner's own time and effort are commonly referred to as **sweat equity**), friends and family, third-party angel investors, and venture capitalists.
- *Stage 4*: After you either start a new business or purchase an existing one, you then must operate and grow the business.
- *Stage 5*: Once the business is in operation and has grown, you will be able to harvest the business through a number of means: by continuing to operate the business and letting it become a **cash cow** (that is, continue operations to generate cash), by **going public** (issuing a public stock offering or IPO), by selling the business, or by allowing it to be acquired by another business. Of course, if the business is not successful, you may be required at this stage to liquidate the business and, if necessary, file for bankruptcy.

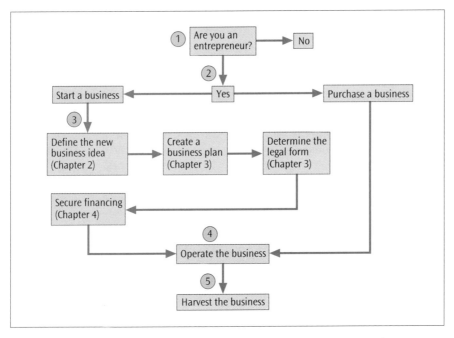

FIGURE 2-1 The Entrepreneurial Process

E-CASE

From Idea to Harvest

The Chicago-area Web marketing company WebPromote was launched by four engineers—Keith Speer, Ken Wruk, Kevin Manley, and John Weiss—whose e-business idea was to create a permissions-based marketing company that sent a weekly newsletter containing advertising messages to newsletter subscribers. In the late 1990s, when the four-year-old e-business's newsletter started generating larger advertising revenues, the founders of WebPromote began searching for funds to expand their e-business, but they didn't have a lot of luck—at least not initially. It seemed that many of the venture capital firms they contacted were concerned that WebPromote lacked the marketing management expertise to succeed.

Enter David M. Tolmie, an experienced marketing professional who met with the founders on behalf of his venture capital firm. Tolmie liked the WebPromote e-business idea, and because he wanted the challenge of running a business, he agreed to join WebPromote as the chief executive officer and president—thereby providing the business the important marketing expertise it needed. As CEO, Tolmie acted quickly to secure a round of venture capital financing; he changed the company name to Yesmail, refocused the company on its permission-based e-mail business, and doubled the company's member base.

continued

The efforts paid off, and in a little more than one year, Tolmie took Yesmail public. Shortly thereafter, Tolmie and the company's founders harvested the e-business's value by selling Yesmail for more than $500 million to the e-business aggregator CMGI, Inc.[10, 11, 12] Today, Yesmail is part of infoUSA and Donnelley Marketing.

After determining that you have the entrepreneurial abilities to take your e-business idea to market and after reviewing the entrepreneurial process, you should next consider some of the major factors that could affect the long-term success of your e-business idea.

FACTORS AFFECTING E-BUSINESS SUCCESS

E-business ideas span the spectrum from selling hard-to-find food products directly to consumers, to auctioning excess oil and gas capacity, to offering online mediation services, to providing Web site hosting services. Successful e-businesses based on viable e-business models and sound business principles will grow, change, and mature; on the other hand, e-businesses based on flawed e-business models or unsound business practices find it difficult, if not impossible, to survive. Many factors can affect the success of a new e-business idea. There are several factors that are unique to doing business online including the network effect, creating innovative marketing ideas, scalability, the ease of entry into electronic markets for you and your competitors, and adaptability to change. Every e-business entrepreneur should carefully consider each of these factors.

The Network Effect

A primary factor in the growth of e-business is the **network effect**. The network effect is the phenomena in which the total value of a product, service, or technology grows as more and more people use the product, service, or technology. A commonly used example that demonstrates how the network effect leads to increasing value is a telephone service. A telephone service or network with a single telephone user has little value to the user or to others because there is no one else to call. However, when you add a second telephone user, the value of the telephone network increases because there are now other people to call. As additional telephone users are added, the telephone service's total value continues to increase.

TIP

The change in the value of a network as it grows, or the network effect, is sometimes called Metcalfe's Law. Bob Metcalfe, an engineer and entrepreneur, was involved in the early development of network technologies at Xerox PARC and is the founder of 3Com Corporation. Metcalfe's Law states that the power of the network is N squared, where N is the number of nodes in the network.[13]

For each Internet user, the value of being online grows as the number of individuals and businesses that are online grows. For global e-business, the network of linked e-businesses and consumers becomes more valuable as more and more e-businesses and consumers participate in the process. Examples of e-businesses and other organizations who have successfully harnessed the power of the network effect include auction sites, such as uBid; professional associations, such as that of the developers of Bluetooth; and B2B technology providers, such as Groove Networks. Figures 2-2 through 2-4 show the Web sites for these organizations.

FIGURE 2-2   uBid

FIGURE 2-3   Bluetooth

FIGURE 2-4 Groove Networks

Take a close look at your e-business idea and decide whether or not the value of the product, service, or technology you plan to offer increases with greater use or distribution over the Internet and the Web. You should also be mindful of the inherent dangers of the network effect on your e-business idea—for example, rapidly circulating negative customer comments or technologies that perform in unexpected and unwanted ways.

As part of the evaluation of your e-business idea, you should identify ways to exploit the network effect by using innovative marketing ideas to promote your e-business.

E-PIONEERS

A Network Effect Backlash

In the late 1990s, an e-business named Third Voice gained worldwide attention by offering Web annotation software that allowed users to post Web versions of "sticky notes" on any Web page. Users simply downloaded the Third Voice browser plug-in software, which enabled them to post comments, view comments by others, and even join discussion groups or communities related to specific Web pages. With this software, users could also import their e-mail address books into Third Voice and use it to generate an e-mail message that included a link to the Web page being viewed. If message recipients didn't have the Third Voice software, a quick click took them to the Third Voice Web site, where they could download it. By enabling people to import their address books, Third Voice immediately created a network of potential users for itself.

continued

What at first seemed like a fun and interesting e-business idea rapidly turned sour, however. Software programmers quickly discovered security problems with the Third Voice software that allowed hackers to access users' data and also enabled them to create fake Web pages. In addition, many Web design professionals and proprietors of e-businesses were outraged that Third Voice users could, in effect, deface Web pages with unflattering and inappropriate comments. While some people championed Third Voice as a new tool for free expression, many others become concerned about threats to user privacy. Before long, Third Voice closed shop, posting instructions on its Web site for how its users could uninstall the Third Voice software from their computers.[14, 15, 16]

Innovative Marketing Ideas

Successful e-business entrepreneurs develop imaginative ways of exploiting the network effect in order to market products and services to new customers. One of the most famous examples of a new e-business that successfully exploited the network effect through the use of an innovative marketing idea is Hotmail.

Hotmail, a wildly successful e-business that offers free e-mail services, almost didn't get launched. As with many e-business startups, Hotmail's founders Sabeer Bhatia and Jack Smith had difficulties getting venture capitalists interested in their idea. When they met with Tim Draper, the founding partner of the venture capital firm Draper Fisher Jurvetson (DFJ), Draper liked the idea of free e-mail but insisted that text be added to the bottom of each outgoing e-mail message encouraging recipients to get their own free Hotmail account. Draper wanted the text to be linked to the Hotmail Web site so that message recipients could click the link, view the features of Hotmail on the Web site, and immediately sign up for the service themselves.

At first, Bhatia and Smith resisted the idea of adding the linked message, considering it to be too much like what today is called spam—unsolicited commercial messages or advertisements that are sent by marketers to a large number of unrelated and uninterested parties. Eager to get financing for their new e-business, however, Bhatia and Smith finally agreed with Draper about including the marketing message link.

Hotmail was quietly launched in 1996 over the July 4th holiday. Because of the holiday and because there was little money to spend on advertising the service, the launch generated little press coverage. But this apparently didn't matter—customers soon began signing up for Hotmail accounts in droves! The first customers to sign up were at colleges and universities. A single user from a school would sign up, and then, the next day, a hundred users from the same school would be signed up. By the end of the week, there would be a thousand users from the same school. In the meantime, word about Hotmail would spread to another school, and this process would begin again.

Before long, users from around the world began signing up, and the propagating process continued at impressive speeds. For example, within six weeks after the first user in India subscribed, India had 100,000 Hotmail subscribers. Less than 18 months after its launch, Hotmail had 12 million worldwide subscribers. And in less than three years, Hotmail had become the fastest-growing media company in history, with more than 30 million subscribers! The network effect enabled Hotmail, the world's first Web-based e-mail service, to spread like wildfire. Shortly after recording 12 million subscribers, Hotmail's founders sold the company to Microsoft for $400 million (Figure 2-5).[17, 18]

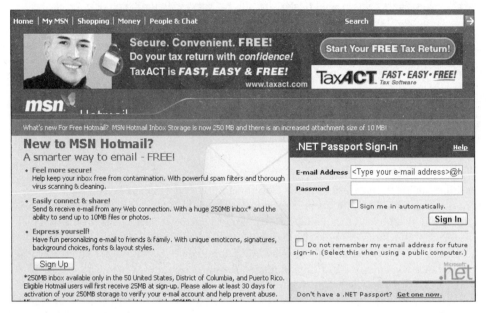

FIGURE 2-5 MSN Hotmail

The astounding spread of Hotmail prompted Tim Draper and his partner at DFJ, Steve Jurvetson, to analyze what was unique about Hotmail's success. They noted that Hotmail spread around the world like an epidemic. When current users sent messages to others containing the Hotmail link, they were "infecting" potential users, who in turn "infected" other potential users. Drawing an analogy to the effects of a sneeze in a crowd, Draper and Jurvetson coined the phrase **viral marketing** to describe Hotmail's innovative method of attracting new customers.[19, 20, 21, 22]

QUOTES ON SUCCESS

"Everybody has good ideas. It is the implementation, which is not trivial. People do not realize that."

Sabeer Bhatia, co-founder of Hotmail

In addition to assessing the network effect and making use of innovative marketing ideas, you must carefully consider how the idea for your e-business will work when your business experiences rapid growth.

Scalability

Scalability refers to the ability of your e-business idea to continue to function well, regardless of how large the business gets. The rapid growth of an e-business can cause problems if your e-business idea does not scale well with unexpected growth. Just ask the founders of AllAdvantage; they not only underestimated the power of the network effect and viral marketing, they also failed to realize how poorly the AllAdvantage e-business idea would scale as a result of rapid growth.

> **TIP**
>
> The term "scalability" is also used to refer to the ability of Web servers to handle increases in traffic loads to a Web site without crashing.

AllAdvantage, along with similar e-businesses such as Spedia and Jotter, tried to put a new twist on generating advertising revenues by paying people to browse the Web. AllAdvantage's e-business idea involved allowing individuals to download a viewing bar in which advertisements targeted to their Web browsing habits would appear as they moved from Web page to Web page. The idea was that advertisers would pay AllAdvantage to display their advertising messages in the viewing bar, while AllAdvantage paid its members up to $.53 per hour to surf the Web and read those messages. Members were encouraged to get others to sign up, and AllAdvantage paid members up to an additional $.10 an hour for the time the new members they had referred surfed the Web. Original members were also paid a bonus if the member they had referred wound up referring someone else, and so on—a classic pyramid scheme.

AllAdvantage had expected to attract 20,000 members in three months. In the first week of operations, 100,000 users enrolled with AllAdvantage; at the three-month mark, the company found itself with one million members. Within a few months, AllAdvantage had 6.7 million registered members, with membership growing at a rate of 15,000–16,000 members per month.

While these numbers may suggest that AllAdvantage was experiencing the same runaway success that Hotmail had enjoyed, the real story is more complicated. While AllAdvantage had no problem building its membership, advertisers resisted the idea of targeting online advertising to individuals browsing the Web. Many advertisers did not yet fully understand how to harness the power of the Internet and exploit the network effect to target customers with their advertising. Because of this resistance, AllAdvantage had trouble selling advertisers on using the AllAdvantage service, and in its first year of operations, the company earned only $14.4 million in revenues. At the same time, with its skyrocketing member acquisitions, AllAdvantage paid out almost $50 million to members. This payout, coupled with operating expenses, caused AllAdvantage to lose more than $100 million in its first year—and from then on, the situation only got worse. Advertisers didn't buy into the AllAdvantage e-business idea quickly enough to match the overwhelming growth in membership. Within two years, AllAdvantage burned through its capital and ceased operations.[23, 24]

Remember! When evaluating your e-business idea, consider its scalability—how well your idea works with unexpected and rapid growth. What will happen to your e-business if demand for your products or services explodes? For example, if your e-business sells products, do you have reliable sources for those products and a strong distribution system that can meet a sudden increase in demand from your customers? If your e-business sells a service, do you have sufficient resources to satisfy a spike in customer orders?

TIP

Many e-businesses—My Points, ClickRewards, and FreeRide, to name a few—attempted to jump on the "pay-to-surf" bandwagon, and only a few still survive. For example, MyPoints and ClickRewards are now reward-based marketing companies in which members earn points toward "free" gifts for Web surfing, answering marketing e-mail, and shopping at specific stores. FreeRide evolved into a marketing company that pays members to take online surveys.

When evaluating your e-business idea, consider how easy it is for you—and your competitors—to enter the marketplace.

Ease of Entry into Electronic Markets

Low-cost Internet technologies have made it easy for many new e-businesses to enter an online marketplace and become competitive. A good example is the C2C auction marketplace, which in recent years has experienced tremendous growth in both the number of auction sites and in auction sales dollars. Three major reasons for the growth of online C2C auction sites are the following.

1. Ease with which consumers can interact with an auction Web site—they need only a computer and Internet access.
2. Availability of Web site auction software that is easily installed and maintained—entrepreneurs don't need tremendous computer expertise to create and maintain a Web site to showcase sellers' items, record bids, and handle buyers' payments.
3. Attractive business proposition—the e-business hosting the auction can earn sales commissions without having to manage, warehouse, or distribute product inventories.

One of the few e-businesses to be profitable from its inception, eBay remains the Web's top auction site despite this intensified competition. Launched as a central location to buy and sell unique items, eBay experienced sales (that is, the value of goods traded on the eBay site) that grew from $347,000 to $47 million within two years. Today, eBay's reported sales in more than ten categories, including automobiles, auto parts, consumer electronics, toys, and collectibles, exceed $1 billion annually. [25, 26, 27]

QUOTES ON SUCCESS

"It was an idea that I had, and I started it as an experiment, as a side hobby basically, while I had my day job."

Pierre Omidyar, founder of eBay

eBay enjoys two business advantages that may be competitive barriers for other C2C auction sites: **first-mover advantage**, or first-to-market advantage, and **name identification**. Because eBay beat everyone else to the Web auction market, the site developed a strong community of buyers and sellers. Thanks to the fact that buyers and sellers are naturally attracted to a Web site where there are already a large number of buyers and sellers—and to the fact that the company was able to fortify its first-mover advantage with aggressive marketing—eBay continues to attract many buyers, who then attract more sellers, who then attract more buyers. Today, eBay reports more than 130 million registered users around the world, including places such as China, India, South Korea, Spain, and Taiwan.[28]

eBay's penetration into the C2C marketplace has become so deep that the name "eBay" is now synonymous with the concept of online auctions. It should be noted that the name also represents a strong online community of already entrenched buyers and sellers. At eBay, buyers and sellers not only enjoy the excitement of hunting for garage sale bargains, but also the thrill of coming into contact and exchanging information with a community of people who share their interests. This type of name identification is the ultimate competitive barrier for new auction sites hoping to break into the C2C auction market.

An e-business' ability to secure first-mover status or name identification does not guarantee that the business will be as successful as eBay. While other e-businesses, such as Dell, Inc., have managed to successfully leverage first-mover advantage and name recognition into a dominant market position, many have not. For example, E*TRADE Financial, one of the first online stock brokerage firms, now jockeys for first place with firms such as Charles Schwab, Fidelity Investments, and others, in an online marketplace that E*TRADE Financial pioneered.

E-CASE

Entrepreneurial Risk-Taking and Vision

As a teenager, Margaret "Meg" C. Whitman's first job was as a snack bar cook and general manager at Valley Ranch in Valley, Wyoming. She made brownies and cookies, and bought other items to sell at the snack bar. When asked by the Detroit News what she learned from that first job, Whitman replied, "You have to work hard to make things work right. It doesn't happen by itself."[29] With a bachelor's degree in economics from Princeton and an MBA from Harvard, Whitman has more than proved this point by tackling a succession of high-profile jobs with companies such as Proctor and Gamble, Disney, and Florists Transworld Delivery (FTD). Whitman has embraced innovative technologies and adopted forward-looking strategies throughout her career. For example, as president and CEO of FTD, she developed FTD's early e-business strategies.

When eBay was a small, three-year-old Internet startup, Whitman took a look and liked what she saw. She quickly recognized that the eBay C2C auction concept was a great e-business idea—it was unique, and it had no brick-and-mortar equivalent. Already very successful in the traditional business world, in 1998 Whitman left her position at FTD and staked her career and livelihood on the new venture by joining eBay as President and CEO (Figure 2-6). From the start, Whitman had the foresight to know that eBay was going to be huge. But no one—not even Meg Whitman—really anticipated just how huge eBay would become in just a few short years. Today, some investment analysts predict that eBay is positioned to control about 75 percent of the global online auction market, which is valued at around $240 billion.[30, 31, 32]

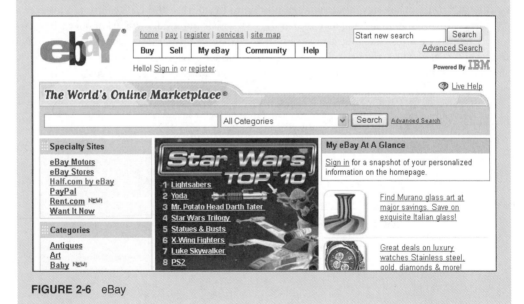

FIGURE 2-6   eBay

You should identify whether your e-business idea has a first-mover advantage or other competitive advantage; if it doesn't, consider the competitive barriers your e-business idea faces in entering the existing online marketplace. Next, think about how quickly your e-business idea can be adapted in an ever-changing business environment.

Adaptability to Change

The e-business marketspace is characterized by rapid knowledge transfer and the need to make decisions quickly. An e-business entrepreneur must act quickly to exploit new ideas and opportunities and to handle new challenges.

The ability to quickly adapt to the marketplace is a hallmark for what may be the most famous e-retailer, Amazon.com. Amazon.com began as an online bookstore. Then, to meet consumer demand for purchasing all kinds of items online, Amazon.com quickly started selling music, toys and games, videos, vehicles, and other items. When the online auction craze hit, Amazon.com promptly launched its own auction site and automatically registered its existing customers so that they could participate in the auctions (Figure 2-7). Next, Amazon.com exploited its expertise in Web technologies by launching a new venture, Amazon Web Services, which resells Web technologies to other e-businesses.

FIGURE 2-7 Amazon.com

Because the rapid pace of change can affect the most fundamental ways in which your e-business idea functions, you must consider how well your e-business idea could respond and adapt to changes in the marketplace. The next step is to then think of ways to exploit the built-in advantages you will have by doing business online.

EXPLOITING E-BUSINESS ADVANTAGES

E-business offers certain inherent advantages. Existing businesses and new e-businesses can exploit these built-in advantages to expand their markets, acquire greater business visibility, maximize customer relationships and increase responsiveness, create new products and services, and reduce costs.

Expand the Market

Because consumers in the e-business marketplace are not bound by the constraints of physical location or time, an e-business has the opportunity to reach a larger market than a traditional brick-and-mortar business. Recognizing this, many successful businesses are revamping their existing business model to add an e-business component that complements their primary value chain. Examples of brick-and-mortar businesses that have successfully modified their business model to incorporate an e-business include Wal-Mart, Costco, and Ticketmaster.

Ticketmaster, the world leader in the market for live event ticket sales, demonstrates how a very successful business idea can add e-business to its business mix. Founded in 1976, Ticketmaster was a pioneer in offering computerized ticketing services at brick-and-mortar ticket outlets. In the mid 1990s, Ticketmaster launched its fledgling Web site, which offered tickets online to events in the Pacific Northwest area. The ease with which a transaction could be completed online was clearly popular with clients and consumers, and within a few months, Ticketmaster was offering tickets online to events across the country.[33]

Next, Ticketmaster was acquired by IAC/InterActiveCorp (formerly USA Network), who combined the Ticketmaster online operations with those of its existing company, CitySearch. Ticketmaster continued to expand its share of the market by actively marketing to consumers via e-mail, holding online ticket auctions, and providing an authorized marketspace where season ticket holders could sell their unwanted tickets. Today, Ticketmaster (Figure 2-8) leads global online and offline ticket sales by combining more than 3,000 brick-and-mortar ticket outlets and 19 call centers with its online operations. In 2004, Ticketmaster sold 98 million tickets across seventeen markets including the U.S.—both online and offline—valued at $5 billion.[34]

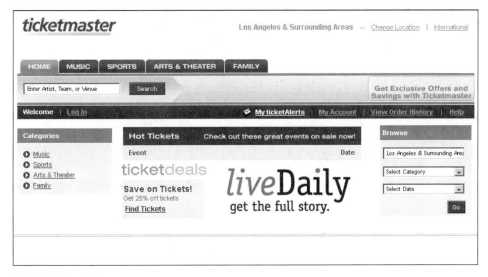

FIGURE 2-8 Ticketmaster

If your e-business idea is based on a successful traditional business model, you should determine whether it takes full advantage of the Internet to expand your market and gain greater visibility for your products or services.

Acquire Greater Business Visibility

The Internet and the Web enable companies to get their name and products or services in front of more potential customers than ever before. In fact, these days, having a presence online has become essential for most companies. In the Web's early days, many companies created Web sites that were little more than company brochures. Today, businesses large and small acquire greater business visibility by hosting useful and informative Web sites. A prime example of this is the way automobile manufacturers such as General Motors, Ford Motor Company, and Chrysler use their Web sites to promote their companies and their vehicles. Figures 2-9 through 2-11 illustrate examples of automotive company Web sites.

Although these major automobile manufacturers sell their products through a network of dealers, potential buyers and current owners can typically find a wealth of information at the company Web sites, including:

- industry and company news
- owners' information about vehicle care, accessories, and warranties
- vehicle images and descriptions
- tools to customize a specific vehicle
- tools to search for nearby dealers
- tools to search dealers' inventories for a specific vehicle
- online gift shops
- calendars of sponsored events

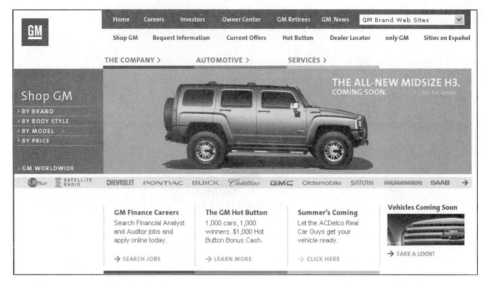

FIGURE 2-9 General Motors

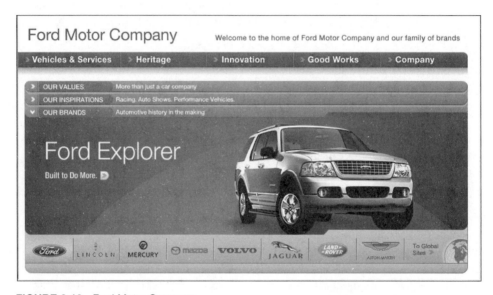

FIGURE 2-10 Ford Motor Company

But simply hosting a Web site is no guarantee that a company will successfully expand the market for its products and services. Market conditions, execution, and timing all play significant roles in this process. One major retailer that tried to harness the power of the Web to gain greater business visibility, but fell very short of the mark, is Federated Department Stores, a chain of brick-and-mortar department stores. In the late 1990s, Federated Department Stores bought Fingerhut Companies, a traditional catalog merchant with

FIGURE 2-11 Chrysler

an online store. Federated planned to leverage Fingerhut's 50 plus years of catalog retailing and logistics expertise, as well as the company's experience with operating an online store, by transforming itself into a major e-retailer and supplier of logistics services (warehousing and distribution) for other e-businesses.[35]

Unfortunately, Federated's timing, as well as the overall market conditions, couldn't have been worse. And, as you will see, the execution of Federated's e-business idea was poor. Shortly after Federated's purchase of Fingerhut in 1999, the U.S. economy weakened. Newly minted e-businesses that had exploded during the "dot.com boom" of the late 1990s were finding it more and more difficult to survive, resulting in a "dot.com bust." In the weakened economy, Fingerhut's online and catalog sales tanked. Fingerhut also had trouble meeting its commitments to its logistics clients, who left in droves.

A major part of the problem for Federated was that Fingerhut's core customer base consisted of low-income credit customers who were granted credit on a purchase-by-purchase basis. To spur sales, Fingerhut began offering revolving credit accounts to its customers—with predictable results. Fingerhut's customers began defaulting at record rates. Within three years, Federated "threw in the towel" and closed Fingerhut, selling its assets to Fingerhut's former CEO and another business partner. Curiously, Fingerhut—the subsidiary—fared better than expected. Reborn as Fingerhut Direct Marketing, Inc., Fingerhut is again open for business as an e-retailer and catalog merchant.[36]

You should consider how your e-business idea can use the Internet and the Web to gain greater business visibility, but always keep in mind that changing market conditions may impact how well you will be able to execute your idea. Next, consider how your e-business idea exploits the Internet and the Web to maximize your customer relationships.

Maximize Customer Relationships and Improve Responsiveness

The Internet is the ultimate tool for communication. E-businesses can use the Internet to build customer loyalty by staying in touch with their customers' needs, building one-on-one relationships, and providing information that enriches their customers' online experience. As consumers become increasingly in control in the new global economy, e-businesses must strive to improve their online responsiveness and maximize the benefits they can offer via the Internet in order to draw and keep customers. These days, e-businesses can create warm customer relationships through a variety of means such as using personalized e-mail, displaying welcome messages, keeping track of customers' interests, and delivering what the customers want. One company that uses its Web site to maximize customer benefits and maintain customer loyalty is Southwest Airlines.

Headquartered in Dallas, Texas, Southwest is famous for its "no frills," low-cost approach to air travel. The company understands that simplicity, low airfares, and customer service are what keep its customers coming back, and the company kept this in mind when developing its Web site. Recognized as the airline industry's leader in capitalizing on the Internet and the Web to boost sales and customer support, Southwest has a simple, easy-to-use Web site. With just a few clicks, a customer can review flight and fare options, book a low-cost flight, print out a boarding pass, book a car and hotel room, and check on flight status. Customers can also enroll in Southwest's Rapid Rewards frequent flyer program at the Web site.

Understanding exactly what its online customers want and then using the power of the Internet and the Web to deliver that is paying off for Southwest Airlines. Today, Southwest Airlines reports that more than 50 percent of its total passenger revenues, about $3 billion each year, are generated by customers booking their flights at Southwest.com (Figure 2-12).[37]

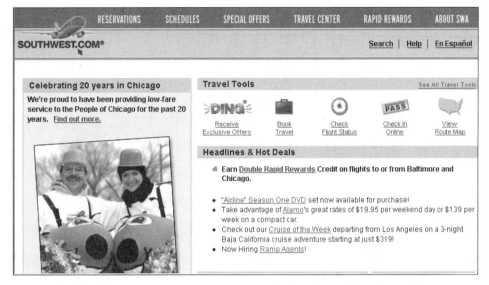

FIGURE 2-12 Southwest.com

E-CASE

A Technology Innovator's Excellent Idea

When W. W. Grainger, an 80-year-old, $5-billion-a-year distributor of electrical merchandise such as power tools, electric motors, and light bulbs, launched its first Web site in the mid-1990s, some analysts thought Grainger was an unlikely e-business candidate. Grainger, which maintained a comfortable lead over its competitors in the sales of machine maintenance and repair supplies, at first seemed to be exactly the kind of traditional intermediary many analysts thought the advent of e-business would destroy.

But Grainger had a history of being a technology innovator in its industry. Grainger's management team understood how a Web site would enable customers to quickly locate product and pricing information. Despite some early problems, Grainger's Web-based sales thrived. Orders placed at the Web site were twice as large on average as orders submitted via the traditional phone- or fax-based methods. Additionally, customers who ordered online were spending 20 percent more annually than they had by ordering via traditional methods.

Grainger continued to fine tune its Web site operations, and before long Forbes magazine named the Grainger site the "Best of the Web" in the "Indirect Procurement Category." B-to-B Magazine named the Grainger Web site to its NetMarketing 100 Top Business-to-Business Web Sites list. Today, instead of browsing through a huge 3,700-page catalog that lists 82,000 products, Grainger's online customers can quickly search for products, get immediate information on their availability, and get up-to-date prices customized for contractual discounts. Grainger reports that 15 percent of its sales are generated at its Web site or through its EDI trading partners (Figure 2-13).[38]

continued

Defining Your E-Business Idea

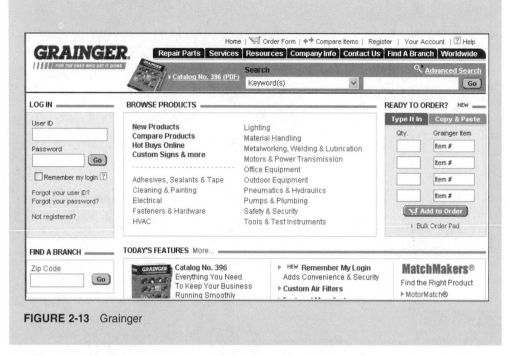

FIGURE 2-13 Grainger

The Internet and the Web create possibilities for all types of new products and services to be bought and sold. To operate a successful e-business, you must be open to the new types of products and services your e-business idea might offer.

Create New Products and Services

The Internet and the Web create an environment that enables existing companies to expand and refine their traditional businesses *and* one that promotes the emergence of new e-business ideas. The list of the new types of e-businesses made possible by the Internet and the Web goes on and on: business applications that can be accessed over the Web, server facilities that offer file backup and storage, firms that provide Web site hosting services to other firms, and more.

One interesting e-business idea is Online Dispute Resolution (ODR). Online Dispute Resolution involves using Internet and Web technologies together with dispute resolution technologies to facilitate dispute mediation, negotiation, and arbitration. A pioneer in the field of Online Dispute Resolution is Cybersettle, Inc.

The idea for Cybersettle was born in the mid-1990s when two New York attorneys, Charles Brofman and James Burchetta, were negotiating a routine auto accident claim. Instead of spending days, weeks, and months in negotiations, the two attorneys chose a simpler method. Each knew what the claim was worth to his client; they wrote that amount down on a piece of paper, and the papers were handed to a court clerk who was told that if the offers were within $1,000 of each other, the attorneys would split the difference and settle the claim. If the difference was greater than $1,000, the clerk was to inform them of this but not divulge the actual amounts, and the attorneys would try settling again. In the case of the auto accident, the claim was settled on the first try.[39] This speedy claim settlement inspired Brofman and Burchetta to think about how similar types of claim settlements could be facilitated using the Internet and the Web. Their idea became the e-business named Cybersettle (Figure 2-14).

Cybersettle uses patented computerized dispute resolution software, which it makes accessible to attorneys and claims professionals at its Web site. The dispute resolution software matches confidential offers and demands in up to three rounds. When the maximum offer is equal to or greater than the minimum demand, the dispute is settled for a monetary figure that is the average of the two—up to the amount of the maximum offer. Cybersettle charges the participants in the dispute settlement fees based on the amount of the settlement.

Cybersettle is now the Official and Exclusive Online Settlement Tool of The Association of Trial Lawyers of America (ATLA). With more than 9,000 claims professionals and 70,000 registered attorneys as clients, Cybersettle has been used to settle an assortment of disputes, ranging from those involving bodily injury to various types of insurance claims. The total value of its settlements is over $700 million.[40]

FIGURE 2-14 Cybersettle

E-CASE

A Brilliant Execution of a Simple Idea

For serial entrepreneur Josh Kopelman, a new e-business idea grew out of an attempt to buy a book online. Unwilling to pay a hefty price for a new book, Kopelman checked out eBay, the auction powerhouse, where he found several copies of the book he was looking for but no bids, which meant he would have to accept a seller's list price. Kopelman quickly realized that the auction model didn't work well for some items, such as used books. Why engage in the bidding process for a used book that could be purchased at any used bookstore for half price or at any flea market for pennies? Following his instincts, Kopelman created an online marketspace for buyers and sellers of used books, CDs, and movies. Since nothing was to be priced at more than half the original price, Kopelman named his online marketspace Half.com.[41]

What happened next is legendary in the annals of startup e-business history. Kopelman and the Half.com marketing team wanted a sure-fire way to get national attention and create greater business visibility for their new startup. They hit upon the idea of renaming a small town "Half.com." The Half.com team looked at several options, and then chose a small Oregon town named Halfway. After meeting with the town's mayor and talking with several of Halfway's residents, Half.com's marketing representative offered to donate computers to the city's elementary school, provide a raffle prize for the County Fair, and donate funds for civic improvements in exchange for the name change. The town accepted and issued a proclamation changing its name to Half.com, Oregon. In January, 2000, Half.com, Inc. launched its Web site during a live broadcast of NBC's Today Show from Half.com, Oregon and Conshohocken, Pennsylvania (Half.com, Inc.'s corporate headquarters), which garnered the company priceless publicity.[42]

At around the same time, Meg Whitman, eBay's CEO, was on the lookout for a way to expand eBay's market by offering fixed price items. Intrigued by what eBay buyers and sellers were saying about Half.com, Whitman took a look and liked what she saw—again! Recognizing that Half.com, Inc.'s business model was complementary to eBay's; she bought Half.com for more than $300 million.[43] After the acquisition, Kopelman worked at eBay for three years, running the Half.com by eBay operation (Figure 2-15). Then, true to his core interests as an entrepreneur, he left and co-founded another e-business startup, TurnTide, a company that develops anti-spam technologies.

continued

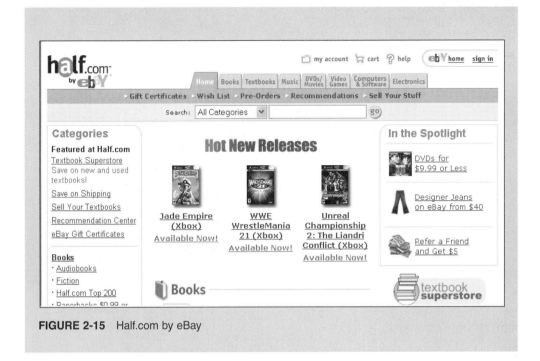

FIGURE 2-15 Half.com by eBay

Conducting business online not only opens up possibilities for new types of products and services, it also offers ways to reduce the costs associated with customer acquisition, transactions, and customer service after the sale.

Reduce Costs

One of the biggest benefits of doing business online is the potential to reduce the costs of running a business. Allowing customers to get quotes as well as place and track orders from a Web site can reduce an e-business' order handling and sales support costs.

Other cost savings can be achieved by providing customer support online. For example, technology companies such as Dell (computers, printers, electronics), Microsoft (computer software), and Cisco Systems (networking products) save many millions of dollars each year by enabling customers to access support services from their Web sites. By making their Web sites the starting point for customer support, Cisco Systems and Microsoft greatly reduce the number of calls to their technical assistance centers, which translates into major cost savings. Figures 2-16 and 2-17 show examples of technology company Web sites.

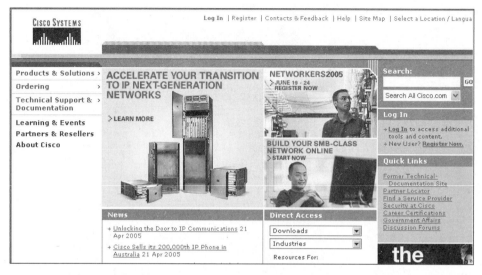

FIGURE 2-16 Cisco Systems

FIGURE 2-17 Microsoft

When evaluating your e-business idea, consider how it might exploit the numerous inherent advantages of doing business online. For example, how might using the Internet and Web help your business expand its market, gain greater business visibility, improve customer relationships, provide new services, and reduce the costs associated with maintaining customer relationships and processing transactions?

E-CASE IN PROGRESS

In the Beginning, the Idea

Dirk Elmendorf and Richard Yoo were students at Trinity University, in San Antonio, Texas in the late 1990s when they started an Internet consulting company named Cymitar Network Systems. Soon, Elmendorf and Yoo were joined by Patrick Condon, another Trinity classmate. The business took off, and it wasn't long before the trio realized that they needed to focus their attention and resources in one specific area: Web hosting. At that time, the Web hosting industry was in its infancy. Elmendorf, Yoo, and Condon believed the field was ripe for a Web hosting provider with a strong emphasis on customer support—and they were right. From that original idea—a Web hosting e-business that emphasized top-level customer support—came Rackspace Managed Hosting (Figure 2-18).

Today, the award-winning Rackspace Managed Hosting has five data centers, revenues approaching $100 million, and clients such as EMI Records, Miller Brewing, Best Buy, and National Geographic. Over the next few chapters, you will learn more about the Rackspace story and how a timely idea became a very successful e-business.[44, 45]

FIGURE 2-18 Rackspace Managed Hosting

. . . IDEAS! IDEAS!

In his free time, Hagar, the lawyer who'd been looking for a midnight snack, began testing his online grocery sales idea with business experts and advisors. He also did his homework. After studying various industries, Hagar came to the conclusion that an online grocery store could operate more efficiently than a regular grocery store because it would not have to deal with the fixed costs associated with running a brick-and-mortar enterprise. He also studied the habits of grocery shoppers and concluded that those shoppers who were already comfortable shopping online were ready for the convenience of online grocery shopping and home delivery. After many months spent doing research, planning, putting together a management team with grocery sales experience, and securing $48 million from private and institutional investors, GroceryWorks.com opened for business in the Dallas area and quickly expanded to Houston. Hagar then helped to engineer a merger between GroceryWorks.com and the grocery giant Safeway.

But GroceryWorks.com—along with other e-businesses that dealt in grocery delivery, such as Webvan and Peapod—struggled with the high cost of operating their businesses and their inability to attract as many customers as they had been expecting. Within a year of its merger with Safeway, GroceryWorks.com was forced to suspend its Texas operations. Ultimately, Safeway partnered with British grocery giant Tesco—a very successful online grocery provider in the U.K.—and resumed online grocery operations under the name Safeway.com. As of this writing, Safeway.com operates in Washington, Oregon, and California.[46, 47, 48, 49]

But what about Kelby Hagar? Ever the entrepreneur, Hagar left GroceryWorks.com after the merger with Safeway to pursue another business idea—surveillance systems for the restaurant industry. Hagar founded a second Dallas-area business, Digital Witness, which provides equipment and software for monitoring brick-and-mortar store operations (Figure 2-19). Digital Witness uses its Web site to promote sales and provide software downloads and technical support to its 400 plus clients, which include Outback Steakhouses, Luby's Cafeterias, and Quick Way stores.[50, 51] Through hard work and an unflagging entrepreneurial spirit, Hagar continues to capitalize on his e-business ideas.

continued

FIGURE 2-19 Digital Witness

Chapter Summary

- An entrepreneur is someone who assumes the risks associated with starting and running a business.

- Entrepreneurial abilities include leadership traits, a high-energy personality, self-confidence, organizational skills, and the ability to act quickly and decisively.

- The entrepreneurial process includes (1) deciding whether you are an entrepreneur, (2) deciding whether to buy or create a business, (3) planning the business, (4) operating the business, and (5) harvesting the business.

- The network effect means that the value of a network to its participants grows as the number of participants grows.

- The nature of the Internet and the Web encourages the development of innovative marketing ideas.

- Scalability refers to the ability of a business idea to continue to function well regardless of how large the business becomes.

- The cost of entry into the online marketspace can be very low for many e-businesses.

- Competitive barriers in the online marketspace include failure to secure first-mover advantage, the lack of name identification, and the lack of customer loyalty.

- The inherent advantages of doing business online include the potential to expand the market quickly, acquire greater business visibility, maximize customer relationships, create new services, and reduce costs.

Checklist

You and Your E-Business Idea:

- ❏ Do you have the energy, self-confidence, organizational skills, and ability to focus on objectives that are needed to start an e-business?

- ❏ Are you a risk-taker—that is, can you make informed but risky decisions and cope with the consequences?

- ❏ If you have to leave your current employment, do you have savings that could provide you with at least six months of personal living expenses? If not, how many hours of your time can you devote to starting your e-business?

- ❏ Does your e-business idea offer a unique or revamped product or service suitable for the online business environment?

- ❏ Does your e-business idea take advantage of positive network effects—in other words, would the value of the product or service you plan to offer increase through faster or wider distribution over the Internet and Web?

- ❏ Is your e-business idea scalable with rapid growth?

- ❏ Can your e-business idea be quickly adapted to accommodate changes in the marketplace and in technology?

❏ Does your e-business idea have a first-mover advantage or other competitive advantage in your market?

❏ Have you identified ways your e-business idea can exploit the inherent advantages of doing business online?

Key Terms

cash cow

entrepreneur

entrepreneurial abilities

entrepreneurial process

first-mover advantage

going public

name identification

network effect

scalability

sweat equity

viral marketing

Review Questions

True/False Questions

1. Low-cost Internet technology discourages the growth of e-businesses by denying easy entry into a market. True or False?

2. First-mover advantage alone guarantees the success of an e-business. True or False?

3. Many traditional businesses are expanding into e-businesses where they are far less restricted by the constraints of physical space and time. True or False?

4. Two advantages of operating an auction e-business are the low barrier to entry and that there is no need to maintain an inventory. True or False?

5. The network effect is the ability of an e-business to continue to function successfully regardless of how large the e-business grows. True or False?

Multiple Choice Questions

1. Which of the following may be a disadvantage to someone planning to start an e-business?

    a. high level of energy

    b. strong belief in the e-business idea

    c. inability to work extended hours most days

    d. support of family members and friends

2. The network effect is the:

    a. advantage of being the first to market.

    b. increased value to each participant as the number of participants increases.

    c. sending of unsolicited messages or advertisements.

    d. ability of an e-business idea to function well regardless of growth.

3. Which of the following factors is not important when evaluating an e-business idea?

   a. The idea provides a new, unique product.

   b. The idea exploits the network effect.

   c. The idea has a first-mover advantage.

   d. The idea is sure to make you famous.

4. Which of the following is not a useful trait for an entrepreneur?

   a. determination to succeed

   b. great organizational skills

   c. hesitancy to take risks

   d. ability to act quickly and decisively

5. Which of the following innovative marketing ideas secured early success for Hotmail?

   a. virtual marketing

   b. spam

   c. viral marketing

   d. scalable marketing

Exercises

1. Using online search tools or other relevant sources, research current e-business trends. Select two e-businesses that are pursuing new ideas, and write a one-page paper describing these ideas, including how each idea exploits the inherent advantages of doing business on the Web.

2. Using online search tools or other relevant sources, research the current status of two e-business examples from this chapter. Then write a one-page paper describing each e-business and its current position in the online marketplace.

3. Review a book from the "For Further Study" list at the end of this chapter or use online search tools to find articles about an e-business entrepreneur of your choice. Then write a one-page paper summarizing the entrepreneur's experiences in creating an e-business.

4. Think about the following successful and not-so-successful e-business ideas you learned about in this chapter: an online magazine, Web "sticky notes," paying people to browse the Web, the hosting of online auctions, and online dispute resolution. Then write a one-page paper describing the e-business factors that led to each idea's success or failure.

5. Using online search tools or other relevant sources, identify several e-businesses in different markets that use Internet and Web technologies to maximize their customer relationships. Then select one of the e-businesses and write a one-page paper describing the e-business and how it uses technology to maximize its customer relationships.

Case Projects

1. You have an idea for an online business but are concerned that you may not have the entre-preneurial abilities to develop the idea into a successful e-business. You would like to know more about what it takes to be an entrepreneur. Use online search tools, your library, and any other helpful resource to research entrepreneurship. Then write a one-page report explaining why you think you are or are not a potential entrepreneur, and whether or not you think you might start your own e-business.

2. You have decided that you are an entrepreneur and that you want to create a new e-business. Considering the current trends in e-business, the current status of the economy, recent changes in technology, and other factors, create a list of five new e-business ideas you might like to pursue.

3. You are the owner of Rob's Concierge Service, a small business that provides personal ser-vices, such as house sitting, dog walking, and deliveries, for busy professionals. You are con-sidering creating a Web site and conducting business online. Write a one-page paper describing how a Web site could help you exploit the inherent advantages of an online business—that is, how it might help you expand your market, create greater visibility for your business, and maximize your relationship with your customers.

Team Project

You and three of your friends love your Aunt Bessie's secret recipe for brownies. You all think it would be a great idea to sell the brownies and other chocolate treats to customers all over the world, and want to create an e-business to do so. Working together with three of your class-mates, use presentation software to create a 5–10 slide presentation that describes this e-business and outlines how the e-business could use the network effect, scalability, and inno-vative marketing to become successful. Make your presentation to a group of classmates who have been selected by your instructor to evaluate your e-business idea.

For Further Study

Here are some resources that might help you in further investigating the topics covered in this chapter.

Student Online Companion

Check out the *Creating a Winning E-Business*, *Second Edition* student online companion Web site for links to the sites discussed in this chapter and to other useful Web sites.

Articles and Books

Ashbrook, Tom. *The Leap: A Memoir of Love and Madness in the Internet Gold Rush.* New York, NY: Houghton Mifflin Company. 2000.

Cohen, Adam. *The Perfect Store*: *Inside eBay.* New York, NY: BackBay Books. 2003.

Dell, Michael and Fredman, Catherine. 2000. *Direct From Dell: Strategies that Revolutionized an Industry.* New York, NY: HarperBusiness.

Ericksen, Gregory K. *Net Entrepreneurs Only: 10 Entrepreneurs Tell the Stories of Their Success.* New York, NY: Wiley. 2000.

Malone, Michael S. *Betting it All: The Entrepreneurs of Technology.* New York, NY: Wiley. 2001.

Reiss, Bob, et al. *Low Risk, High Reward: Starting and Growing Your Business with Minimal Risk.* New York, NY: The Free Press. 2000.

Stevenson and Roberts, et al. *New Business Ventures and the Entrepreneur, 5th Edition.* New York, NY: McGraw-Hill Irwin. 1999.

End Notes

[1] *Fulfilment & Logistics.* "How an E-grocery Became the Darling of Dallas." www.elogmag.com/magazine/01/1-feature3.shtm. April 2000.

[2] *Digital Witness.* "About Us." www.digitalwitness.net/team_kh.html. 2005.

[3] Reiss, Bob, et al. *Low Risk, High Reward: Starting and Growing Your Business with Minimal Risk.* New York, NY: The Free Press. 2000.

[4] Stevenson and Roberts, et al. *New Business Ventures and the Entrepreneur, 5th ed.* New York, NY: McGraw-Hill Irwin. 1999.

[5] All Things Considered. "Failure." *National Public Radio.* www.npr.org/templates/story/story.php?storyId=1079633. July 17, 2000.

[6] Allis, Sam. "Making a Success Out of Failure." *The Boston Globe.* www.failuremag.com/news_boston_globe.html. July 14, 2000.

[7] Andres, Clay. "Born Loser, and Proud of It." *The Westchester County Times.* www.failuremag.com/_news_westchester.html. July 2000.

[8] *Failure Magazine.* "Fact Sheet." www.failuremag.com/fact.html. 2005.

[9] Williams, E. E. and Napier, H. Albert. *Preparing an Entrepreneurial Business Plan.* Chicago: T & NO Book Company. 2004.

[10] Fitzpatrick, Michele. "Twists, Turns, Triumphs Mark the Life of Start-Up Yesmail." *Chicago Tribune*, 8. *Chicago Tribune* Archives. www.chicagotribune.com/. July 10, 2000.

[11] Rose, Barbara. "Yesmail Swept Up in Online Consolidation Wave." *Chicago Business.* chicagobusiness.com/cgi-bin/mag/article.pl?article_id=13873&bt=David%20M.%20Tolmie&searchType=all. January 3, 2000.

[12] Little, Darnell. "CMGI In Accord With Web Marketer Yesmail.com Agrees to Sale; Stock Climbs." *Chicago Tribune*, *Chicago Tribune* Archives. www.chicagotribune.com/. December 16, 1999.

[13] Wikipedia. "Metcalfe's Law." en.wikipedia.org/wiki/Metcalfe's_law. May 10, 2005.

[14] Oakes, Chris. "Third Voice Rips Holes in Web." *Wired Magazine.* www.wired.com/news/print/0,1294,20636,00.html. July 9, 1999.

[15] Oakes, Chris. "The Web's New Graffiti?" *Wired Magazine.* www.wired.com/news/print/0,1294,20101,00.html. June 9, 1999.

[16] *Ebituaries: Home of Dead Dot-Coms.* "Third Voice Home Page." www.ebituaries.whirlycott.com/viewsite.php?site=90. November 6, 2001.

[17] *BusinessWeek Online*. "Could Anyone Have Thought Up Hotmail? Excerpts from *The Nudist on the Late Shift*." www.businessweek.com/smallbiz/news/coladvice/book/bk990903.htm. September 3, 1999.

[18] Bronson, Po. "HotMale: Sabeer Bhatia Started His Company on $300,000 and Sold It Two Years Later for $400 million. So, is He Lucky, or Great?" *Wired Magazine*. www.wired.com/wired/archive/6.12/hotmale.html. December 1998.

[19] *San Francisco Chronicle*. "Draper Fisher Jurvetson On the Record: Tim Draper." sfgate.com/chronicle/ontherecord/. March 13, 2005.

[20] Jurvetson, Steve and Draper, Tim. "Viral Marketing." Draper Fisher Jurvetson. www.dfj.com/cgi-bin/artman/publish/steve_may00.shtml. May 1, 2000.

[21] Microsoft Corporation. "MSN Hotmail Continues to Grow Faster Than Any Media Company in History." www.microsoft.com/presspass/features/1999/02-08hotmail.asp. February 8, 1999.

[22] Ransdell, Eric. "Network Effects." *Fast Company*, 208. www.fastcompany.com/magazine/27/neteffects.html. September 27, 1999.

[23] Kirby, Carrie. "Pay-to-Surf Not Paying Off for Web Sites." *San Francisco Chronicle*. www.sfgate.com/cgi-bin/article.cgi?file=/chronicle/archive/2000/07/12/BU90515.DTL. July 12, 2000.

[24] Gimein, Mark. "Meet the Dumbest Dot-Com in the World," *Fortune*, 46; 142(2). July 10, 2000.

[25] Rudi, Corey. "Profiting From Online Auction Sites." *Entrepreneur.com*. www.entrepreneur.com/article/0,4621,311268,00.html#. October 6, 2003.

[26] *Internet Retailer*. "Auto Parts Exceed $1 Billion in Sales at eBay." www.internetretailer.com/dailyNews.asp?id=11944. May 11, 2004.

[27] *Hoover's*. "eBay Inc. Fact Sheet." www.hoovers.com/ebay/--ID__56307--/free-co-factsheet.xhtml. 2005.

[28] Ibid.

[29] *Detroit News*. "Meg Whitman." www.detnews.com/menu/stories/32861.htm. January 22, 1996.

[30] *eBay*. "The Company Executive Team: Meg Whitman." pages.ebay.com/aboutebay/thecompany/executiveteam.html#Whitman. 2005.

[31] Krawcheck, Sallie. "Builders & Titans: Meg Whitman, A New Kind of Auction Hero." *Time Magazine*. www.time.com/time/2005/time100/. April 18, 2005.

[32] *Forbes.com*. "eBay's 'Theoretical Fair Value' Seen at $105." www.forbes.com/home/markets/2005/01/25/0125automarketscan13.html. January 15, 2005.

[33] *The Internet Archive Wayback Machine*. "Searched for www.ticketmaster.com." web.archive.org/web/*/ticketmaster.com. 2005.

[34] *Ticketmaster*. "About Us." www.ticketmaster.com/h/about_us.html?tm_link=tm_home_i_abouttm. 2005.

[35] Odell, Patricia. "Federated Gives Up." *DIRECT*. directmag.com/mag/marketing_federated_gives_2/. February 1, 2002.

[36] *Hoover's*. "Fingerhut Direct Marketing, Inc." www.hoovers.com/fingerhut/--ID__10827--/free-co-factsheet.xhtml. 2005.

37 Southwest Airlines New Releases. "Southwest Airlines Now Offers Internet Boarding Pass; Added Convenience From the Country's Most Successful Airline Web Site." *Southwest.com.* phx.corporate-ir.net/phoenix.zhtml?c=92562&p=irol-newsArticle&ID =531949&highlight=online%20sales. February 5, 2004.

38 *W. W. Grainger.* "Company Information." www.grainger.com/Grainger/static.jsp?page=about.html. 2005.

39 Springsteel, Ian. "E-Legal Activities." *CIO Magazine Case Files: E-Business Models.* www.cio. com/archive/031500/legal.html. March 15, 2000.

40 *Cybersettle.* "About Us." www.cybersettle.com/about/about.asp. 2005.

41 Cohen, Adam. "Less Hassle, By Half." *Time Magazine.* corp.half.com/pressfiles/time_ magazine_02_05_01/time_magazine_02_05_01.html. February 5, 2001.

42 Ibid.

43 Jhaveri, Amish. "Kopelman Offers Insights on Entrepreneurship." *The Wharton Journal,* The Wharton School University of Pennsylvania. www.whartonjournal.com/news/2005/01/24/News/ Kopelman.Offers.Insights.On.Entrepreneurship-838689.shtml. January 24, 2005.

44 Rackspace Managed Hosting Press Releases. "Rackspace Reports Rapid Growth and Continues Climb to $100 Million Mark." www.rackspace.com/aboutus/listings.php?hidelistings=1&detail=1176. September 30, 2004.

45 Eisner, Adam. "Rackspace Managed Hosting Off to Great Start in 2001." *Web Host Industry Review.* www.thewhir.com/features/rackspace.cfm. May 3, 2001.

46 Habal, Hala and Bounds, Jeff. "Suit: GroceryWorks Deteriorating," *Dallas Business Journal.* www.bizjournals.com/dallas/stories/2001/03/12/story1.html. March 12, 2001.

47 *Fulfilment & Logistics.* "Safeway.com Rides Again—With Tesco's Help." www.elogmag.com/magazine/17/safeway.shtml. February 2002.

48 Barrow, Becky. "Tesco to Click With Net Shoppers in US." *The Daily Telegraph.* millennium-debate.org/tel26jun012.htm. June 26, 2001.

49 Patsuris, Penelope. "E-Groceries Still Have a Shelf Life." *Forbes.com.* www.forbes.com/2001/07/10/0710webvan.html. July 10, 2001.

50 Hall, Cheryl. "Surveillance Firm Helps Restaurants Reduce Theft and Improve Service." *The Dallas Morning News.* www.dallasnews.com/sharedcontent/dws/bus/columnists/chall/stories/ 102404dnbushall.9f1ee.html. October 23, 2004.

51 *Digital Witness.* "About Us." www.digitalwitness.net/team_kh.html. 2005.

CREATING AN E-BUSINESS PLAN

FROM BRIDESMAID TO E-BUSINESS . . .

Jenny Lefcourt and Jessica DiLullo met at Stanford Business School in the late 1990s, where they were M.B.A. candidates. Lefcourt, who received her B.S.E. from the University of Pennsylvania's Wharton School, and DiLullo, who graduated with a B.A. in Economics from Stanford University, were experienced managers before beginning the M.B.A. program.

Lefcourt and DiLullo soon discovered two important facts about each other. They both wanted to create a startup business, and they shared the same business idea—an online bridal registry. They wanted to create an easy way for people to register for a broad selection of gifts and for givers to send those gifts.

continued

Lefcourt and DiLullo decided to create a business plan for their online bridal registry e-business idea and enter the plan in an entrepreneurship contest at Stanford. In the contest, a panel of judges reviewed business plans through several rounds, eventually awarding a prize of $25,000 in funding. Relying on their real-world experience as consumers (both had been bridesmaids many times, and DiLullo was engaged to be married), Lefcourt and DiLullo began to prepare their plan. Working out their ideas, the two of them would spend long hours in the Stanford library constantly asking each other, "Would you register [for gifts] that way?" If the answer was "No!" they went back to the drawing board. The hard work they put in to creating, reviewing, and revising their business plan paid off. After the first round of judges' reviews, Lefcourt and DiLullo pulled their business plan out of the competition—and for good reason![1, 2]

BUSINESS PLAN ORGANIZATION

In Chapter 2, you learned about entrepreneurship and how to evaluate an e-business idea. After you determine your interest and abilities as an entrepreneur and evaluate your e-business idea, the next step is to create a plan for your new e-business. Business planning is an ongoing process of setting goals for a business and then determining the strategies that would best accomplish those goals. A **business plan** is a formal business planning document that:

- identifies the business and its mission
- names the key players on the management team
- describes the products or services to be offered
- provides an analysis of the current marketplace in which the business will operate
- identifies customers and competitors
- determines the resources necessary for profitable operations
- sets a timetable for profitability

For a startup e-business, an initial business plan is used to evaluate the feasibility of the new e-business, to seek initial funding for operations, and to guide the management team in operating the e-business. Since business planning is an ongoing process, however, a business plan is always a "work-in-progress" to be updated and maintained over the entire life of the e-business.[3]

TIP

Strategic business planning—determining a company's long-term goals and then identifying the actions required to reach those goals—is an ongoing process for any existing business. For a startup e-business, creating a business plan is usually part of its strategic planning process. Therefore, business plans are sometimes called strategic plans.

Creating and developing a business plan forces you to take a sharp look at your e-business idea; the process requires a good deal of thought, time, and effort. Fortunately, a comprehensive business plan can help you not only find and exploit the hidden strengths of your e-business idea, but can also help identify and fix hidden weaknesses.

QUOTES ON SUCCESS

"It's no mystery that one of the essential keys of launching, operating, and growing a successful company is sound business planning. However, too many entrepreneurs carry critical planning records in their heads and never transfer the ideas to paper. This is one of the critical mistakes owners can make in running a business."

William Clark, former Executive Director of FastTrac™

You can get assistance in creating a business plan for your e-business idea from a number of non-profit, governmental, and educational Web sites, such as those of the Service Corps of Retired Executives (SCORE®) and the Small Business Administration (SBA). These and similar Web sites provide information about business planning and offer examples of formal business plan documents. While much of the information offered at these sites is directed toward brick-and-mortar businesses, most of the tips and examples are also appropriate for developing a business plan for an e-business. Other sites, such as the Canadian government's E-Future Centre site, provide business plan guides and tips directed specifically at e-business startups. Figures 3-1 through 3-3 provide examples of Web sites with business planning information.

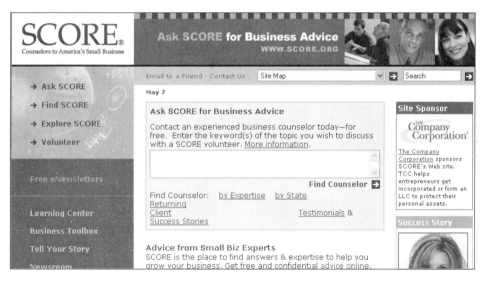

FIGURE 3-1 SCORE

FIGURE 3-2 SBA

FIGURE 3-3 E-Future Centre

Several e-business Web sites—such as BizPlanIt and Bplans.com—sell business plan preparation services and programs containing business plan tips and templates (Figures 3-4 and 3-5). They also offer free tips and business plan examples.

FIGURE 3-4 BizPlanIt

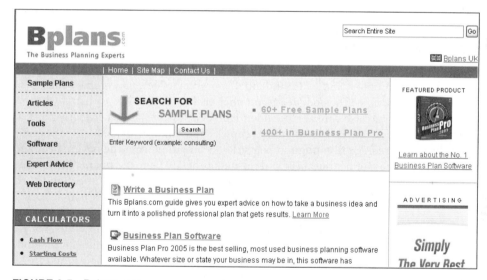

FIGURE 3-5 Bplans.com

Although there are no hard and fast rules for the arrangement of the components of a business plan, a plan usually includes some or all of the following items:[4]

- a cover sheet and a title page
- a table of contents
- an executive summary
- a description of the business

- a vision and/or mission statement
- information on products or services to be offered
- analyses of the e-business's overall industry, targeted customers, and competition
- operational, financial, and managerial plans for a three- to five-year planning period
- identification of critical risks
- an exit strategy

The arrangement of pages in a business plan and individual page formatting, such as headings and subheadings, are a matter of style and will vary from plan to plan. While it is very important to have a professional-looking and error-free business plan document, the real goal is to create a description that helps a reader clearly understand your startup e-business idea and when and how you expect it to become profitable.

QUOTES ON SUCCESS

"When I read a business plan, I look to find the answers to three questions. Is there really an opportunity here? Can these people pull it off? Will the cash flow?"

Dr. Ray Smilor, President of Beyster Institute at the Rady School at UC San Diego

Cover Sheet and Title Page

A business plan's **cover sheet** usually includes the title of the document, the preparer's name, the plan copy number, and a "Confidential" notation. When you prepare multiple copies of a business plan, it is a good idea to number each copy and keep a list of who has each copy. You should also consider confidentiality; each plan should be clearly marked as confidential, with a notation that additional copies should not be made.[5]

An inside cover sheet, or **title page**, repeats the information from the cover sheet and can include appropriate contact numbers and the names of key team members. The inside cover sheet may also include the name of the person to whom the copy of the plan is assigned. If you can afford it, you may consider adding a custom cover for a bound business plan.

The business plan examples in this chapter are for a fictitious startup e-business, WildCountryTreks.biz, a company that offers custom backcountry backpacking tours and wilderness survival training. Figure 3-6 and Figure 3-7 illustrate the cover sheet and the title page, respectively, for the WildCountryTreks.biz sample business plan.

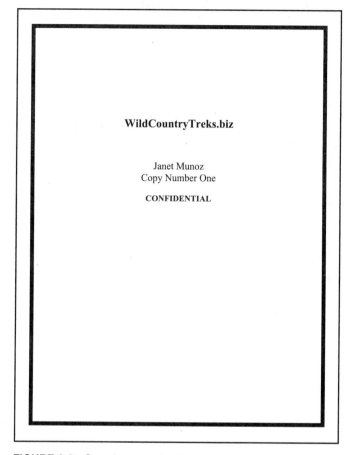

FIGURE 3-6 Sample cover sheet

WildCountryTreks.biz

Janet Munoz
Dave Winston
Richard Wang
(505) 555-2356
info@wildcountrytreks.biz
Copy Number One Assigned to Jason Elliott

CONFIDENTIAL
Do not make copies of this document

FIGURE 3-7 Sample title page

Table of Contents

A business plan's **table of contents** should list all the major sections and subsections of the business plan by page number, allowing the reader to locate key information quickly. It may be helpful to maintain a table of contents checklist as you prepare different sections of the plan. You can then review the checklist when preparing the final table of contents.

The final table of contents should not be prepared until the rest of the business plan is finished. Remember to double-check the final table of contents to look for common mistakes such as missing sections or subsections and incorrect page numbering. Obviously, a table of contents with errors will not help your readers locate information quickly, and will give them a negative feel for your business before they've even read your plan! Figure 3-8 illustrates the sample table of contents for the WildCountryTreks.biz business plan.

TABLE OF CONTENTS

FIGURE 3-8 Sample Table of Contents

E-CASE

An Evolving E-Business Plan

In the mid 1990s, three Chicago-area entrepreneurs—Christian Crone, Russ W. Rosenzweig, and Robert Hull—pooled together their business experience and resources to create a new e-business startup named Round Table Group. Their e-business idea was to create a virtual matchmaking service between academic experts and businesses that needed quick answers to very specific questions. The original Round Table Group business plan envisioned clients from Fortune 500 companies submitting simple, targeted questions via e-mail. Then the Round Table Group would forward these questions to its roster of pre-screened academic experts in the field to see which expert wanted to provide the answer. The expert who agreed to answer the question worked directly with the client company and set his or her own fees; Round Table Group would then charge the client company a "matchmaking fee."[6]

Crone, Rosenzweig, and Hull soon found that what the Round Table Group's business clients really wanted wasn't part of their original business plan. Instead of short, targeted answers to specific questions, clients wanted consulting help on projects that might last for months. In addition, many prospective clients wanted litigation consultation, another service area that was not part of the original business plan. The Round Table Group's management moved quickly to revamp its business plan and give its clients what they wanted. Today, Round Table Group matches more than 10,000 academic, industry, and government experts in several fields (including political and social policies, biotechnologies, environmental policies, and international trade policies) with clients needing consulting services. Round Table Group also provides expert services to attorneys and investment managers, and manages a corporate learning and speakers' bureau. The company's revenues have grown in each year of operation, and in both 2003 and 2004, *Inc. Magazine* named Round Table Group to its list of the 500 fastest-growing private U.S. companies (Figure 3-9).[7]

continued

FIGURE 3-9 Round Table Group

QUOTES ON SUCCESS

"About 70 percent of the work [client requests for services from the Round Table Group] is litigation consulting. We weren't even contemplating lawyers; they weren't part of the business plan."

Russ Rosenzweig, co-founder of Round Table Group

Executive Summary

An **executive summary** is a condensed version of your complete business plan, and is considered by many to be the most important section of the plan. It enables investors, bankers, and other interested parties to quickly read and understand your e-business idea without having to wade through the entire business plan document. Typically, investors, bankers, and other sources of funding have more business plans and proposals than they have time to evaluate; therefore, they use the executive summary page of business plans to quickly weed out those plans in which they have no interest.[8]

QUOTES ON SUCCESS

"A business plan needs to get to the point quickly since most investors will make a decision within 10 seconds whether to contact the entrepreneur or not. One of the biggest turn-offs to an investor is a paragraph that starts with 'The Internet is growing at a tremendous rate.' You need to get to the beef of the message and get to it quickly."

Bill Reichert, venture capitalist and Managing Director of Garage Technology Ventures

Keep your executive summary short—one page is best, but if a longer summary is required, keep it to no more than three pages. A reader should be able to scan your executive summary quickly and understand how your e-business idea works. Use clear and concise language to describe your e-business idea, and use action words to get the reader's attention and generate excitement. Emphasize the reasons why your e-business idea will be successful, and support these reasons using concrete facts and descriptions. You will probably want to include brief information on:

- your staff and management team
- the marketplace in which your e-business will operate
- any competitive advantages your e-business will enjoy
- a summary of financial projections

For best results, save writing the executive summary until you have worked out the details in the remainder of your business plan. This will enable you to summarize the plan's key points more effectively. Figure 3-10 depicts the Executive Summary for the WildCountryTreks.biz business plan.

You can use either a bulleted list style or narrative paragraphs in your executive summary. You can also choose to position your executive summary either before or after the table of contents, but remember: no matter which style you use or where you position your executive summary in relation to the table of contents, your executive summary is the first—and perhaps only—part of your business plan that most investors or other interested parties will read.[9]

Vision and Mission Statements

Some business plans include a vision and/or mission statement. Although closely related, a vision statement and a mission statement are often defined differently. A **vision statement** can be defined as a formal statement of a business's hopes and desires, and it is written to inspire, guide, and encourage its employees toward achieving the business's long-term goals. A **mission statement** is a formal statement that explains—in a few words—the overall purpose of a business.

Executive Summary

WildCountryTreks.biz (WCT) begins online operations in January, 2007 to offer customized wilderness backcountry backpacking tours to clients worldwide. WCT also offers wilderness survival training to individuals or groups sponsored by corporations and government agencies.

WCT's projected total sales for the first year of operations are $1,500,000. Sales are projected to increase 15% annually for the next three to five years. WCT expects to break even in the first year and be profitable in the second and remaining years.

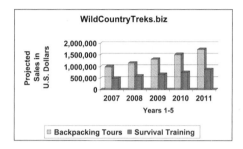

The market for both customized wilderness backpacking tours and wilderness survival training is growing at a rate of 15 percent annually. WCT has only two major competitors in the customized backcountry wilderness backpacking tour and wilderness survival training marketplace. Because each of these major competitors has only a 20 percent share of the market, WCT believes the marketplace is underserved. Our total focus on flexibility in meeting client needs and providing ongoing client service gives WCT a competitive advantage in this marketplace.

WCT's potential clients are 1) adventure-oriented individuals, 2) groups interested in backpacking tours to rugged, out-of-the-way, and unspoiled locations, and 3) individuals and groups for whom wilderness survival training is critical. With its roster of 200 highly experienced professional tour guides and survival trainers, WCT is prepared to offer specialized services to each of these three potential client markets.

With a top management team and advisory board, an easy-to-use Web site, and a strong marketing plan, WCT is poised to exploit the wilderness backpacking tours and survival training marketplace.

FIGURE 3-10 Sample Executive Summary

QUOTES ON SUCCESS

"Good business leaders create a vision, articulate the vision, passionately own the vision, and relentlessly drive it to completion."

Jack Welch, former Chairman and CEO of General Electric

Many businesses have vision and/or mission statements, but define them in other ways, such as "corporate identity statements" or "statements of goals, values, and objectives." The terms "purpose," "aspirations," "ideals," and "direction," along with other similar phrases, are often used interchangeably with the terms "vision" and "mission." In this text, we use the term mission statement to broadly define statements about a company's philosophy and goals.

Before you attempt to write a mission statement for your startup e-business, you should examine published examples of various mission statements to see how they reflect company philosophies. Many businesses and other organizations, such as those noted below, publish their mission statements on their Web sites, for example:

> "Google's mission is to organize the world's information and make it universally accessible and useful."
> *Google*

> "At Microsoft, we work to help people and businesses throughout the world realize their full potential. This is our mission. Everything we do reflects this mission and the values that make it possible."
> *Microsoft® Corporation*

> "To grow a profitable airline, that people love to fly and where people love to work."
> *Virgin Atlantic Airways Ltd.*

> "To provide services for students and other writers in the University community."
> *University of Miami College of Arts and Sciences, Writing Center*

When writing a mission statement, you can choose to target a single audience or a combined audience made up of employees, customers, competitors, shareholders, the government, the press, and the general public. You can also choose to focus your mission statement on a specific theme such as customer support, quality, leadership, or innovation.

Crafting an effective mission statement that makes your e-business's purpose clear to others might take time. You will probably need many drafts, as it may not be easy to identify the few meaningful words that clearly express your message to your targeted audience. Thinking very carefully about your e-business idea, goals, and objectives before you begin is critical to drafting an effective mission statement. You might also find that it is helpful to brainstorm with interested others—family members, friends, mentors—as you search for just the right words for your mission statement.[10]

TIP

Mission statements that are too wordy and too general can be meaningless. For funny examples of how not to write a mission statement, check out the Mission Statement Generator in the Games section at Dilbert.com.

E-CASE IN PROGRESS

Rackspace Managed Hosting

As e-businesses evolve, their mission statements will probably also evolve. For example, here is the original mission statement for Rackspace Managed Hosting, the managed hosting provider introduced in Chapter 2:

"To profitably enable e-businesses around the world to succeed by providing a scalable, productized, outsourced, internet server platform for their businesses."

In the beginning, Rackspace simply rented Web servers to its clients—providing no other hosting services. This arrangement was similar to that of a traveler arriving at the airport and renting a car. After renting the car, the traveler is completely responsible for everything else—getting directions to his or her destination, driving the car, parking the car, and returning the car after the trip.

Today, Rackspace provides its clients with all the Web hosting services they need, and thus the business now functions more like a limousine or taxi service that provides the traveler with a complete transportation package—vehicle, driver, pick up at the curb, and drop off at the destination. Rackspace has transformed itself from renting servers to running the servers and being a world-class service provider.

Just as Rackspace's e-business has evolved, so has its mission statement. Rackspace's current mission statement reflects this evolution from equipment renter to service provider:

"To be recognized as one of the world's great service companies along with names like Lexus, Nordstrom, The Ritz-Carlton and _____."

The blank line is deliberate—indicating Rackspace's long-term goal of adding its name to this list of great service companies.

Business Description

The **business description** portion of your business plan should also be brief and include only pertinent information. It should provide the reader with a summary of your e-business's background and business idea. You may also include information about the legal form of your e-business, when and where your e-business was formed, its history to date, key personnel, and future goals. Figure 3-11 illustrates a portion of a sample Business Description section.

> **Business Description**
>
> WildCountryTreks.biz (WCT) is a privately-owned provider of guided backpacking tours and wilderness survival training. Created in September, 2006 as a sole proprietorship, WCT matches individuals and groups seeking adventure travel to unspoiled jungle, desert, and mountain wilderness locations with professional backcountry guides. WCT also provides wilderness survival training programs for government- or corporate-sponsored individuals.
>
> **Our mission**:
>
> *To share the joy of unspoiled wilderness with people who love exciting adventures.*
>
> **Management team**:
> Janet Munoz, President
> Dave Winston, Vice President, Tours
> Colonel Richard Wang, U.S. Army (Ret.), Vice President, Training
> Judith Estes, Call Center Manager
> Leslie Astor, Web Site Operations
> Lynn Washington, Administrative Manager
>
> **Contact information**:
> WildCountryTreks.biz
> 302 West Sunset Drive
> Santa Fe, New Mexico 87503
> (505) 555-2356
> E-mail address: info@wildcountrytreks.biz
> Web site: www.wildcountrytreks.biz

FIGURE 3-11   Portion of a sample Business Description section

Products or Services

Your business plan's **products** or **services section** is one of the most important sections in the plan. In this section, you provide a description of the products or services your e-business offers and the anticipated sales that each product or service is expected to generate. When developing this section, be sure to provide enough detail about each product or service to enhance understanding, but not so much that a reader might be confused.

If your e-business sells products, you should describe each product in terms that make clear the product's benefits to potential customers. You might decide that high-quality graphics would be helpful in promoting your products. If so, you can include them in your business plan's appendix or in a separate catalog. In either case, you would then add a statement in the products section that refers the reader to the appendix or states that a product catalog is available upon request. If your e-business is selling a service, describe the service: what it is, how it works, what makes it unique, and how potential customers can benefit from it. Figure 3-12 illustrates a sample Products or Services section.

> **Services**
>
> WildCountryTreks.biz (WCT) is focused on two business areas: guided backpacking tours and wilderness survival training.
>
> **Guided Backpacking Tours**
>
> WCT has a roster of over 200 professional backcountry wilderness tour guides located around the world. Professional backcountry guides can register to become part of the WCT roster at our Web site. However, all guides are thoroughly vetted before being added to our roster. A typical WCT backcountry guide has more than five years experience in desert, mountain, or jungle wilderness backpacking; is certified by the Worldwide Backcountry Guides Association; and holds one or more first aid certifications.
>
> All backpacking tour packages are customized to each client's specifications and include roundtrip airfare and ground transportation to and from the tour's starting point. Food and all necessary equipment are also provided as part of the tour package.
>
> Because each backpacking tour is customized to the client's specific needs, tour package prices vary. Factors built into the pricing include backpacking location, duration of the tour, food, equipment rental, and tour guide fees. WCT adds a 15 percent service fee to the total cost of each tour.
>
> **Wilderness Survival Training**
>
> Each wilderness survival training course is designed to meet the client's specific needs for desert, mountain, or jungle survival training. WCT charges the client a fee for developing the course curriculum and for providing the course trainer. The typical WCT wilderness survival trainer is an ex-Navy Seal or ex-Army Ranger or ex-Army Special Forces member. Fees vary according to the length of the course and the depth of the curriculum. The client is responsible for all costs related to the location, transportation, equipment, and supplies need for the course.

FIGURE 3-12   Portion of a sample Products or Services Section

Marketplace Analysis

A business plan's **marketplace analysis** should include information about the specific industry of which your e-business is a part. It should also include a description of your e-business's targeted customers and competitors. You can choose whether to include an overview of the marketing and sales strategies you will employ to reach your targeted customers.

Industry Information

An **industry** includes all the businesses that make or sell similar or complementary products or services. For example, airlines, hotels, and travel agencies are all part of the travel industry. Auto dealers, tire retailers, and auto maintenance businesses are all part of the auto industry. Computer hardware manufacturers, software developers, and silicon chip manufacturers are all part of the computer industry. The industry information in your business plan can include industry size, marketplace trends, technology trends, anticipated labor shortages or other personnel issues, changes in government regulations, or other relevant topics that might affect the success of your e-business.

When putting together your industry information, be sure to use verifiable data that has been gathered from recognized sources such as government agencies, industry trade associations, or qualified studies from reliable organizations. Use charts and graphs to enhance the reader's understanding of the industry data you include. Charts and graphs should be large enough to be readable, but not so large that they detract from the overall format of the plan.[11]

Targeted Customers

In addition to providing industry information, the marketplace analysis portion of your business plan should identify your e-business's potential customers, changes in consumer preferences, shifts in consumer demographics, and new consumer-oriented technologies or other advantages your e-business can exploit to attract customers. Your plan should also define the characteristics of your targeted customers that will affect your e-business's success, including age, gender, income level, education, social class and lifestyle, geographic location, and buying habits. The targeted customers portion of your marketplace analysis should clearly explain how your products or services are a good fit with the characteristics of your targeted customers. Some would argue that this section is the most important part of the marketplace analysis.

Competitors

The marketplace analysis section of your business plan should emphasize your e-business's specific competitive advantages and strengths. It must also present a careful analysis of competitive issues—number of competitors, ease of entry into the marketplace, and so forth—that will affect your e-business's success. Each of your major competitors should be identified by name, and you should include a summary of each competitor's products and marketplace strengths and weaknesses. Charts depicting various competitors' market share can be helpful.

Marketing and Sales Strategies

The marketplace analysis section should also provide an overview of your e-business's major marketing objectives and strategies. You should describe in detail the steps required to accomplish a business's marketing objectives—in other words, strategies and tactics that lead directly to sales and cash flow. Your marketing strategies section should therefore give details about the following:

- the features and benefits of the products or services your e-business offers
- prices and pricing strategies
- how the products or services will be promoted
- how the products or services will be distributed

Your business plan might also contain a detailed marketing plan and budget. You will learn more about how to create a detailed marketing plan in Chapter 6. Figure 3-13 illustrates a sample portion of a Marketplace Analysis.

Marketplace Analysis

According to the travel industry research company *AdventureTravelStats*, the market for customized backcountry wilderness backpacking tours is growing at a rate of 15 percent annually. The reasons for this growth include:

- Growing interest among all age groups in protecting and savoring the natural world.
- Increasingly affluent worldwide travel client base.
- Rising expectations for more adventure and excitement in vacation travel.

Potential Client Base for Wilderness Backpacking Tours

WCT's potential client base for its backcountry wilderness backpacking tours are adventure-oriented individuals and groups interested in backpacking tours to rugged, out-of-the-way, and unspoiled locations. According to *BackPacker Today* magazine, the typical clients for WCT's backcountry wilderness backpacking tour services are affluent and well-educated males or females between the ages of 21 and 55 who live in major metropolitan areas around the world.

Total Market by Gender	Total Market by Annual Income

Male Female	$50,000 $50-100,000 $100,000+

Potential Client Base for Wilderness Survival Training

The uncertain worldwide political climate has led to more and more corporations and government agencies being interested in ensuring that their employees who travel on business are prepared for unexpected events, including the need to survive in a desert, mountain, or jungle environment until rescued.

FIGURE 3-13   Portion of a sample Marketplace Analysis

Operational Plan

Traditionally, the **operational plan** section of a business plan describes a business's physical location and equipment, as well as the manufacturing or service actions necessary to get the business's products and services to customers. Business location information can include headquarters location, branch office locations, warehouse space, and manufacturing space. Any competitive advantages inherent in the business's locations should be noted in the operational plan section, and a layout of the facilities should be provided in the appendix. An overview of significant vehicle, computer, office equipment, or manufacturing equipment needs—including purchase or lease costs—should also be supplied.[12]

An e-business's operational plan should incorporate these traditional elements where applicable. But an e-business operational plan should also focus on the operations that are critical to the specific e-business model or models being followed. For example, if the e-business is a B2C online store selling physical products, then information about product distribution, such as the location of distribution centers or the need for a fleet of specially designed delivery trucks, should be included. In contrast, an e-business that sells a service and thus requires a large customer support staff, might emphasize call center locations and the availability of a labor pool (from which to hire customer support personnel) near those locations.

An e-business's operational plan should also include information on how the e-business's Web site is integrated into overall operations. Adding a diagram to the appendix of the Web site elements and a description of each element can help readers better understand the Web site operations. You can also include a screenshot of the e-business Web site and information about the internal or outsourced personnel needed to host, create, and maintain the Web site. Summary information about any special technologies, hardware, and software you will use—and what, if any, advantages this use gives you in the marketplace—can also be included. Figure 3-14 illustrates a portion of a sample Operational plan.

Operational Plan

WildCountryTreks.biz (WCT) clients contact us using Web-based application forms or through our call center. Client inquiries are immediately assigned to an account executive: a backpacking tour or wilderness survival specialist, as appropriate. The account executive immediately contacts the client to discuss the client's needs in more detail.

Each WCT account executive is responsible for:

- All client support activities before, during, and after the sale
- Matching the client with the appropriate backpacking guide or trainer
- Coordinating all aspects of a backpacking tour with the tour guide and the client, including air and ground transportation, food and equipment, and other client requests
- Coordinating wilderness survival curriculum development with the trainer and the client

Web Site Operations

WCT is outsourcing its Web site development to Creative Webs, Inc., a Santa Fe, New Mexico-based Web developer. See Exhibit A for a mockup of several Web pages and our Web-based forms. WCT is outsourcing its Web site hosting to Rackspace Managed Hosting, a San Antonio, Texas-based managed hosting provider.

FIGURE 3-14   Portion of a sample Operational Plan

Financial Plan

The **financial plan** section of your business plan explains how your e-business idea, goals, and strategies translate into profits. It should also show when your e-business will hit the break-even point after initial funding or investment. You may choose to include a number of financial documents in your plan, but there are three items that your financial plan section must include: a pro forma balance sheet, a projected income statement, and a planned cash flow statement.

Pro forma (or projected) financial statements include income, expense, and cash flow data based on current assumptions of future events. A **pro forma balance sheet** indicates the value of the e-business's assets, liabilities, and ownership or equity at a specific point in time (Figure 3-15). A **projected income statement** shows the e-business's anticipated revenues, expenses, and net profit or loss over a specific period of time (Figure 3-16). A **planned cash flow statement** illustrates what cash the e-business plans to receive over a specific period of time and how it will spend that cash (Figure 3-17). A business plan generally includes monthly financial statement projections for 12–24 months in addition to the three-to-five year summary statements shown in Figures 3-15 through 3-17.

Pro Forma Balance Sheet

	FY 2007	FY 2008	FY 2009	FY 2010	FY 2011
Assets					
Short-term Assets					
Cash	$ -	$ -	$ -	$ -	$ -
Accounts Receivable	-	-	-	-	-
Other Short-term assets	-	-	-	-	-
Total Short-term Assets	$ -	$ -	$ -	$ -	$ -
Fixed Assets	$ -	$ -	$ -	$ -	$ -
Less Accumulated Depreciation	-	-	-	-	-
Total Long-term Assets	-	-	-	-	-
Total Assets	$ -	$ -	$ -	$ -	$ -
Liabilities and Equity					
Accounts Payable	$ -	$ -	$ -	$ -	$ -
Other Short-term Liabilities	-	-	-	-	-
Long-term Liabilities	-	-	-	-	-
Total Liabilities	-	-	-	-	-
Equity	-	-	-	-	-
Total Liabilities and Equity	$ -	$ -	$ -	$ -	$ -

FIGURE 3-15 Pro Forma Balance Sheet example

Projected Income Statement

	2007	2008	2009	2010	2011
Sales					
Backpacking Tours	$ -	$ -	$ -	$ -	$ -
Wilderness Survival Training	-	-	-	-	-
Total Sales	$ -	$ -	$ -	$ -	$ -
Direct Sales Costs					
Tour Guide Fees	$ -	$ -	$ -	$ -	$ -
Tour Food and Equipment Costs	-	-	-	-	-
Survival Trainer Fees	-	-	-	-	-
Curriculum Development Fees	-	-	-	-	-
Total Direct Sales Costs	$ -	$ -	$ -	$ -	$ -
Gross Margin	$ -	$ -	$ -	$ -	$ -
Gross Margin %	%	%	%	%	%
Operating Expenses					
Promotion and Advertising	$ -	$ -	$ -	$ -	$ -
Web Site Hosting and Operations	-	-	-	-	-
Employee Expenses	-	-	-	-	-
Rent and Utilities	-	-	-	-	-
Insurance	-	-	-	-	-
Travel	-	-	-	-	-
Miscellaneous	-	-	-	-	-
Total Operating Expenses	$ -	$ -	$ -	$ -	$ -
Profit Before Interest and Taxes	$ -	$ -	$ -	$ -	$ -
Interest Expense	-	-	-	-	-
Taxes	-	-	-	-	-
Net Profit	$ -	$ -	$ -	$ -	$ -
% of Total Sales	%	%	%	%	%
Per employee	$ -	$ -	$ -	$ -	$ -

FIGURE 3-16   Projected Income Statement example

Planned Cash Flow Statement

	12/31/2007	12/31/2008	12/31/2009	12/31/2010	12/31/2011
Cash Inflow					
Cash on Hand	$ -	$ -	$ -	$ -	$ -
Cash from customers	-	-	-	-	-
Other cash	-	-	-	-	-
Cash Inflow	$ -	$ -	$ -	$ -	$ -
Cash Outflow					
Expenses	$ -	$ -	$ -	$ -	$ -
Asset Purchases	-	-	-	-	-
Estimated Tax Payments	-	-	-	-	-
Other	-	-	-	-	-
Cash on Hand 12/31/XX	-	-	-	-	-
Cash Outflow	$ -	$ -	$ -	$ -	$ -

FIGURE 3-17   Planned Cash Flow Statement example

A financial plan may also include a statement describing the financial assumptions used to generate the numbers used in the plan, a break-even analysis (Figure 3-18) that identifies the amount of sales needed to cover fixed and variable expenses, and a standard financial ratios analysis (Figure 3-19) illustrating how well the e-business will perform compared to other companies in the same industry. You may include an investment proposal, some-times called a "deal plan," that lays out to investors how much capital you need and in what form, the anticipated return on investment (ROI) calculations, and a timeline for profitability. If your e-business has a financial history, you may also include historical financial statements.

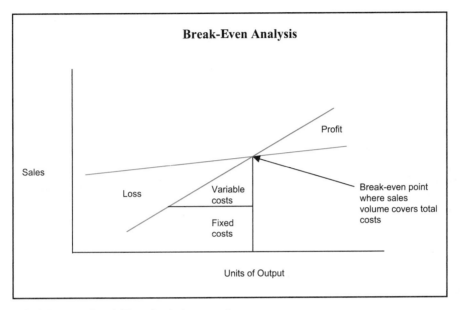

FIGURE 3-18   Break-Even Analysis example

Ratio Analysis

	FY 2007	FY 2008	FY 2009	FY 2010	FY 2011
Profitability Ratios					
Gross Margin					
Net Profit Margin					
Return on Assets					
Return on Equity					
Liquidity Ratios					
Current Ratio					
Quick Ratio					
Solvency Ratios					
Debt to Equity					
Debt to Assets					
Efficiency Ratios					
A/R Turnover					
Collection Days					
A/P Turnover					
Total Asset Turnover					
Other Ratios					
Assets to Sales					
Acid Test Ratio					
Asset Turnover					
Sales to Equity					
Sales per Employee					

FIGURE 3-19   Ratio Analysis example

Management Plan

Because many investors base their investment decisions on the strength of the management team behind a new venture, a strong management team is critical for a startup e-business. The management plan section of your business plan should therefore focus on your management team, including any outside advisors, mentors, and consultants. This section should provide information about the people on the management team who are or who will be involved in the day-to-day operations of your e-business. Typically, the information about team members is organized into brief narrative descriptions of each person's title, duties and responsibilities, previous related experience, previous successes, and education. You may, in addition, choose to add a detailed resume for each top team member in the business plan's appendices.

Part of the planning process for any startup business is determining what work will be done in-house by employees and what work will be outsourced to professionals outside your company. Outsourcing allows you and your management team to concentrate on your e-business's core activities, while other professionals focus on areas in which you and your team lack experience or skills. Examples of activities that businesses outsource include:

- getting advice from an attorney on the legal form of the e-business, the content of the business plan, the wording of confidentiality agreements, and contractual obligations
- having a professional accountant prepare or review your pro forma financial statements
- hiring technology professionals to create, maintain, and host your e-business Web site

Outside advisors can be very important to a startup e-business because they add experience and background that might be missing from the management team. An impressive **Advisory Board** of experienced outside advisors can add credibility to an e-business startup. If your e-business has an Advisory Board, be certain to include the members' names and other pertinent information. Figures 3-20 and 3-21 show examples of the types of information you can include in your Management Plan section.

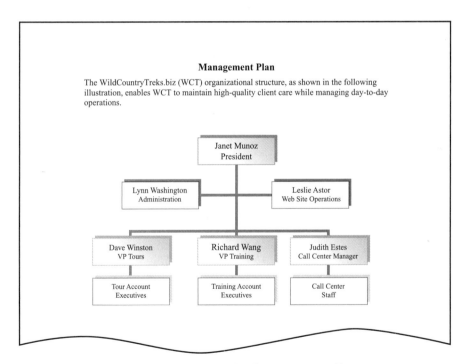

Management Plan

The WildCountryTreks.biz (WCT) organizational structure, as shown in the following illustration, enables WCT to maintain high-quality client care while managing day-to-day operations.

FIGURE 3-20 Management Plan organizational structure example

Management Plan

The WCT Management Team

Janet Munoz	Janet is a graduate of the Wharton School of the University of Pennsylvania. She also holds a Bachelor of Science degree in Information Technology from the University of New Mexico, where she graduated with honors. Janet is an avid backpacker and a charter member of the Worldwide Backcountry Guides Association.
Dave Winston	Dave has more than 15 years' experience in the travel tour industry, including positions with American Airlines, American Express Travel, and Travelocity. Dave is a graduate of the University of Illinois with a Bachelor of Business Administration degree.
Richard Wang	Richard recently retired from the U.S. Army after 25 years as an Army Ranger. His last assignment was at the Combined Forces Survival College, where he directed curriculum development and training. Richard is a graduate of the United States Military Academy at West Point.
Judith Estes	Judith is a graduate of the University of New Mexico with a degree in Business Administration. As a student, Judith worked for various call centers in the area. Most recently, Judith was manager of a local technology company's call center.
Leslie Astor	Leslie graduated from the New Mexico Technical College in with an Associate Degree in Web Site Operations and Web Design. Leslie will oversee all outsourced Web site design and operations.
Lynn Washington	Lynn has been an administrative manager for more than 10 years, most recently with a local Santa Fe art gallery. Lynn holds an Associate Degree in Accounting from New Mexico Technical College.

The WCT Advisory Board

James Delmar
Vice President World Travel Group

Rickie Rodriguez, Certified Public Accountant
Rodriguez and Capshaw, LLP

Avery Jamison, Attorney at Law
Jamison and Jamison, LLP

FIGURE 3-21 Management Team and Advisory Board example

E-CASE IN PROGRESS

Rackspace Managed Hosting

Because business planning is an ongoing process, the documents generated from the business planning process are dynamic, changing as new opportunities, threats, and business needs arise. In the late 1990s, when Elmendorf, Yoo, and Condon—the three original Rackspace Managed Hosting founders—first pitched their e-business idea to Graham Weston and Morris Miller, two successful San Antonio-based entrepreneurs, the trio had not yet created a formal business plan, just a spreadsheet showing potential profitability.

The spreadsheet analysis illustrated how Rackspace could successfully and profitably follow a B2B subscription e-business model. The analysis also indicated that the payback period for providing a client with a Web server was only five months. In addition to these promising projections, Weston and Morris—like the original founders—were convinced that a focus on customer service and support would keep customer attrition low and encourage new customer growth. Low customer attrition, new customer growth, and the five-month payback per server projection made the Web hosting e-business idea very attractive—so attractive that Weston and Morris agreed to provide seed money for Rackspace and to add their entrepreneurial experience to the Rackspace management team.

Fortunately, the Rackspace management team's expectations that the company would experience low customer attrition and new customer growth panned out. Before long, it became apparent that Rackspace would need to raise additional funds to support its rapid growth. But before the Rackspace management team approached venture capitalists about investing in the company, the team took the time to prepare a more formal business planning document with components similar to those illustrated in this chapter.

Issues Analysis and Critical Risks

An **issues analysis** or **risk assessment** identifies both threats and opportunities an e-business encounters from outside influences such as economic changes, impending product innovations, technological advancements, environmental issues, government regulations, and any barriers to entry into the market. The analysis should take into account all potential losses as well as any legal factors, staffing concerns, changes in competitor market share, new technologies, or other issues that could have a negative impact on the e-business. The analysis should also describe the business's internal strengths or weaknesses, such as a lack of managerial expertise among the company's principals. After you have determined which issues are most significant, you must prepare a one-page issues statement that lists these potential problems and the contingency plans you have developed to resolve them.[13]

Exit Strategies

Like an issues analysis, a realistic **exit strategy** provides owners and investors with some understanding about how they will get their money back from a new venture; it is an essential feature of a business plan. Your business plan should in fact include a handful of exit strategies that your e-business can execute in order to enable your investors (and you) to recover their business investments. Some possible exit strategies that your e-business's management team might consider include going public with an IPO (initial public offering), being acquired by another company, selling the company (or any part of it) to other individuals, and merging with an existing company to form a new company.[14]

Appendices

The appendices to a business plan can include items that provide additional details to the plan, such as Web site schematics and Web page descriptions; pictures of products or details about services; marketing materials; management team resumes; details about order fulfillment locations and equipment; copies of relevant legal documents, such as leases, contracts, patents, partnership agreements, incorporation agreements, and licenses; and any other documentation necessary to support the information included in the other sections of the plan.

In addition to describing an e-business's goals and operations, a business plan should also identify the way the e-business is legally organized.

LEGAL FORMS OF ORGANIZATION

One of the critical issues in the planning process for an e-business startup is determining the legal form of organization the new business should adopt. This legal form is noted in a business plan, often in the business description section, and is supported by documentation in the appendices. The most common legal forms of organization that businesses adopt are a sole proprietorship, a partnership, and a corporation.

The simplest form of organization is a sole proprietorship. A **sole proprietorship** is a business started and operated by an individual; in terms of tax and legal liability, the owner (proprietor) and the business are considered to be one and the same. This means that a sole proprietorship does not pay taxes as an entity separate from the owner. Also, legal claims against the sole proprietorship can be satisfied against an individual owner's personal property as well as his or her business assets.[15]

A **partnership** is a legal business entity that consists of two or more owners. A **general partnership** consists of multiple co-owners of a for-profit business. Partnership arrangements can be complicated, so a written partnership agreement that defines all the details of the partnership arrangement is very important. Topics that should be covered in the partnership agreement include:

- each partner's capital contribution to the partnership
- how the partnership's profits and losses are allocated between the partners
- what partnership salaries and drawings against profits are allowed
- the management responsibilities allocated to each partner
- what happens to the partnership when a partner withdraws, retires, becomes disabled, or dies
- how the partnership can be dissolved

In a general partnership, tax and legal liabilities flow through to the individual partners, just as with a sole proprietorship. Profits are distributed to the partners according to the partnership agreement, and each individual partner pays his or her own taxes. Legal claims against the partnership can be pursued against any of the partners for any amount, without regard to the amount of partner's investment in the partnership or share of its profits. Individual partners are responsible for the full debts of the partnership and an individual partner can bind the entire partnership to a business deal.[16]

A **limited partnership** has both a general partner and limited partners. In a limited partnership, a general partner assumes management responsibility and unlimited liability for the partnership. The **limited partners** have no management participation and are legally liable only for the amount of their capital contribution plus any specifically accepted debt.[17]

Another common legal form for a business to adopt is that of a corporation. A **corporation** is a legal and taxable entity separate from its owners and managers. A common type of corporation is a "C" corporation. The owners of a **"C" corporation** are its shareholders. Unlike a sole proprietor or the partners in a partnership, a corporation shareholder's legal liability is limited to the extent of his or her investment in the business. A corporation is also a taxpaying entity, and its earnings are actually taxed twice: once as income to the corporation, and again when shareholders pay taxes on dividends issued by the corporation.[18]

Two other forms of organization are the "S" corporation and the limited liability company. An **"S" corporation** gives its owners partnership tax status and corporate liability protection. An organization must meet very restrictive conditions to qualify for S corporation status. A **limited liability company (LLC)** combines the limited liability feature of a corporation with the tax status of a partnership or sole proprietorship. Like a partnership, an LLC may have an operating agreement that distributes profits and losses in many ways, not just in proportion to the owners' capital contributions.[19, 20]

When planning your new e-business's legal form, keep in mind who the owners are and who the potential investors will be. For a small group of individual owners and investors, forming a partnership, an S corporation, or an LLC might be most appropriate. If venture capital or other professional investors are solicited, if there is a need for the flexibility of stock options in incentive stock plans, or if there is a plan to take the business public, forming a corporation is a more suitable course of action. In any case, when preparing your business plan, you should consult with attorneys and accountants to make certain that you understand the details and ramifications of your state's laws and the different legal and taxable forms your e-business can take.

A business plan should also include a description of the business partnerships which will play an important role in the business's future success.

E-BUSINESS PARTNERSHIPS

For any business—including an e-business—achieving success depends strongly on the business's relationships with its strategic business partners, such as its suppliers and distributors. Partnering, building alliances, and collaborating with others offer a faster way to get an e-business's products and services sold and distributed. An e-business can have a variety of strategic business partners who provide support for a wide range of business activities, including marketing, sales assistance, technology support, and financial services. As an e-business grows and evolves, old partnerships may dissolve and new partnerships may take their place. For startup e-businesses, there are two general categories of partnerships: partnerships that help the e-business build market awareness, and partnerships that assist in the actual e-business operations, such as customer service, or shipping and transportation.

For example, the key strategic partners for Rackspace Managed Hosting include its equipment and software providers, such as Dell, Inc., Microsoft Corporation, and Red Hat, Inc., as well as a number of Web designers who comfortably recommend Rackspace to their clients based on the managed hosting e-business's dedication to customer support as emphasized by its business slogan—Fanatical Support™.

Startup e-businesses are not the only businesses that benefit from e-partnerships. New e-business partnerships have been formed by traditional brick-and-click retailers who find new and different ways to use the Internet to create and exploit new strategic partnerships. One example of a traditional brick-and-click business forming an e-partnership is the story of PartsAmerica.com, an e-business partnership that provides broader and more efficient distribution of auto parts and accessories.

The original PartsAmerica.com partners were Advance Auto Parts and CSK Auto, two major auto parts retailers who became e-partners in order to offer automotive parts and delivery options to both the B2B and B2C automotive marketplace. At the time the initial e-partnership was formed, Advance Auto Parts was ranked second in the Aftermarket Business Auto Chain Report Top 50 and had more than 1,600 stores in 37 states (primarily in the East, Midwest, Puerto Rico, and the Virgin Islands). CSK Auto was ranked third in the same report, with 1,120 stores in 17 states, primarily in the western United States. Today, the PartsAmerica.com partnership also includes Checker Auto Parts, Schuck's Auto Supply, and Kragen Auto Parts. By partnering together as PartsAmerica.com, these five companies benefit from being able to cover 50 states, with a total of 3,000 local stores for parts pickup, delivery, and returns, and having an inventory of more than $1 billion. PartsAmerica.com customers benefit from broader product options, same-day or overnight delivery, and in-store order pickup and returns across the country.

To identify appropriate strategic partnerships, take a careful look at the information already included in your business plan, in particular your product and service offerings and your targeted customers. Careful analysis might reveal areas in which your e-business could benefit from various partnerships. Ask for recommendations from mentors and advisors, and keep a close monitor on your competitors' partnering activity. Once you have identified potential partners with whom you might make mutually advantageous strategic partnership arrangements, you can begin the process of contacting them. When you enter into a strategic partnership, remember to formalize the arrangements you make in a contract containing clearly identified expectations for both partners, a specific period of time for the contract obligations, and an exit clause. Remember also to include a reference to these key partnerships in your business plan as part of the startup phase of your e-business.

A credible, professional business plan that describes your e-business idea; illustrates your products or services and how they will be marketed; shows how your e-business operations can make money; and lists its management team, advisors, and key partners, can be a map to success for your new e-business. In the next chapter, you learn about ways to fund your startup e-business.

One of the judges during the first round of the Stanford business school's entrepreneurial contest was Dave Wharton, an associate at the venture capitalist firm Kleiner Perkins Caufield & Byers. Wharton thought the online bridal registry was such a good idea that he contacted Lefcourt and DiLullo after the first round of judging. Following a series of conversations, Lefcourt and DiLullo decided to pull their plan out of the contest. They were getting serious about it and didn't want to circulate their idea too widely. Kleiner Perkins funded Lefcourt and DiLullo's online gift registry idea, and the e-business Della & James, named for the two characters in O. Henry's story "The Gift of the Magi," was born. Lefcourt and DiLullo left Stanford at the end of that first year to enter Della & James in the estimated $45 billion-per-year bridal registry market.[21]

During the first year after funding, Lefcourt and DiLullo concentrated on creating partnerships with retailers and hiring staff. By the time the Della & James Web site was launched in 1999, the e-business had 50 employees, including CEO Rebecca Patton, formerly a senior vice president with E*TRADE Group. Della & James' retail partners included Crate & Barrel, Neiman Marcus Group, Williams-Sonoma, Dillard's, Recreational Equipment, and Gump's, a well-known San Francisco specialty retailer.[22, 23]

In 2000, Tiffany & Co. acquired a 5 percent stake in Della & James, and the e-business, whose name was soon simplified to Della.com, became the exclusive wedding registry partner for Tiffany's. Shortly after partnering with Tiffany's, Della.com merged with the WeddingChannel.com (Figure 3-22). Today, customers can use the WeddingChannel.com's patented registry system to search more than 1.5 million gift registries. As of 2005, WeddingChannel.com customers had used these gift registries to purchase more than $100 million in gifts.[24, 25, 26]

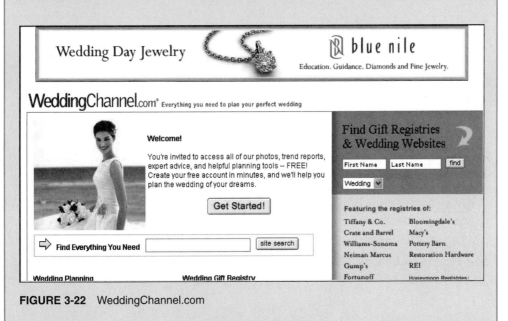

FIGURE 3-22 WeddingChannel.com

Chapter Summary

- A business plan is used to check the feasibility of a startup business, to seek funding for a business, and to guide the business operations after startup.

- A business plan should include some or all of the following components: a cover sheet and title page, a table of contents, an executive summary, a business description, information on products or services, and a marketplace analysis with information about the industry, potential customers, and competitors.

- A business plan should also include a marketing plan, an operational plan, a financial plan, an issues statement, and a formalized exit strategy.

- The appendices to a business plan include items such as resumes, pictures of products, descriptions of services, legal documents, and other supporting documentation that provides additional detailed information for the plan.

- The business plan business description should include a notation about the legal form of organization of the e-business. The most common legal forms of organization are a sole proprietorship, a partnership, and a corporation. A limited liability company combines features of partnerships and corporations.

- A business plan should also note any key business partnerships and alliances with suppliers or distributors and so forth that the e-business has developed.

Checklist

Your E-Business Plan:

- ❏ Demonstrates overall professionalism.
- ❏ Is neatly prepared, assembled, and bound.
- ❏ Is clearly and concisely written.
- ❏ Has been professionally edited for organization, clarity, spelling, and grammar.
- ❏ Begins with a cover sheet and a title page with contact information and a proprietary warning against reproducing the plan.
- ❏ Is numbered so that you can account for all copies of the plan.
- ❏ Contains a table of contents and an executive summary.
- ❏ Contains some or all of the following components, as appropriate: a business description, a summary of products or services, a marketplace analysis, a marketing plan, an operational plan, a financial plan, a management plan, a survey of critical risks, a formalized exit strategy, and appendices.
- ❏ Uses verifiable data from reliable sources.
- ❏ Has been reviewed by attorneys and accountants as necessary and contains any appropriate legal notations relating to the plan's contents and financial projections.

Key Terms

Advisory Board

business description

business plan

"C" corporation

corporation

cover sheet

executive summary

exit strategy

financial plan

general partnership

industry

issues analysis

limited liability company (LLC)

limited partners

limited partnership

marketplace analysis

mission statement

operational plan

partnership

planned cash flow statement

pro forma balance sheet

products section

projected income statement

risk assessment

"S" corporation

services section

sole proprietorship

table of contents

title page

vision statement

Review Questions

True/False Questions

1. A startup company's business plan is used only to solicit funding from investors. True or False?

2. It is not necessary to number multiple copies of a business plan. True or False?

3. An Executive Summary page is the least important part of a business plan. True or False?

4. A "C" corporation is a taxable entity separate from its owners and managers. True or False?

5. Information about key strategic partnerships should be included in a business plan. True or False?

Multiple Choice Questions

1. A business plan's executive summary:
    a. allows plan readers to locate specific plan contents.
    b. provides a miniature version of the complete plan.
    c. provides details about products or services.
    d. is a "to do" list for promoting products and services offered.

2. A sole proprietorship:
    a. consists of multiple co-owners.
    b. is not a separate taxable entity.
    c. has both a general partner and several limited partners.
    d. provides an owner personal protection from legal claims against the business.

3. An issues analysis identifies:

    a. outside threats and opportunities.

    b. marketing strategies for pricing, promoting, and distributing the e-business's products and services.

    c. the e-business's business model.

    d. targeted customers' buying habits.

4. A mission statement:

    a. describes business activities that will be outsourced to consultants.

    b. identifies outside advisors who bring additional background and experience to the e-business.

    c. briefly explains the overall purpose of the e-business.

    d. projects how and when the e-business can be profitable.

5. In order to explain how you and your investors will get your money back from a startup e-business, your business plan should include a formalized:

    a. marketing plan.

    b. appendix.

    c. table of contents.

    d. exit strategy.

Exercises

1. Using online search tools, locate at least two e-businesses that publish their mission statements on their Web sites. Write a short paper in which you describe each e-business and quote each mission statement. Include a brief analysis of each mission statement, identifying the mission statement's theme or focus and its target audiences. Based on your analysis, explain why you think each of the mission statements is or is not effective at explaining the business's purpose.

2. Using Bplans.com or similar e-business Web sites that provide examples of business plans, review at least three sample business plans. Then create the cover sheet, title page, and table of contents for a business plan whose e-business idea is based on the B2C e-retailing model.

3. Using online search tools or other relevant resources, search for information about Amazon.com to find answers to the following questions:

    a. What basic industry research did Jeff Bezos, the founder of Amazon.com, do about the book-selling industry? How did this information inform his early business planning for an online bookstore?

    b. What types of strategic partnerships has Amazon.com entered into?

    c. What business model or models does Amazon.com follow today? Has the Amazon.com business model evolved? If yes, how?

    d. Do you think Amazon.com's overall mission or vision has evolved over time? Why or why not?

4. Using online search tools or other relevant sources, identify two e-businesses and their key strategic partners. Describe each e-business and its partnerships, and explain how each partner benefits from the arrangement.

5. Using links on this text's student online companion, tour the following business planning sites: the Small Business Administration (SBA), the Service Core of Retired Executives (SCORE), the Small Business Advancement National Center at the University of Central Arkansas (SBANC), and E-Future Centre. Create a table of features for each site you review, then report back to your class on the Web site features and how an individual starting a new e-business can benefit from them.

Case Projects

1. You have been thinking about creating a new e-business to sell your custom needlework designs, needlework supplies, and the completed handcrafted needlework projects you accept on consignment. You decide it's time to get serious about your idea. Draft a business description section for your e-business's business plan and include the types of e-partnerships that would benefit your e-business.

2. You and three close friends are all freelancers: a technical writer, a Web designer, and a computer programmer. Each of you would like a better way to make contact with potential clients. Over dinner, the three of you decide that creating an online freelance talent "matchmaking service" that matched freelancers with clients who needed help on specific projects would be a great e-business idea. But before you go any further, you want to know if other e-businesses are currently offering this type of online service. Using online search tools or other relevant sources, identify e-businesses that currently offer freelance talent matchmaking services; then draft a brief portion of a marketplace analysis that describes the major competitors in the marketplace. If you need to see examples of industry or marketplace analysis sections, use links on this text's student online companion to review sample business plans.

3. You are the product manager for an exclusive line of cosmetics and fragrances, sold in shopping mall kiosks, which is targeted to teenage girls. You are considering submitting a proposal to your management team to encourage the company to begin selling its products online. To prepare the proposal, you need to learn more about the online buying habits of teenage girls. Using Web sites and other relevant sources, research the online buying habits of teenage girls; then draft the targeted customer portion of a marketplace analysis for your proposal.

Team Project

You and several associates are planning to start a new e-business that sells Web site design services to other businesses. Your first step is to draft the e-business's description, the business concept, and a mission statement. Working together with your assigned team members, define your e-business's concept and description, and then draft its mission statement. Make certain that the mission statement:

1. reflects the e-business's purpose

2. is succinctly worded

3. uses meaningful and understandable language

4. addresses a clearly defined target audience or audiences

As a group, present the e-business concept and mission statement to your classmates and your instructor. After your presentation, ask your classmates to evaluate your mission statement. They should address the clarity of your wording, try to identify your target audience(s), and appraise the effectiveness of the mission statement in conveying its message to the targeted audience(s).

After your mission statement has been evaluated, prepare, at the direction of your instructor, a complete business plan for the e-business, including (but not limited to) the following components: cover sheet, title page, executive summary, services section, marketplace analysis, and operational, financial, and management plans. Limit the business plan to 30 pages or less.

For Further Study

Here are some resources that might help you in further investigating the topics covered in this chapter.

Student Online Companion

Check out the *Creating a Winning E-Business, Second Edition* student online companion Web site for links to the sites discussed in this chapter and to other useful Web sites.

Articles and Books

BizPlanIt.com "Ten Painless Steps to Start and Finish Your Business Plan." www.bizplanit.com/free/articles/ten_painless_steps.html. 2001.

Berry, Tim. "Business Plan Essentials." *Entrepreneur.com.* www.entrepreneur.com/article/0,4621,320299,00.html. March 14, 2005.

Napier, H. Albert. "Moneymakers: A Business Plan Provides a Clear Path to Success." HoustonChronicle.com. www.chron.com/cs/CDA/ssistory.mpl/business/3174449. May 10, 2005.

Williams, Edward E., et al. *Preparing an Entrepreneurial Business Plan.* New York, NY: T&NO Book Company. 2004.

End Notes

[1] Millard, Elizabeth. "Higher Earning: Business Plan Contests at Leading Graduate Schools Create an Embarrassment of Riches." *Business 2.0.* www.business2.com/content/magazine/vision/1999/08/01/11414. August 1, 1999.

[2] Ratnesar, Romesh. "Start Me Up: The Post-Techie World has New Players: Women, Immigrants and even Republicans, Women Entrepreneurs Doing It for Themselves." *Time Magazine.* www.time.com/time/archive/printout/0,23657,992065,00.html. September 27, 1999.

[3] Sherman, Andrew J. "Business Planning: Building an Effective Business Model." *entreworld. org.* www.entreworld.com/Content/EntreByline.cfm?ColumnID=393. May 6, 2005.

[4] *National Association of Women Business Owners* and *Women's Business Institute* as reported at *Business.gov.* www.business.gov/phases/launching/write_business_plan/index.html. "Essential Elements of a Good Business Plan for Growing Companies." 2005.

5 PriceWaterhouseCoopers. "Business Plan Basics." www.pwcglobal.com/extweb/industry.nsf/docid/936CBF75BD75BD2585256AC6005D8B10. 2001.

6 Gaines, Sallie. "3000 Scholars at Your Service." *Chicago Tribune*. www.roundtablegroup.com/about/article.cfm?ID=25. August 26, 1999.

7 Press Release. "Inc. Magazine Names Round Table Group to Inc. 500." *Inc. Magazine* as reported by Round Table Group. www.roundtablegroup.com/about/release.cfm?ID=65. October 28, 2004.

8 PriceWaterhouseCoopers. "Business Plan Basics." www.pwcglobal.com/extweb/industry.nsf/docid/936CBF75BD75BD2585256AC6005D8B10. 2001.

9 *BizPlanIt: The Virtual BizPlan*. "Executive Summary: Business Plan Basics." www.bizplanit.com/vplan/execsum/basics.html.

10 *Leader to Leader Institute*. "How to Develop a Mission Statement." www.pfdf.org/leaderbooks/sat/mission.html.

11 *United States Small Business Administration (SBA)*. "Essential Elements of a Good Business Plan for Growing Companies." www.sba.gov/managing/strategicplan/guide.html#mkt_analysis. 2005.

12 CCH Business Owner's Toolkit. "Planning Your Business Operations." 2005. www.toolkit.cch.com/text/P02_5511.asp. 2005.

13 SBA Online Women's Business Center. "Marketing Plan Components: Competitor & Issues Analysis." www.onlinewbc.gov/docs/market/mk_mplan_competitor.html. 2001.

14 Lavinsky, Dave. "Documenting the Exit Strategy in Your Business Plan." Ezine@rticles. www.onlinewbc.gov/docs/market/mk_mplan_competitor.html. April 2, 2005.

15 *FindLaw for Small Business*. "Sole Proprietorships." www.smallbusiness.findlaw.com/business-structures/sole-proprietorship.html. 2005.

16 *FindLaw for Small Business*. "Partnerships." www.smallbusiness.findlaw.com/business-structures/partnership.html. 2005.

17 *FindLaw for Small Business*. "Limited Partnerships." www.smallbusiness.findlaw.com/business-structures/business-structures-overview/business-structures-overview-types(1).html. 2005.

18 *FindLaw for Small Business*. "Corporations." www.smallbusiness.findlaw.com/business-structures/corporations.html. 2005.

19 *FindLaw for Small Business*. "S Corporation Facts." www.smallbusiness.findlaw.com/business-structures/corporations/corporations-s-corp-facts.html. 2005.

20 *FindLaw for Small Business*. "Limited Liability Companies (LLCs)." www.smallbusiness.findlaw.com/business-structures/llc.html. 2005.

21 Bazdarich, Colleen. "An E-Commerce Company Is Born: Della & James Goes After the Bridal Registry Market." CBS MarketWatch. weddings.della.com/about_guest/pc_cbs_6-22-99.asp. June 22, 1999.

22 Feurstein, Adam. "Taking the Wraps Off Next Internet Battle: Gift Giving." *BizJournals.com*. www.bizjournals.com/sanfrancisco/stories/1999/06/21/newscolumn2.html. June 18, 1999.

23 Bounds, Wendy. "Several Major Retailers Say 'I Do' to Wedding-Registry Web Site." *The Wall Street Journal*. weddings8.della.com/about_guest/pc_wsj_6-9-99.asp. June 9, 1999.

[24] Press Release. "WeddingChannel.com and NBC Get Hitched." *WeddingChannel.com* wedding. weddingchannel.com/press_release/pr_nbc.asp. April 23, 2004.

[25] Anders, George. "Della & James to Receive $45 Million From Investors." *The Wall Street Journal*. www.weddings8.della.com/about_guest/pc_wsj_9-23-99.asp. September 23, 1999.

[26] Press Release. "WeddingChannel.com and della.com Merge to Create World's Definitive Marketplace for Weddings." *WeddingChannel.com*. www.zoominfo.com/. April 27, 2000.

CHAPTER 4

GETTING YOUR E-BUSINESS OFF THE GROUND

LEARNING OBJECTIVES

In this chapter, you will learn to:

- Describe the financing issues associated with an e-business startup
- Discuss the role of informal investors in an e-business startup
- Identify issues important to venture capital investors
- Pitch your e-business idea to investors
- Discuss the advantages and disadvantages of business incubators

CONSULTING AN ONLINE GURU . . .

At first, Kannan Srinivasan, a professor at Carnegie Mellon University in Pittsburgh, Pennsylvania, was just looking for a way to save money when he decided to hire a moonlighting computer technician to set up his new computer system. But this was the late 1990s, and e-business was booming. Like many other Internet-savvy professionals, Srinivasan wondered if people like him, who found themselves with an immediate need for temporary professional help, might appreciate the convenience of finding such help from an online service. He soon began toying with ways this idea could be turned into an e-business—an online marketplace that would bring together employers with short-term projects and self-employed or moonlighting professionals looking for work.

continued

Serendipitously, Srinivasan ran into a former student, Inderpal Guglani, a business planner with FedEx. The two men began discussing the idea, and it wasn't long before they pooled their resources (about $60,000), and Emoonlighter, an online freelance talent marketplace, was born.[1] Emoonlighter courted employers by providing special services, such as helping them prepare the necessary tax documents associated with hiring contract employees, and it charged moonlighting professionals a fee for each job they secured through the Emoonlighter marketspace.

In the beginning, Emoonlighter made a ripple rather than a splash in the existing online freelance talent marketplace. At the time, the industry already consisted of at least ten e-business competitors, including Ants.com, Elance.com, and the San Francisco-based Guru.com, who were backed by more than $250 million in total venture capital funding.[2] With limited funds, two employees, and an office in the basement of Guglani's home, what chance did Emoonlighter have to succeed against its well-funded rivals?

STARTUP FINANCING

As an entrepreneur starting a new e-business, you must be prepared to invest time, effort, and your own money to get your new e-business off the ground. As you learned in a Chapter 1, the hard work you put in to building your e-business creates value in the business in the form of sweat equity. But in addition to this sweat equity, you must be prepared to invest cash in your e-business startup. Starting an e-business can involve a myriad of basic operational activities such as building a Web prototype, doing market research, paying rent, and leasing office equipment. To pay for these and other various expenses, entrepreneurs of startups often find themselves drawing on personal savings, mortgaging personal assets, or obtaining a personal loan. Some creative entrepreneurs manage to avoid borrowing money and raising equity financing by finding unique and inventive ways to acquire resources for their startups—this type of self-funding is known as **bootstrapping**. Many entrepreneurs, however, find that they can't finance the start up of their new business on their own, and therefore raise money by turning to two major groups of investors: informal investors and venture capitalists.

Informal Investors

Informal investors include friends and family members who invest in a business, and angel investors, which you'll learn about in a moment. According to a July, 2004 report by the Global Entrepreneurship Monitor (GEM) at Babson College, informal investors in the U.S. contributed more money for startups and growing businesses—more than $100 billion—than professional investors, making informal investors the "lifeblood of U.S. entrepreneurship."[3] The GEM report also notes two other important findings:

- Almost 20 percent of adults surveyed reported that they had invested in someone else's private business in the prior three years.
- More than half of all informal investing is made by relatives.

Many entrepreneurs discover that they can successfully tap their network of friends and family for a solid round of funding that helps cover a startup e-business's legal fees, Web site prototype development, and other out-of-pocket expenses until additional funding is secured. Another benefit of financing from family and friends is that family and friends who know and trust you are likely to stand by you during tough times in the startup process, such as when building your customer base takes longer than projected. This is because they are largely investing in you rather than in your e-business idea.

QUOTES ON SUCCESS

"We weren't betting on the Internet [with a $300,000 investment]. We were betting on Jeff."

Jackie Bezos, mother of the founder of Amazon.com, Jeff Bezos

But there is a downside to accepting startup money from family members and friends: the risk of jeopardizing personal relationships with business misunderstandings. For example, you may think of the $5,000 Uncle Phil gives you for your new e-business as a loan or gift, while he may think he is buying equity in your business and is entitled to a portion of its future earnings. To avoid misunderstandings that might strain a personal relationship, follow a few simple rules:[4, 5]

- Make certain that your family member or friend understands that an e-business startup is a risky proposition. Find out how that person would feel if you weren't able to give the money back.
- Don't let a family member or friend invest any money that he or she cannot afford to lose.
- Provide the same marketing material to a family member or friend that you would provide to any investor, enabling the person to make an informed investment decision about your business proposition.
- Deal with a family member or friend the same as you would deal with a stranger and make it clear you're doing so for the benefit of both of you. Keep the funding transaction "at arm's length," and require legal documentation for the transaction.

Funding from your network of family members and friends may be the easiest money you will ever find, so this is a great way to obtain enough money to get started. After tapping into your family and friends network, your next step in financing your startup e-business is to seek funding from angel investors.

The term "angel investor" originally referred to wealthy investors in Broadway theatrical productions who swooped down and saved shows from going bankrupt. Now **angel investor** commonly refers to an individual with money and time who enjoys the excitement of investing in the early stages of a new business and is not averse to taking risks. According to the Center for Venture Research at the University of New Hampshire, angels invested more than $20 billion in startups in 2004—20 percent more than the previous year. Also, the rate of angel investment is expected to continue growing in the future, at about 10 percent a year.[6] Unlike friends and family investors, who typically invest in the entrepreneur, angel investors are primarily interested in the e-business idea. A typical angel investment can range from a few thousand to a few hundred thousand dollars.[7]

Perhaps one of the more well-known examples of angel investing involves Andy Bechtolsheim, a co-founder of Sun Microsystems, and two young Stanford University graduate students, Larry Page and Sergey Brin. In the late 1990s, Page and Brin were developing a new type of search engine which they called "BackRub." When they decided to start a company in order to commercialize their search engine, they went looking for investors and managed to snag an early morning meeting with Bechtolsheim. Bechtolsheim was in a hurry and had no time to listen to a detailed pitch for the new e-business, so Page and Brin gave him a quick demo. Bechtolsheim quickly grasped the new search engine's potential and right on spot wrote out a check for $100,000 made payable to Google, Inc., the name of Page and Brin's new company. Bechtolsheim's investment clearly paid off—especially when Google went public in August, 2004 with an IPO valued at about $1.67 billion.[8, 9]

QUOTES ON SUCCESS

"My whole strategy in dealing with companies is that if I can't figure out what they're doing in a 15-minute phone call, I'm not interested. If it makes sense, then I'll listen to more. I invest in companies that make sense to me, companies that are doing things that need to be done. If I like the idea, other people probably do too."

Andy Bechtolsheim, co-founder of Sun Microsystems and angel investor

Although many angels operate alone, some are now participating in investment clubs (also known as syndicates, associations, forums, or networks) through which they can combine their investments with other angels and thus spread their investment risk. An investor who wishes to join an **angel investment club** must qualify as an **accredited investor** under federal securities law, which means he or she must have a minimum net worth of $1 million, or an individual income of at least $200,000 per year, or a household income of $300,000 per year in each of the two most recent years.[10] Additionally, each angel investment club has its own membership criteria, and usually includes a financial commitment to the club itself.

One well-known angel investment club is Band of Angels founded in 1994 by Hans Severiens. Severiens, a former executive and venture capitalist, realized the advantages of pooling his resources together with other angel investors. In addition to minimizing the risk of an investment by spreading it out over a number of investors, a network of investors would offer a variety of expertise that individual investors could tap in order to quickly assess whether a new business was a good investment. Severiens brought together several Silicon Valley entrepreneurs and retired high-tech executives to create Band of Angels, which today has more than 100 members. They meet monthly to review presentations from three new entrepreneurs whose business plans have gone through a rigorous pre-screening process. Focusing on high-tech companies, Band of Angels members invest individually in the range of $300,000 to $750,000. As of 2004, Band of Angels members had invested more than $100 million in 151 companies (Figure 4-1).[11]

FIGURE 4-1 Band of Angels

E-PIONEERS

Of Angels and E-Mail

One high-profile investment made by Band of Angels members was a $4 million investment in Sendmail, Inc. Sendmail has its origins in an e-mail server software program that was released to the open source community in 1981 by Eric Allman, who at the time was a graduate student at the University of California at Berkeley. The software, which was free, made it possible to forward e-mail from one network to another. The Sendmail software became so popular that Allman began spending much of his time providing support to Sendmail users, including corporate users such as AOL and Earthlink. After giving free support to Sendmail users for a few years, Allman decided that there was a business opportunity in selling packaged Sendmail software with full service and support. In 1998, he teamed up with Greg Olsen, a Cornell graduate, to form Sendmail, Inc.

In July 1998, Allman and Olsen raised $6 million from angel investors: the $4 million from Band of Angels members, and another $2 million from the co-founders of Sun Microsystems, Andy Bechtolsheim and Bill Joy.[12] Today, Sendmail, Inc. customers include major companies from a variety of industries, such as Fidelity Investments (financial services), Johnson & Johnson (healthcare), Starbucks (retail), and the Executive Office of the President (U.S. government).[13] The Sendmail e-mail server and security software runs on 40 percent of all Internet e-mail servers and is used to deliver 70 percent of all global e-mail messages (Figure 4-2).[14]

continued

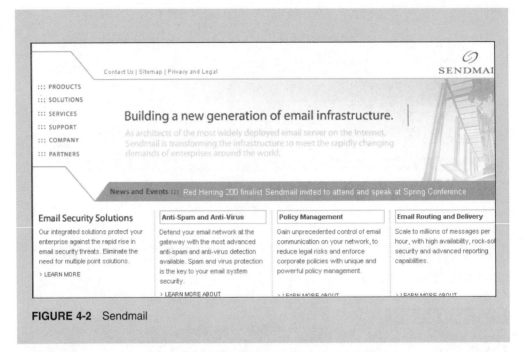

FIGURE 4-2 Sendmail

One way to find angel investors is to ask family members, friends, and business associates for referrals to angel investors in your area. Another way to locate information on angel investors is to search the Web; a good place to start is the Active Capital Web site (Figure 4-3). Active Capital is a spin off of the Angel Capital Electronic Network (ACE-Net), an online venue for entrepreneurs looking for angel investors and angel investors looking for investment opportunities. ACE-Net was originally created in the late 1990s through a collaborative effort by the Small Business Administration and various academic and non-profit communities. Active Capital, which was formed in 2004, is allowed to match entrepreneurs and accredited investors across state boundaries as a result of protective rulings by the Security and Exchange Commission (SEC) and state regulators.[15] The Active Capital Web site and databases are managed by the Office of Technology Transfers and Commercialization (OTTC) at California State University at San Bernardino.

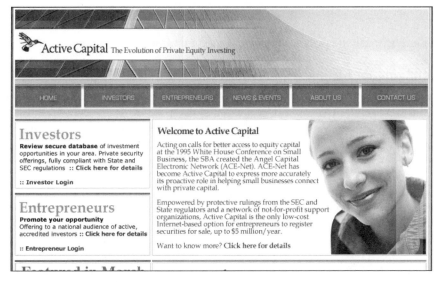

FIGURE 4-3 Active Capital

When presenting your e-business idea to an angel investor, remember that even though how you present yourself is important (and that it does help to have management experience on your e-business's team), the potential investor is most interested in how your e-business fills a marketplace need. Angel investors can provide what might be the first significant funds available to you and can usually provide experience and advice. Individual angels generally invest a relatively small amount in each new business that catches their interest; investment angel clubs or networks may be able to provide a larger amount of startup funds.

E-CASE IN PROGRESS

Rackspace Managed Hosting

In 1999, soon after Graham Weston and Morris Miller brought their management expertise and investment dollars to Rackspace Managed Hosting, business was booming. Before long, Weston and Miller needed to find additional funding to pay for the company's expansion into European markets—and they did. In January 2000, Rackspace secured $6.3 million in investments: $2.5 million from members of the Rackspace management team, and the rest from two angel investor groups, Teton Capital and Isom Capital. In April 2000, Rackspace received an additional round of funding in the amount $11.5 million from three venture capitalists: Sequoia Capital, Norwest Venture Partners, and Red Hat, Inc. Altogether, Rackspace managed to secure more than $27 million in venture capital funding from 1999 until February, 2001, when the e-business became profitable and no further equity funding was necessary.

After you secure startup seed money from your personal capital and investments from informal investors such as family, friends, and angels, you should attempt to solicit your next round of funding from one or more venture capital investors.

Venture Capital Investors

A **venture capitalist (VC)** firm is a professional investment company that provides funds for startup businesses, generally in exchange for an equity position in the new business. Venture capital firms raise hundreds of millions of dollars in funding for new businesses from sources such as endowments, insurance companies, and pension funds. Some prominent VCs include Kleiner Perkins Caufield & Byers, Murphree Venture Partners, and Sigma Partners. According to The MoneyTree™ Survey by PricewaterhouseCoopers, Thomson Venture Economics, and the National Venture Capital Association, VCs were on track in 2005 to fund $20-$23 billion in investments.[16] But VCs can provide a new e-business with more than just money. These firms can also offer extensive knowledge about specific industries and access to important contacts.

VCs come in many flavors: from traditional partnerships of two or more wealthy individuals to professionally managed investment groups, from government-sponsored investment companies to the corporate funding programs of high-tech companies that consider funding technology startups a way to extend their own research and development interests.[17] In general, VCs invest anywhere from $250,000 to $1,500,000 in the early stages of a new business, and may be prepared to offer several million dollars over the course of the business's life. Examples of VCs and the successful e-business startups they have funded include Draper Fisher Jurvetson (DFJ) and Hotmail; Kleiner Perkins Caufield & Byers and Google; the SBA's Small Business Investment Companies (SBIC) program and America Online; and Motorola Ventures and Web portal developer Epicentric. Figures 4-4 and 4-5 illustrate the Web sites of two of these VCs.

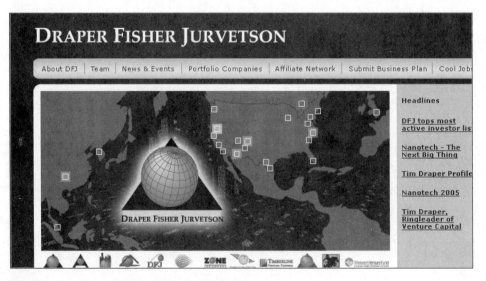

FIGURE 4-4 Draper Fisher Jurvetson (DFJ)

FIGURE 4-5 Small Business Investment Companies (SBIC) Program

When seeking VC funding, you should strive to find venture capital firms whose interests would fit well with your e-business's focus and its investment needs. Often VCs invest in a portfolio of complementary businesses, each of which adds value to the overall portfolio. Many VCs focus their investments in a specific industry, or they participate in specific funding rounds, such as seed or startup money, first round funding, or growth funding for new product development by established companies. Also, VCs often limit their investments to businesses in a specific geographic area. Therefore, it is a good idea to begin by researching the VCs in your area.

For each VC that might make a good match with your e-business, you should find out about the VC's principal investors, its management team and staff, and its investment strategies and history before you approach the company for funding. The VC firms you target through your research probably receive hundreds of business plans each year, and they give serious consideration to only a few of these plans—those with an exciting e-business idea backed by sound market research and an experienced management team.[18] One of the best ways to approach a VC is to arrange a face-to-face meeting with a principal through a source the VC knows and trusts—such as an accountant or attorney or other business professional—or to have your business plan submitted to the VC through the trusted source.[19]

QUOTES ON SUCCESS

"In the past, entrepreneurs started businesses. Today they invent new business models. That's a big difference, and it creates huge opportunities."

John Doerr, partner at Kleiner Perkins Caulfield & Byers

While there are numerous advantages to obtaining VC investments, you should be aware of some significant downsides. In exchange for receiving VC funding for your e-business, you can expect to give up a large stake in your e-business's equity. The typical equity percentage is in the 30-40 percent range, but depending on your situation, this can vary greatly—anywhere from 10 percent if you acquire growth funding for an already established company, to as much as 90 percent for a startup. You can also expect to have one or more of the VC principals on your board of directors, because a VC investor firm will want to be actively involved in any strategic business decisions that could change your e-business's direction or deplete its resources.

TIP

One of the strategies that entrepreneurs and VC investors implement in order to recover their investments in a new e-business is to take that business public. Going public is not a simple matter; therefore e-businesses that are serious about going public should consult with their attorneys about the ramifications of a public stock offering, including how to file the appropriate documents. For a brief overview of the IPO process, check out the Initial Public Offerings Web page at the CCH Business Owner's Toolkit Web site.

Whether you are soliciting funding from your friends and family network, angel investors, or venture capital firms, you need to be ready to pitch your e-business to investors.

PITCHING YOUR E-BUSINESS TO INVESTORS

The first meeting with a potential investor of any type is a sales meeting. Your immediate objective is to get the potential investor excited about your e-business idea and interested in pursuing more extensive discussions that might lead to financing. When pitching to an angel or VC investor, the first meeting is likely to be brief, about one hour in length. Come prepared, be on time, and bring everything you need with you.[20]

At a first meeting, you should present a brief pitch document rather than your entire business plan (although you should be ready to provide the plan upon request). A **pitch document** is a short marketing document about one to three pages in length based on the Executive Summary portion of your business plan. The pitch document should briefly highlight a market need, how your e-business meets this need, what potential profits you expect, and how your e-business's management team can make it all happen.

QUOTES ON SUCCESS

"I've never heard a pitch that was too short. A pitch can't be too short because a good one will motivate listeners to ask questions that extend it."

Guy Kawasaki, Managing Director and Chairman of Garage Technology Ventures

Your pitch document should be the basis of your verbal presentation or pitch. You can supplement your pitch visually by creating an easy-to-understand slide show presentation using Microsoft PowerPoint or a similar tool. But keep the slide presentation brief—no more than 10-15 slides. If you plan to make a slide show presentation, remember to bring a laptop computer, a projector, an extra projector bulb, and an extra copy of your slide show on portable media such as a diskette or flash drive. Be sure to bring printed copies of your slide show just in case your equipment malfunctions.[21]

Your pitch to investors, supported by your pitch document and slide show, should:

- define your product or service
- identify who will buy your product or service and how much they will pay for it
- describe your key industry competitors, and describe how your product or service is differentiated from those of your competitors
- explain how much it will cost to provide the product or service
- make clear when the investor(s) can expect your e-business to be profitable and thus earn a return on the investment
- illustrate the planned exit strategies both for the investor(s) and for your e-business principals
- specify how much money you are looking for and how you will spend it

Expect potential investors to ask questions during your pitch in order to determine how well you understand your e-business, your target market, your competitors, and critical marketplace issues. Also be prepared to field questions about any potential risks or problems associated with your e-business idea, and point out how your e-business is positioned to handle those risks or problems. If you do not have an answer to a particular question, don't try to pretend that you do. Simply acknowledge that you need to do further research on the issue and move on.

E-PIONEERS

Zipping Along to an IPO

It is critical to prepare in advance for each meeting with potential investors by having your handout materials organized and reviewed, by practicing your presentation, and by trying to anticipate the possible questions so that you can give decisive answers. If you think all this preparation is "overkill," just ask Scott Kucirek, co-founder and Executive Vice President of ZipRealty, about his experiences pitching a new startup to investors.

Kucirek and his partner, Juan Mini, began planning to start their own e-business while they were working their way through business school at the University of California at Berkeley in the late 1990s. By the time they graduated in 1999, Kucirek and Mini had secured $1.7 million from Vanguard Venture Partners and a group of angel investors led by Barrington Partners.[22] By August 1999, their new e-business—an online real estate brokerage business named ZipRealty.com—was up and running. Later that year, ZipRealty secured $16 million in a second round of financing; this involved the two original investors but was led by a venture capital firm named Benchmark Capital.[23]

continued

But Kucirek and Mini learned a few hard lessons along the way to getting that financing. Their first lesson was about being careful and being brief. The original business plan for ZipRealty.com contained 40 pages of text, with 20 pages of financial statements and projections. After scheduling a series of presentations to various investors, Kucirek and Mini decided to revise some of their handout material, but in their haste, they mixed up the revised pages. Unfortunately, this mix-up was discovered on the way to an actual presentation, and so they had to stop, fix the errors, and then re-copy and re-bind the handout material. Then, things went from bad to worse. During their presentation, Kucirek and Mini noticed that there were inconsistencies in the name of their proposed e-business on several of their charts. Not surprisingly, no one expressed interest in investing in ZipRealty after that presentation. After this disastrous start, Kucirek and Mini managed to pare their business plan down to 20 pages and developed procedures for making sure their material was carefully reviewed before future presentations.[24]

Another lesson Kucirek and Mini learned was the need to practice their presentation and try to anticipate questions from potential investors. The day before one investor meeting, Kucirek and Mini were unexpectedly asked to bring along two other members of their management team who had real estate experience and could discuss why buyers and sellers would use ZipRealty's services. Unfortunately, the busy ZipRealty.com management team hadn't set aside enough time prior to the meeting to consider the various questions that might come up and practice their answers. During the meeting, the potential investors asked questions that the management team members were unable to answer; the presentation went nowhere. Continuing to learn from their mistakes, Kucirek, Mini, and the others on the management team made certain that they practiced their pitch before each future meeting.[25]

Through luck, perseverance, and the ability to learn from their mistakes, the ZipRealty team members managed to "get their act together" and eventually began receiving serious interest from investors such as Vanguard Venture Partners, Barrington Partners, Benchmark Capital, and Pyramid Technology Ventures. By May 2001, ZipRealty was profitable in some of its metropolitan markets.[26] By 2003, ZipRealty had doubled its annual revenues to $34 million, and it received another infusion of capital from its VCs.[27] And in November 2004, ZipRealty (Nasdaq: ZIPR) inaugurated a $69 million IPO.[28] With more than 1,000 employees, ZipRealty now represents buyers and sellers in 13 metropolitan markets in 10 states and Washington, D.C., and the e-business generates more than $60 million a year in revenues (Figure 4-6).[29]

continued

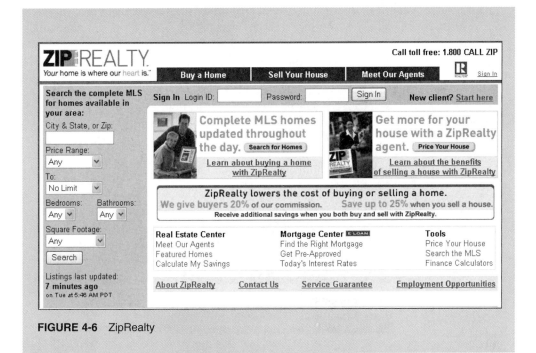

FIGURE 4-6 ZipRealty

QUOTES ON SUCCESS

"Investors don't like improvised presentations, we found. The VCs asked questions we hadn't anticipated. We ended up making references to an *Austin Powers* movie and giving long-winded, inconsistent answers. Surprise! No dough from them either. That was the last time we went anywhere unrehearsed."

Scott Kucirek, co-founder and Executive Vice President of ZipRealty

During your pitch, be certain to differentiate yourself and your management team from the competition by describing how your team's background and experiences position your e-business for success. Be enthusiastic. Give examples that show a real commitment to your e-business, and that create the feeling that your e-business idea is a viable, exciting opportunity for your investors. The value of learning as much as you can about your potential investors before meeting with them should not be underestimated; use your research to establish a good fit between your e-business idea and their existing portfolio of businesses.

TIP

You should consider developing a one- or two-minute **elevator pitch**—a quick explanation of your e-business idea that you can use to interest a potential investor when time is a premium, such as during a brief encounter on an elevator.

If an angel investor or VC firm is interested in your e-business, you might receive a term sheet. A **term sheet** is a list of the major terms or conditions of a proposed financing arrangement being offered by the investor, and is used to start negotiations for the investment deal. The term sheet will include a valuation of your e-business. For example, if the investor proposes to invest $200,000 in your e-business for 10 percent equity, then the investor values your e-business at $2 million. Term sheets can also include demands for a certain class of stock, automatic buyouts in case of an acquisition, seats on your e-business's board of directors, and other contingencies. Because term sheets can be very complicated, you should have them reviewed by an attorney experienced in negotiations with investors before agreeing to any investment deal. You can find an example of a term sheet and other investor-related legal documents at the National Venture Capital Association (NVCA) Web site illustrated in Figure 4-7.

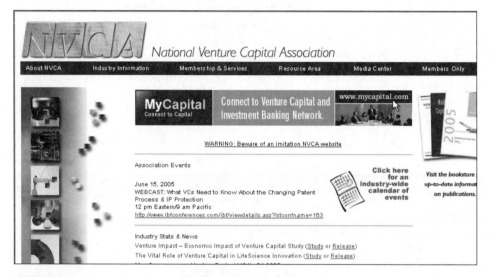

FIGURE 4-7 National Venture Capital Association (NVCA)

> **TIP**
>
> Before soliciting equity funding you should consult with a qualified attorney who has experience in investments and securities. Federal or state securities laws may require you to have an attorney prepare an offering document, known as a private placement memorandum. A **private placement memorandum** discloses the benefits and risks of an investment in your e-business to potential private investors.

Pitching your e-business to investors can garner tremendous gains in funding and, in the case of VC firms, valuable expertise and contacts; but it is an involved process that requires patience and perseverance, and getting good results can sometimes depend on sheer luck and timing. Another way to help get your e-business off the ground is with a business incubator.

Business incubators, sometimes called **business accelerators**, are non-profit organizations (usually sponsored by local governments, businesses, and universities) or commercial businesses that nurture startup businesses by offering fee-based business development and administrative-support services or by providing such services for a percentage of ownership equity in the startup company. The average startup business remains part of a business incubator's portfolio of companies for about three years. As you will learn shortly, the services that non-profit and commercial business incubators offer are various, but generally include office space, telecommunications hookups, reception and conference room areas, computer networking facilities, Web site design and hosting, clerical support, mentoring, and referrals to potential investors. An alternative way that entrepreneurs can get access to such business-support services is through self-incubation, whose advantages and disadvantages will be detailed at the end of this section.

Non-Profit Business Incubators

Approximately 90 percent of the nearly 1,000 business incubators in the U.S. are non-profit organizations.[30] A **non-profit business incubator** is often a cooperative venture between a university and the local community designed to stimulate and support the growth of local business startups. For example, the Austin Technology Incubator (ATI) at the University of Texas at Austin provides startup technology companies with a variety of services, such as business consulting, arranging for CEO mentors, providing advisory review boards, offering referrals for financing, and helping with market research (Figure 4-8). ATI-incubated companies also have access to office space and furniture, conference rooms, and telecommunications services. Along with the support of the University of Texas, ATI receives funding from the City of Austin and the Greater Austin Chamber of Commerce, and charges fees of the startups that participate in its incubation program. Additionally, startup companies grant ATI a one-percent ownership equity.[31] Similar business incubators or accelerators supported by local government, area businesses, and universities are the Advanced Technology Development Center (ATDC) at the Georgia Institute of Technology, the Houston Technology Center, and the Illinois Technology Enterprise Center (ITEC) at Northwestern University Evanston Research Park. Figures 4-9 through 4-11 illustrate the Web sites for these non-profit business incubators.

An interesting private non-profit business incubator is the Women's Technology Cluster (WTC), which incubates women-led startup companies and promotes entrepreneurship for women (Figure 4-12). Initially funded by Catherine Muther, a former vice president of corporate marketing for Cisco Systems and now president of the Three Guineas Foundation, WTC now gets support from a number of government, philanthropic, and business organizations, such as The Kapor Foundation, The Kauffman Foundation, Citibank, Wells Fargo, Pillsbury Winthrop LLP, and the City and County of San Francisco.[32] Women-owned startups incubated by WTC include MsMoney.com (a financial services Web portal for women), Quantum Intech, Inc. (a provider of stress management programs and training), and Remedy Interactive (a provider of risk management technologies and consulting services).[33]

FIGURE 4-8 Austin Technology Incubator (ATI)

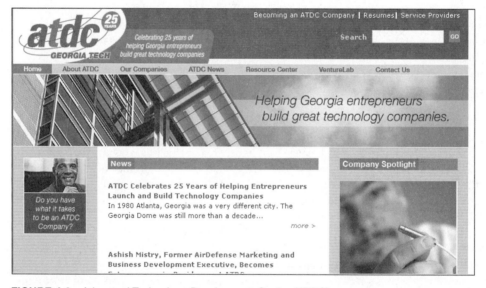

FIGURE 4-9 Advanced Technology Development Center (ATDC)

TIP

If you're an entrepreneur needing funding for the early-stages of your startup e-business, you may qualify for state-funded bridge grants (short-term funding grants intended to help you keep afloat while you are looking for other more substantial financing), or other assistance from state or academic organizations. When seeking funding, you should check with your state agencies and local universities for economic development programs to see what kind of assistance is available in your area.

FIGURE 4-10 Houston Technology Center

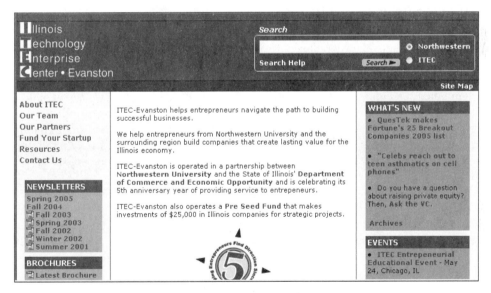

FIGURE 4-11 Illinois Technology Enterprise Center (ITEC)

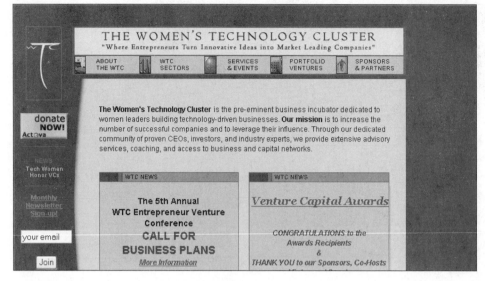

FIGURE 4-12 Women's Technology Cluster (WTC)

Commercial Business Incubators

Commercial business incubators are commercial businesses that typically charge fees and require a large share of ownership equity for providing startup companies with business-development services. One of the first commercial business incubators on record is the Batavia Industrial Center (BIC). In the 1940s and 50s, Massey-Harris Ferguson, Ltd. was a heavy equipment manufacturer that occupied approximately 850,000 square feet of building space in Batavia, New York. The company was the area's largest employer until 1956, when it closed its Batavia factory, leaving thousands of people out of work and a large manufacturing complex sitting empty and unused.[34] In 1959, the Mancuso family purchased the empty complex and charged one of its family members, Joseph, with the responsibility of putting people back to work.

When Joseph Mancuso was unable to find a single employer large enough to fill the entire complex, he named the building the Batavia Industrial Center (BIC) and started renting space to startup and small businesses. Then he did whatever he could to help the startups and small businesses grow. One of the first businesses in the BIC was a Connecticut chicken hatchery, or incubator. Playing on the incubator idea, Mancuso began describing the BIC as a "business incubator." Under the direction of Joseph Mancuso, who eventually came to be known as the "father of business incubation," and his son Tom, more than 1,100 businesses have gotten their start at the BIC.[35] Today the BIC is managed by the Mancuso Business Development Group (Figure 4-13).

During the economic and dot-com boom of the late 1990s, many new commercial or for-profit business incubators, such as Idealab, eCompanies, and Cambridge Incubators, formed with the hope that they could accelerate the growth of hot new e-businesses that could quickly—sometimes within three to twelve months—be spun out on their own and start to generate huge profits for investors. It was not uncommon for many of these new commercial business incubators—some of whom called themselves **Internet accelerators**—to demand up to 60 percent or more equity in an e-business startup *and* to charge these startups monthly

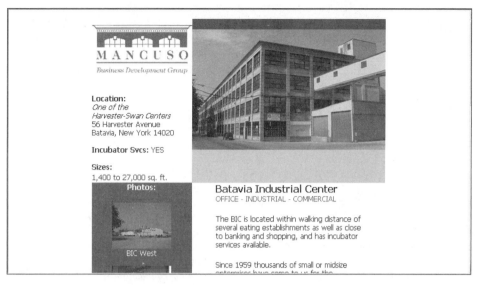

Location:
One of the
Harvester-Swan Centers
56 Harvester Avenue
Batavia, New York 14020

Incubator Svcs: YES

Sizes:
1,400 to 27,000 sq. ft.

Photos:

BIC West

Batavia Industrial Center
OFFICE - INDUSTRIAL - COMMERCIAL

The BIC is located within walking distance of several eating establishments as well as close to banking and shopping, and has incubator services available.

Since 1959 thousands of small or midsize

FIGURE 4-13   Batavia Industrial Center (BIC)

fees for using their services.[36] The economic downturn in 2000-2001 that led to the dot-com bust also wreaked havoc on many new commercial business incubators who focused on start-up e-businesses. Some, like Digital Hatchery and i-Incubator, shut down. Others, such as Idealab and eCompanies, are still active, maintaining substantial portfolios of incubating companies. Critics of commercial business incubators suggest that giving up so much equity makes the cost of incubation too high for entrepreneurs. But proponents counter that some entrepreneurs require more guidance and thus may be more willing to give up a measure of independence and equity in exchange for that guidance. Such entrepreneurs like the low-risk option of starting a business that commercial business incubators offer.

E-PIONEERS

Incubating Veritas Medicine

In the late 1990s, Stephen Knight, a licensed physician and MBA graduate from the Yale School of Organization and Management, knew he had a good e-business idea. Knight's idea was to create a Web-based database of clinical drug trials that would be available to both patients and pharmaceutical companies. Pharmaceutical companies run clinical trials on drugs being developed, but often the details of those trials are kept secret for competitive reasons. Many people are willing to participate in these trials to get free medical examinations and free medications or to benefit from cutting-edge research, but they do not know how to find out about them. Knight reasoned that by providing a comprehensive Web-based database of trial sites, along with crucial medical information, both patients and pharmaceutical companies could win. Patients could locate and enroll in the trials easily, and the pharmaceutical companies could fill their trials faster, thereby bringing their drugs to market sooner and saving millions of dollars in the process. But

continued

Knight, who had just become president and chief operating officer of a medical supply firm, needed someone else to get the e-business off the ground; so he called on a friend from Yale Medical School, Robert Adelman. Adelman, a board-certified physician with a consulting and entrepreneurial background, agreed to become chief operating officer for the e-business venture, now named Veritas Medicine.[37]

At first, after several months of trying, Knight and Adelman were unsuccessful in securing venture capital funding; then, as part of a last-ditch effort, they pitched their e-business idea to Cambridge Incubator (CI). By early September 1999, the Veritas team had signed a deal with CI and a venture capital firm, SeaFlower Ventures, that provided seed money, office space and furniture, phones, a T-1 line, and a computer network. This deal transformed Veritas Medicine from an entrepreneur's dream into an operating company. As part of the deal, CI would also provide Veritas Medicine with access to the professional services that e-business startups need, such as access to Web developers, lawyers, public relations and marketing specialists, and human resources specialists, as well as introductions to other funding sources. Joining the business incubator jump-started Veritas Medicine and put it weeks, perhaps even months, ahead in the startup process.[38] But the deal between Veritas Medicine and CI was expensive. In exchange for providing $834,000 in seed money and access to facilities and professional services, CI and SeaFlower Ventures got 51.22 percent ownership of Veritas Medicine. In addition to its equity position, CI charged Veritas Medicine about $19,000 a month for the infrastructure and professional services supplied during incubation.[39]

Was the expensive jump-start worth it? Perhaps so. Today, the award-winning Veritas Medicine is a leading recruiter for participants in the pharmaceutical industry's clinical trials; it has more than 20 employees, and more than a dozen part-time physician researchers.[40] Following revelations in 2004-2005 about the very serious side effects of some popular drugs, the pharmaceutical industry is under pressure to conduct broader and more comprehensive clinical trials. Veritas Medicine is poised to exploit that need by providing the pre-qualified participants that these more extensive clinical trials will require (Figure 4-14).[41]

continued

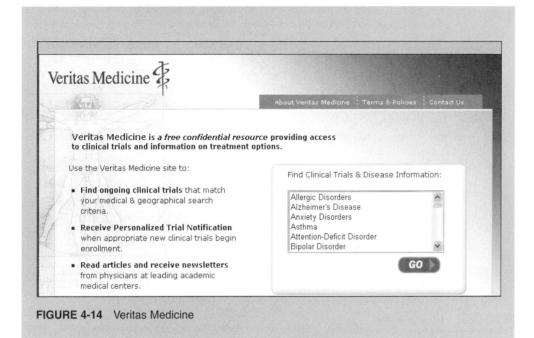

FIGURE 4-14 Veritas Medicine

QUOTES ON SUCCESS

"The Internet is the most competitive marketplace in history, and we needed a leg up in terms of resources. We knew we had competition, so for us speed was more important than money. On day one, we walked into prime office space in an area with a 0 percent occupancy rate. On day two, our T-1 line was installed. On day three, the computer network was set up. Phones and support personnel began arriving on day four. By the end of nine months, we had a staff of 70 and were hitting our stride while our main competitor was still working out of his apartment."

Robert Adelman , co-founder of Veritas Medicine and principal of OrbiMed Advisors

For many entrepreneurs, starting a new e-business under the wing of a non-profit or commercial business incubator may be most appealing because it offers a "one-stop solution" to many startup problems. But before making a decision to incubate your e-business startup, you should consider the following questions:[42, 43]

- Who are the business incubator's principals and staff? What entrepreneurial experience do they have? What do they do to stay up-to-date with economic issues and trends in business incubation?
- What is the business incubator's track record? Find out how long the incubation program has been operating. Check out the organization's formerly incubated companies to see how well they are doing; ask the principals about their incubation experience.

- What specifically will the business incubator do to help your e-business? The startup services provided by business incubators vary. You should know up front which services are provided by the business incubator and which services you must provide on your own.
- How easy will it be to leave the incubator program? Find out how long, on average, former clients stayed in the incubator.
- In addition to services, does the business incubator offer seed money, or are there venture capital funds linked to the incubator? It might be worth joining the incubator simply to get access to seed money or first-round venture capital financing.
- How much will it cost you in cash and equity to participate? Remember that the value from a commercial incubator's services must warrant giving up a large equity position in your e-business. In making this decision, you should determine whether it makes financial sense for your e-business to pay for the incubator's services (if services are part of the deal) as well as give up a large equity position.

After a careful review of the advantages and disadvantages of incubation, you can decide if joining a business incubator is the right thing for your startup e-business. One alternative to participating in a business incubation program is self-incubation.

Self-Incubation

For entrepreneurs, one of the primary advantages of participating in a business incubator is the ability to share ideas and discuss startup problems with other entrepreneurs whose companies are in the incubator's portfolio. Another way to enjoy the advantages of communicating with other entrepreneurs of startups is through **self-incubation**—that is, by participating in a members-only group that focuses on entrepreneurs. The Portland, Oregon group Starve Ups is an example of such a group.

Starve Ups was started in 2000 by two young entrepreneurs who, in the course of attending a number of meetings and lectures on entrepreneurial topics, found that talking with their peers in the parking lot afterward was the most important part of the experience.[44] Membership in Starve Ups is composed of no more than 18 companies whose founders meet on a regular basis to discuss issues related to running their businesses. At these meetings, no topic—from pricing to profit margins—is off limits. But for Starve Ups members, it isn't just the conversation that's helpful. Members also share practical expertise and access to contacts, sell or barter products and services among themselves, and even borrow equipment from other members when necessary.[45]

To avoid becoming just another business networking organization, Starve Ups keeps its membership roster lean, with a maximum of 18 companies, and is very selective about admitting new members. Of the more than 90 companies that have applied for membership since its founding, most have been turned away for various reasons. For example, membership in Starve Ups is open to only those companies that offer their customers a distinct product in addition to customer support services. Has membership in Starve Ups been beneficial to its companies? It seems so. As of this writing, all seven founding companies of Starve Ups are not only still in business, they have become profitable.[46]

QUOTES ON SUCCESS

"When you go to [entrepreneurship] events, and there are 300 to 400 people in the room, it's great to network and talk to people. But in Starve Ups, you can talk about real issues with people you trust and literally lean on, and get valuable feedback."

Josh Friedman, CEO of Eleven Wireless Inc. and member of Starve Ups

. . . CONSULTING AN ONLINE GURU

Although competing in a marketplace full of well-funded rivals, Emoonlighter had, as it turned out, one very big advantage—Inderpal Guglani and his "conservative, tight purse strings, no-nonsense approach" to doing business. After operating on a shoestring budget for the first three years, Emoonlighter managed to garner $400,000 in funding from the angel investors at Fairview Funds LP in August 2000.

With the additional funding and the ever-frugal Guglani at the helm, Emoonlighter concentrated on becoming profitable fast—and its plan worked. Guglani's tight rein on expenses and focus on profitability helped Emoonlighter stay afloat during the economic turndown and dot-com bust of 2000-2001. By November, 2001, many of Emoonlighter's marketplace rivals were struggling financially, looking for more funding, scrambling to change direction, or going out of business. But Emoonlighter had grown to five employees and—by holding down costs and generating a steady stream of fee revenues—was making a profit.[47, 48, 49]

According to the U.S. Department of Labor, in 2001 almost 10 million U.S. workers described themselves as being self-employed. Many of these were professionals who had been newly laid-off as a result of the unpleasant side effects of the economic downturn and dot-com bust, and had subsequently turned to self-employment to find work.[50] By managing its resources carefully, Emoonlighter survived this difficult economic period and was then poised to exploit this pool of newly available self-employed professionals.

By this time Guglani had become concerned that the name "Emoonlighter" was a bit too unprofessional for the types of self-employed professionals his employer clients needed. A name change was in order. In an interesting twist of fate, one of Guglani's major marketplace rivals, Guru.com, had blown through its $63 million in VC funding without becoming successful, and its assets were for sale. In late 2003, using a bank line of credit for financing, Emoonlighter acquired some of Guru.com's assets, including its URL, name, and client base, after which point Emoonlighter became the new Guru.com (Figure 4-15).[51] Today, Guru.com boasts a database of more than 400,000 self-employed professionals, 30,000 employer clients, and 40,000 employer-posted projects. And Guglani runs it all with fewer than 10 employees![52]

continued

FIGURE 4-15 Guru.com

QUOTES ON SUCCESS

"Pursuing an innovative idea on a young platform like the Internet presents many unknowns. The trick is to identify the bad ideas at a low cost, so one can avoid them. A rigorous process of hard work and patience has brought us to this point—not one good idea or decision."

Inderpal Guglani, co-founder of Emoonlighter, now President and CEO of Guru.com

Chapter Summary

- In addition to sweat equity, an entrepreneur starting a new e-business should expect to invest personal funds in the e-business by drawing on savings, mortgaging assets, or taking out a loan.

- An entrepreneur's network of friends and family can be an important source of seed money for a startup e-business.

- An angel investor is a wealthy individual who enjoys investing in startup businesses and is not afraid of taking some risk.

- Venture capital firms can raise hundreds of millions of dollars to fund new businesses, but in return they usually expect an equity percentage, a seat on the board of directors, and some management control over the businesses in which they invest.

- The first meeting with potential investors is really a sales meeting in which the entrepreneur's objective is to excite interest in the e-business idea. For investor meetings, an entrepreneur should be on time, be prepared, and keep it brief.

- A pitch document is a short marketing document that highlights the key facets of an e-business idea.

- Before meeting with potential investors, entrepreneurs should practice their presentation and try to anticipate the questions they might be asked.

- To succeed in getting funding for their startup e-businesses, entrepreneurs must learn from each presentation they make, and use the resulting feedback to refine their pitch document and verbal presentation for future meetings.

- Business incubators are non-profit organizations (often sponsored by the government or a university) or private companies that nurture startup businesses.

- Commercial business incubators can provide a startup company with a wide range of services, but in exchange may demand a large stake, sometimes 60 percent or more, in the startup's equity.

- To enjoy the benefits of incubation without relinquishing equity, some entrepreneurs prefer self-incubation, which involves participating in small groups that are composed of fellow entrepreneurs and that foster meaningful exchanges of information and resources among peers.

Checklist

Pitching Your E-Business:

❑ Be sure that your pitch or marketing document is short and to the point. It should highlight a need in the marketplace, describe how your e-business idea will fill that need, forecast how profitable the e-business could be, list any important partnerships you've cultivated, outline all the critical risks you've anticipated, and explain why you and your team can overcome those risks to turn this idea into a successful business. If you are making a slide show presentation, keep it less than 15 slides.

❑ Be prepared to pitch your e-business idea under different circumstances: with your pitch document handout, using a slide presentation, or by making a quick three-minute elevator pitch.

❑ Do your homework and be ready to answer questions about the e-business. For example, when asked, you should be able to describe the benefits your e-business provides to potential customers, the size of its target market, how fast the target market is growing, and the basis for the e-business model.

❑ Try to anticipate all the questions that a potential investor might pose, but if you are asked a question that you don't know the answer to, resist the urge to "wing it" with an improvised answer. Instead, simply state that you need to do more research and move on.

❑ When pitching your e-business idea, be enthusiastic and demonstrate your commitment by relating any personal investment (time, effort, money) you have made in the e-business.

❑ Be able to describe how the management team's experience positions the e-business for success.

❑ Practice! Practice! Practice! Don't forget that pitching your e-business idea to others will help you refine it.

❑ Bring everything you need with you to the pitch meeting, including printed copies of electronic presentations, an extra copy of the electronic presentation file on portable media, and your own computer and projector if you plan to use them. Be ready to be able to deliver your pitch if your equipment fails to operate properly or if your electronic media is unusable.

❑ Be prepared to pitch your e-business idea to potential investors a number of times. Listen to their feedback and use it to refine your presentation for future meetings.

Key Terms

accredited investor

angel investment club

angel investor

bootstrapping

business accelerators

business incubators

commercial business incubators

elevator pitch

informal investors

Internet accelerators

non-profit business incubators

pitch document

private placement memorandum

self-incubation

term sheet

venture capitalist (VC)

Review Questions

True/False Questions

1. A term sheet is a list of the major points of a proposed investment deal. True or False?

2. An angel investor is typically an individual with money and time who enjoys the excitement of investing in startup businesses and is not averse to taking risks. True or False?

3. An accredited investor must have a minimum net worth of $2 million. True or False?

4. Venture capital firms can be a good source of first- and second-round funding for a new startup e-business. True or False?

5. Commercial business incubators generally charge an incubated company service fees as well as an equity position in the company. True or False?

Multiple Choice Questions

1. Sweat equity is the value added to a startup business by:

    a. friends and family investors.

    b. commercial business incubators.

    c. entrepreneurs.

    d. angel investors.

2. Informal investors include:

    a. non-profit business incubators.

    b. angel investors.

    c. commercial business incubators.

    d. venture capitalists.

3. A private placement memorandum is:

    a. a negotiating document sent to an entrepreneur by a potential investor.

    b. used to secure a personal bank loan.

    c. part of the business plan for a new e-business startup.

    d. an offering document prepared by a qualified attorney that discloses the benefits and risks of an investment.

4. Which of the following is a disadvantage to an entrepreneur attempting to obtain startup funding from a member of his or her family and friends network?

    a. It may be the easiest money an entrepreneur can find.

    b. Friends and family members are investing in the entrepreneur rather than in the e-business idea.

    c. Losing Aunt Ruth's nest egg may upset family relationships.

    d. Each friend and family investor must have a minimum net worth of $1 million.

5. One way to enjoy the advantages of interacting with other startup entrepreneurs—such as being able to share ideas, discuss problems, and exchange products and services—without giving up any equity in your e-business or paying fees for business-development services is to become part of a:

   a. non-profit business incubator's portfolio.

   b. self-incubation group of entrepreneurs.

   c. commercial business incubator's portfolio of companies.

   d. None of the above.

Exercises

1. Using links located on this text's student online companion Web site, research the financing information available to an entrepreneur at the Small Business Administration's Web site, especially on the page that discusses the SBIC program. Then write a brief summary of how you could use this information to help find financing for a new e-business.

2. Visit the National Venture Capital Association's Web site. Review and print the different types of legal investment documents provided there, including a term sheet.

3. Select three venture capital firms discussed in this chapter, and use links located on this text's student online companion Web site, online search tools, or other relevant sources to learn more about each firm. Write a one-page paper that compares and contrasts each firm's investment focus. Give examples of the types of successful e-businesses in which each has invested.

4. Using online search tools or other relevant sources, research angel investment clubs, VCs, and non-profit or commercial business incubators that focus their investment efforts on women and minority entrepreneurs. Summarize your research on three to five funding sources in a presentation of 10 to 15 slides, and present your findings to a group of classmates.

5. Using the link for Starve Ups provided at this text's student online companion Web site, learn more about the way this entrepreneurial self-incubating group works. Then, using online search tools or other relevant sources, find similar groups in your area. Write a brief summary of what you find and present the information to your classmates.

Case Projects

1. Suppose you and three classmates have just graduated from college and have a great idea for a new e-business. Write a brief description of a possible e-business idea that follows the e-business model of your choice. Next, using online search tools or other relevant resources, investigate non-profit and commercial business incubators in your area. Then join your three classmates in a roundtable discussion of the pros and cons of incubating your e-business with one of the business incubators. Make a decision for or against joining the business incubator, and list the reasons for your choice.

2. You are considering starting a new e-business. Write a description of the e-business idea using the e-business model of your choice. Then create an outline of the steps necessary to develop equity in the startup, get seed money, and solicit first-round funding. Using online search tools or other relevant resources, identify appropriate funding sources in your area for each stage of funding.

3. You and three classmates have developed a new wireless technology that will make it easy for C2C e-businesses to make transactions over wireless devices, and the three of you want to start a new e-business to market this technology. When each of you discuss your e-business idea with your respective families, you discover that you all have family members who are willing to lend you money to get started—in fact the father of one of your classmates is so enthusiastic about the idea that he has offered to cash out his retirement plan. Working with your classmates, discuss the advantages and disadvantages of accepting your family members' offers, and then make a decision about whether you should accept these offers. Write a brief explanation justifying your group's decision.

Team Project

You and three classmates have just developed new software that enables B2B exchanges to process buy and sell transactions much faster than they can with the tools they currently use. You and your classmates want to create a new e-business to sell this software to B2B exchanges, but you need seed money to get started. A professional associate of one of your parents has secured a third-party introduction to a group of angel investors, and you are meeting with the investors next week to pitch your e-business idea. Now you and your team must prepare for the meeting.

Create a one-page pitch document and, using Microsoft PowerPoint or another presentation tool, develop a five- to ten-slide presentation that highlights the selling points of your e-business idea. (Aside from the facts explicitly stated above, you are free to elaborate on your e-business idea and make any useful assumptions.) Draw up a list of questions that you anticipate your potential investors might ask, and prepare short answers for these questions. Don't forget that you will probably be asked to explain how your new software works, what the current trends are in the B2B exchange marketplace, and how well your team can successfully manage the new e-business.

Using your pitch document and your slide show, pitch your e-business idea to a group of angel investors that consists, in this case, of your classmates and instructor. Your team will have 45 minutes to an hour to make your presentation and to answer any questions about your e-business idea that are posed by your potential investors.

For Further Study

Here are some resources that might help you in further investigating the topics covered in this chapter.

Student Online Companion

Check out the *Creating a Winning E-Business, Second Edition* student online companion Web site for links to the sites discussed in this chapter and to other useful Web sites.

Articles and Books

Ashbrook, Tom. *The Leap: A Memoir of Love and Madness in the Internet Goldrush*. Boston, MA: Houghton Mifflin Company. 2000.

Kawasaki, Guy. *The Art of the Start*. New York, NY: Portfolio Hardcover (Penguin Group). 2004.

End Notes

[1] Shropshire, Corilyn. "Lessons for a Guru: Small Online Tech Outsourcing Firms Survives by Watching Pennies, Buys Giant Rival Guru.com." *Pittsburgh Post-Gazette*. www.post-gazette.com/pg/04001/256944.stm. January 1, 2004.

[2] Kersting, Jonathan. "Moonlighting Guru: Inderpal Guglani Takes a Sustainable Route to Building a Business." *Pittsburgh Technology*, *TEQ Magazine*. news.pghtech.org/teq/teqstory.cfm?ID=1103. December 2003.

[3] Bygrave, William. "Founders, Family, Friends, and Fools." *United States GEM 2003 Report* as reported by BusinessWeek Online. www.gemconsortium.org/document.asp?id=355 and www.businessweek.com/smallbiz/content/sep2004/sb2004093_9929_sb014.htm. September 3, 2004.

[4] Robbins, Steven. "Asking Friends and Family for Financing." *Entrepreneur.com*. www.entrepreneur.com/article/0,4621,293174,00.html. October 1, 2001.

[5] Ennico, Cliff. "Accepting Money from Friends & Family." *Entreprenuer.com*. www.entrepreneur.com/article/0,4621,299420,00.html. May 6, 2002.

[6] Ganapati, Priya. "Angel Funding Grew in 2004." *Inc.com*. pf.inc.com/criticalnews/articles/200503/angels.html. March 29, 2005.

[7] Gomes, Lee. "Angel Investors Return, and They are Serious." *Startup Journal: The Wall Street Journal Center for Entrepreneurs*. www.startupjournal.com/financing/trends/20050412-gomes.html. April 12, 2005.

[8] Sherman, Chris. "Happy Birthday, Google!" *SearchEngineWatch*. searchenginewatch.com/searchday/article.php/2160731. September 8, 2003.

[9] Kopytoff, Verne. "Google IPO." *San Francisco Chronicle*. www.sfgate.com/cgi-bin/article.cgi?file=/chronicle/archive/2005/05/06/BUC200GOOGLE.DTL&type=tech. May 6, 2005.

[10] Ewing Marion Kauffman Foundation. "Join an Angel Group." givingback.kauffman.org/cwp/appmanager/givingBack/givingBackDesktop?_nfpb=true&_pageLabel=givingBack_investingInCompanies_opp&title=Join%20an%20Angel%20Group. 2005.

[11] "Band of Angels: Who We Are, Deal Statement, and FAQ Pages." www.bandangels.com. 2005.

[12] "Sendmail Pits Angel Investors against VCs." *Red Herring*. www.redherring.com/PrintArticle. aspx?a=5887&sector=Archive. November 1998.

[13] Sendmail, Inc. "Who We Serve." www.sendmail.com/company/customers/. 2005.

[14] Sendmail, Inc. "Who We Are." www.sendmail.com/company/whoweare/. 2005.

[15] "Welcome to Active Capital." www.activecapital.org/. 2005.

[16] The MoneyTree™ Survey by PricewaterhouseCoopers, Thomson Venture Economics and the National Venture Capital Association. "Venture Capital Investing Settles Back to $4.6 Billion in Q1 2005." www.nvca.org/pdf/MoneyGTreeQ1-05.pdf. 2005.

[17] Hosmer, LaRue Tome. "A Venture Capital Primer for Small Business." *U.S. Small Business Administration*. www.sba.gov/library/pubs.html#fm-5. 2005.

[18] Ibid.

[19] "Venture Capital 101." *Blumburg Capital*. www.sba.gov/INV/vc101.pdf. 2002.

[20] Ibid.

[21] Kawasaki, Guy. "The Art of Pitching." *Forbes.com*. www.forbes.com/ceonetwork/2004/09/10/0910gkbook.html. September 10, 2004.

[22] Sinton, Peter. "Campus Creations Venture Capital Firms Investing in Student Startups." *San Francisco Chronicle*. www.sfgate.com/cgi-bin/article.cgi?file=/chronicle/archive/1999/05/26/BU49146.DTL. May 26, 1999.

[23] Roussel, Tara. "Smart Startup: ZipRealty Has Graduated Into Major Venture Funding." *San Francisco Business Times*, 14(18), 32. December 10, 1999.

[24] Kucirek, Scott. "Raising Money Is Like Theater: You Better Have Your Act Together: ZipRealty's Team Tripped, Goofed, and Rambled Before They Got It Right." *BusinessWeek Online: Net Journal*. www.businessweek.com/smallbiz/news/coladvice/diarynj/nj991008.htm. October 8, 1999.

[25] Ibid.

[26] Press Release. "ZipRealty Raises $8 Million: New Investor Pyramid Technology Ventures Joined by Previous Investors Benchmark Capital, Vanguard Ventures, and Barrington Partners." www.ziprealty.com/about_zip/news/article_detail.jsp?id=110&type=press. June 11, 2001.

[27] Copeland, Michael. "Home Sweet Deal: How is ZipRealty Closing So Many Sales? Killer Discounts and User-Friendly Agents." *Business 2.0*. www.ziprealty.com/about_zip/news/article_detail.jsp?id=256&type=news. June 23, 2004.

[28] Press Release. "Recent IPO Included Directed Shares Program for Customers Who Bought or Sold a Home with ZipRealty." www.ziprealty.com/about_zip/news/article_detail.jsp?id=258&type=news. December 14, 2004.

[29] "ZipRealty, Inc. Fact Sheet." *Hoover's*. www.hoovers.com/ziprealty/--ID__132807--/free-co-factsheet.xhtml. 2005.

[30] Campbell, Mary. "Small Business Builder: Incubators; Incubators Offer Nurturing Environment for Small Business." *ABCNews*. abcnews.go.com/Business/print?id=88065. September 30, 2004.

[31] *Austin Technology Incubator*. www.ic2.org/main.php?a=2&s=4. 2005.

[32] Zich, Janet. "The New Face of Philanthropy." *Three Guineas Fund*. www.3gf.org/news_sb. html. February 2001.

[33] "Portfolio Company Industries." *The Women's Technology Cluster*. www.wtc-sf.org/portfolio_industry.html. 2005.

[34] "Founders Awards: Joseph Mancuso." *National Business Incubators Association*. www.nbia. org/resource_center/what_is/founders_awards/mancuso.php. 2004.

[35] Phipps, Jennie L. "Incubators Bring Out the Best in a Business." *Bankrate.com*. www.bankrate. com/brm/news/biz/Biz_ops/19990607.asp. June 7, 1999.

[36] Norman, Jan. "Ready to Hatch." *The Orange County Register*. www.ec2.edu/main/000124a. html. January 24, 2000.

[37] Singer, Thea. "Inside an Internet Incubator." *Inc. Magazine*. www.inc.com/magazine/20000701/19547.html. July 2000.

[38] Ibid.

[39] Ibid.

[40] Hendrickson, Dyke. "Internet Campaigning, Pharmaceutical Style." *Mass High Tech: The Journal of New England Technology*. www.masshightech.com/displayarticledetail.asp?art_id=65320. April 5, 2004.

[41] Hendrickson, Dyke. "Biomed Rounds: Bigger is Better: Veritas Set for Broader Drug Trials." *Mass High Tech: The Journal of New England Technology*. www.masshightech.com/displayarticledetail.asp?art_id=67669. January 24, 2005.

[42] Newton, David. "Pros and Cons of Incubator Funding." *Entrepreneur.com*. www.Entrepreneur. com/article/0,4621,319712,00.html. January 24, 2005.

[43] National Business Incubators Association (NBIA). "Tips for Entrepreneurs." www.nbia.org/resource_center/entrepreneurs_tips/index.php. 2005.

[44] Williams, Geoff. "Guiding Light." *Entrepreneur Magazine*. www.wired.md/news/coverage_entrepreneur_july22.shtml. July 22, 2003.

[45] Lindquist, Christopher. "I'm OK, You're Still in Business: Starve Ups, a Self-Help Group for Entrepreneurs, Keeps the Dot.com Dream Alive." *Darwin Magazine*. www.darwinmag.com/read/070102/starve.html. July 2002.

[46] Earnshaw, Aliza. "Advice Comes From the Trenches." *The Business Journal of Portland*. tampabay.bizjournals.com/cincinnati/moneycenter/story.html?id=2575. May 29, 2005.

[47] Kersting, Jonathan. "Moonlighting Guru: Inderpal Guglani Takes a Sustainable Route to Building a Business." *Pittsburgh Technology, TEQ Magazine*. news.pghtech.org/teq/teqstory.cfm?ID=1103. December 2003.

[48] Desmarais, Martin. "Emoonlighter Illustrates Elusive Dot-Com Perseverance." *Indus Business Journal*. www.indusbusinessjournal.com/global_user_elements/printpage.cfm?storyid=389054. March 1, 2003.

[49] Shropshire, Corilyn. "Lessons for a Guru: Small Online Tech Outsourcing Firm Survives by Watching Pennies, Buys Giant Rival Guru.com." *Post-Gazette*. www.post-gazette.com/pg/04001/256944.stm. January 1, 2004.

50 U.S. Department of Labor. As reported by AARP "Free Agents: The World of Independent Contractors." www.aarp.org/money/careers/selfemployment/a2004-06-14-freeagents.html. 2001.

51 Guzzo, Maria. "eMoonlighter Buys Guru, A Rival Freelance Job Site." *Pittsburgh Business Times*. www.bizjournals.com/pittsburgh/stories/2003/06/30/daily13.html?t=Printable. July 1, 2003.

52 "Guru.com Facts." www.guru.com/aboutGuru.pdf. 2005.

OPERATING YOUR E-BUSINESS

MONEY FROM E-MAIL. . .

In late 1998, two young men—Peter Thiel, a hedge fund manager and attorney, and Max Levchin, an engineer—had an idea for an e-business: developing and providing security software for personal digital assistants (PDAs) and other handheld devices. After spending a few months building this e-business, which they named FieldLink, Inc., Thiel and Levchin abandoned the security software idea when they realized that there was a greater consumer need that was being neglected by the marketplace—the need to quickly and easily send money from person to person. Thiel and Levchin believed that consumers would find this ability to transfer money useful for a variety of daily transactions, ranging from splitting dinner checks to paying for online auction purchases. They also believed that the Internet was the perfect medium for these kinds of person-to-person payments. By the summer of 1999, FieldLink, Inc. had become Confinity, and Thiel and Levchin's focus had changed from building an e-business that developed security software to one that developed "money-beaming" software for handheld devices.[1]

continued

Like Thiel's and Levchin's original e-business idea, the Confinity e-business idea called for the company to focus on the wireless marketplace. PDA and cell phone users who wished to be able to transfer or exchange money could open a free account with Confinity that would be funded with their respective credit cards. Using their handheld device's infrared (IR) port to connect with other handheld devices, customers would then be able to "beam" money from their accounts to another person's Confinity account. Confinity's plan was to make its money "on the float"—that is, by earning interest on the money sitting in customer accounts.[2]

As suggested in earlier chapters, the late 1990s were the "rock and roll" days of the dot.com boom, and so Confinity had little trouble finding investors, including Nokia Ventures (the venture capital arm of the mobile communications giant Nokia) and Deutsche Bank. In July 1999, Confinity demonstrated its money-beaming software at a press party by having Nokia Ventures "beam" $3 million in startup funds to Confinity through Thiel's PDA.[3] With the popularity of handheld devices growing both in the U.S. and around the world, Confinity was counting on the network effect and viral marketing to make its person-to-person payment system a success. But the dot.com bust was right around the corner. What did the future have in store for Thiel, Levchin, and Confinity?

STARTUP AND OPERATIONAL CHALLENGES

In Chapters 3 and 4 you learned about planning for your new e-business and finding ways to fund its startup. The next step in the entrepreneurial process is to begin operating your e-business. Gearing up for operations requires considering a number of issues; for example, what are some of the legal issues that might be involved with starting and operating your e-business? When is it time to move into commercial office space? What do you need to know about hiring employees? How will your customers pay for the products and services they buy? What technologies are appropriate for your e-business, and who provides them? How will you coordinate your Web site operations with accounting, product distribution, and other "back office" operations? Must you maintain all the skills, technologies, and operational systems your e-business needs in-house, or can you save money by outsourcing some of these e-business needs? Finding good answers to these operational questions may be critical to the success of your new e-business. In Chapter 6, you will examine another important operational issue: how to market your e-business's products and services. But first, this chapter will introduce you to some of the other diverse operational challenges that your startup e-business is likely to face, and will suggest ways to deal with those challenges.

LEGAL ISSUES

Establishing an early relationship with an attorney experienced in business startups and, if possible, with e-business startups, is important for an e-business entrepreneur. An experienced attorney can help you develop your business plan, formalize the legal structure of your business, and establish a company valuation for investors. An established attorney may also be able to provide leads to angel investors and venture capitalists. In addition to providing such services, your attorney should participate in any negotiations with investors and review all legal documents related to securing investments, such as term sheets.

You should obtain legal assistance not only for startup business planning and investor activities, but also for many ongoing business activities. As with a brick-and-mortar business, an e-business is subject to state and federal laws and regulations. The legal issues associated with an e-business can be more complicated, however, because business activities on the Web often move across traditional jurisdictional boundaries, and because the growth in Web-based business is itself leading to changes in laws that are likely to affect your e-business.

When dealing with issues such as copyrights, trademarks, employee benefits and compensation, retirement plans, and personnel policies, you should seek advice from your attorney. Understanding and negotiating marketplace issues such as taxation, content liability, and information privacy may also require the advice of an experienced attorney.

Copyrights and Trademarks

A **copyright** is a form of legal protection for original work, including writing, drama, music, art, and other intellectual property—whether the work is published or unpublished. A copyright takes effect the moment the work is created. Virtually everything on the Web is protected by copyright law, whether the copyright statement is visible or not. As a matter of form and best practice, however, an e-business's Web site should contain a copyright notice. While the fair use concept built into copyright law allows for the limited use of copyrighted material under certain circumstances, you should not include any text, graphics, audio, or other material on your e-business Web site without obtaining permission from the copyright holder.[4, 5] Both the U.S. Copyright Office Web site (Figure 5-1) and the private Copyright Web site provide good information on copyright issues.

FIGURE 5-1   U.S. Copyright Office

A **trademark** is a distinctive symbol, word, or phrase that a business uses to identify its products and to distinguish them from those of other businesses. Similarly, a business's services are distinguished from those of other businesses by distinctive words, phrases, or symbols, which are referred to as **service marks**. Trademarks and service marks make it easier for a consumer to identify the source of a product. For example, the word "Nike" and the Nike "swoosh" symbol identify shoes and other products manufactured by Nike. Using a trademark to heighten consumer awareness of a product or service is called branding. You learn more about branding in Chapter 6.

When you develop a trademark or service mark for your e-business, you should take steps to register it with the U.S. Patent and Trademark Office. The U.S. Patent and Trademark Office (Figure 5-2) and the International Trademark Association Web sites are good sources of information about trademark issues.

FIGURE 5-2 U.S. Patent and Trademark Office

Content Liability and Customer Information Privacy

An e-business may be exposed to potential liabilities for defamation, libel, copyright infringement, obscenity, indecency, and other issues on the basis of the content and hyperlinks on its Web site. To protect against potential lawsuits, you should have your attorney draft an appropriately worded disclaimer of liability and post this disclaimer on your e-business's Web site so that it is easily accessible to viewers.

One of the challenges for any e-business is the collection and analysis of customer information. Many potential customers may be hesitant to purchase products or services online or to provide valid information for fear that the information will be misused or sold to others. Your e-business should have a clear policy on how it handles and secures the information it collects at its Web site. Additionally, your e-business should take steps to make customers aware of its information privacy policies before it collects the information, including informing customers how this information will be secured from possible unauthorized or fraudulent use.

One way an e-business can build customer confidence about information privacy is to become a TRUSTe licensee. TRUSTe is a non-profit organization that advocates disclosure of information privacy policy, informed user consent, and consumer education about information privacy issues.[6] E-businesses that are licensed to display the TRUSTe online seal, or "trustmark," on their Web sites have agreed to make customers aware of what personal information is being gathered, with whom it will be shared, and how it will be used. These companies have also agreed to provide customers with options that allow them to control the way their information is disseminated. TRUSTe (Figure 5-3) licensees include e-businesses such as Yesmail, Classmates.com, and eBay.

FIGURE 5-3   TRUSTe

Taxation

The taxation of business conducted on the Web is one of the toughest issues facing government. Sales taxes are the largest single source of revenue for many state and local governments. These governments fear that increased sales on the Web (as opposed to in-state brick-and-mortar businesses) will continue to cause a serious decline in their sales tax revenues—a decline that must be made up from other sources, such as increased taxes on property and income. As of this writing, e-businesses are subject to the same existing sales taxes, property taxes, and corporate income taxes as the brick-and-mortar businesses in their state, but they are not required to collect sales taxes from customers unless the e-business has a physical presence in the customer's state. Instead, buyers are responsible for paying their state's applicable sales taxes, although few, if any, do so. Although the Internet Tax Freedom Act, originally passed by Congress in 1998 and extended to 2006, has placed a moratorium on new Internet taxes, there are several ongoing initiatives by state and local government officials to find an efficient method of collecting sales taxes on Web-based purchases.[7]

It is important that your e-business's management team stay informed about changes in tax law. You should certainly seek the advice of an experienced tax accountant when dealing with tax issues for your e-business. It also bears repeating that you should establish a relationship with an experienced attorney who can help you sort through the legal and tax issues your e-business faces.

In addition to legal issues, another operational challenge your e-business will probably need to resolve is the issue of physical office space. If, like most startups, your e-business occupies temporary office space, chances are you will eventually outgrow this space. When that happens, it's time to look into suitable commercial office space.

COMMERCIAL OFFICE SPACE

Many startup e-businesses begin life at the entrepreneur's kitchen table and then quickly expand to a spare room that serves as temporary office space. An arrangement like this might work sufficiently for some time, but when you reach the point at which your employees have to share desks (and sometimes chairs), when a growing stacks of manila folders serves as your "file cabinet," and when extension cords and surge protectors fill every electrical outlet in the room, it's time to consider commercial office space. Perhaps even more important than these internal needs are considerations related to your e-business's interactions with the public—in other words, you should consider relocating to commercial office space when your e-business needs appropriate meeting spaces for clients and business associates.

If your e-business happened to have a need for large facilities, such as a call center or product distribution center, you should have already evaluated those needs and outlined them in your business plan. In the case of many e-business startups, however, the details about commercial office space needs may not be a significant consideration in the original business planning process. If this is true for you, it is important that you develop a detailed facilities plan before you begin searching for commercial office space. Effectively planning your search for commercial office space will help you avoid common pitfalls such as moving to a poor business location, leasing space that doesn't easily allow for your business's future growth, or occupying facilities that turn out to be inappropriate for clients, business associates, and employees. Asking and answering the following questions will help you create a facilities plan and select appropriate office space:

- How much space does your e-business need now, and how much will it need in the near future?
- How will you divide the space into offices or work areas?
- What are your electrical wiring requirements?
- What telecommunications services and office equipment does your e-business need?
- Will there be a need for additional space for break rooms, reception areas, and conference rooms?
- Will the physical location of the facilities offer access to adequate parking and to nearby businesses that employees could patronize, such as restaurants and childcare centers?
- What major roads and public transportation are nearby?
- Is there a nearby pool of qualified prospective employees?
- What type of security requirements does your e-business have?
- What property damage and liability insurance requirements does your e-business have?

A facilities plan should also include a budget for the office space facilities plus any other costs associated with the facilities, such as utilities, security, insurance, and parking fees. Any estimated costs that may be incurred during the leasing process, such as legal fees, real estate brokerage fees, and general contractor fees (if modifications to commercial space are necessary), should also be part of your facilities plan budget.

After developing your facilities plan, you may choose to look for opportunities to save money on office space. One way to do that is to look for office space with a non-profit business incubator. Another way is to share office space with other businesses, so that the costs of common areas such as reception areas, conference rooms, and break rooms can be split among the businesses. Subletting space from an existing business is yet another good way to save money. Alternatively, you may, as some startup e-businesses do, find ways to trade services with strategic business partners in exchange for office space. Lastly, an additional way to downsize your office space requirements is to consider providing options that allow some of your employees to telecommute.[8]

It is a good idea to work with an experienced and reliable commercial real estate broker who can help you locate commercial office space and help negotiate the lease on the selected space. It is also a good idea for your attorney to review all lease agreements.

One of the major reasons your startup businesses might find it necessary to move from a temporary home office into commercial space is to accommodate the additional employees that your e-business needs to grow.

HIRING EMPLOYEES

The key to success for any business is its people. In the case of startup e-businesses, finding key people—that is, employees who are critical to your e-business operations—can be difficult. Knowing whom to hire first may be confusing, as the employee needs of individual e-business startups can vary. Generally speaking, however, your first key hire may need to be the chief information or technology officer. Even if you plan to outsource most of your e-business's technology needs, it is critical for your management team to include someone who understands e-business technologies and who can help design and monitor the development of your e-business's systems and their operations. The next type of hiring you might do involves building your e-business's technology staff with—depending on your e-business's specific needs—employees who can serve as software engineers, Web developers, product managers, and network administrators.[9]

Because maintaining an accurate picture of your e-business startup's financial position and **burn rate** (the rate at which your e-business is using its cash reserves) is critical, your next hire is likely to be a chief financial officer. Your chief financial officer is responsible for all your e-business's accounting and financial activities and may also play a key role in securing additional financing. In addition to a chief financial officer, you should consider hiring an experienced marketing professional to begin working on business development. After you secure another round of financing, it may be time for your e-business to concentrate on hiring employees for the marketing and customer support teams. Lastly, you should remember that as the number of employees grows, you will probably need to hire someone to handle the e-business's human resources functions.[10]

QUOTES ON SUCCESS

"What characteristics do I look for when hiring somebody? That's one of the questions I ask when interviewing. I want to know what kind of people *they* would hire."

Jeff Bezos, founder of Amazon.com

Leads on finding qualified and talented employees can come from many sources. Investors, boards of advisors or directors, family, friends, strategic business partners, and current employees are all great sources for new employee leads. When hiring senior managers or technology workers, you may consider using a professional recruiting firm. Professional recruiters, also known as headhunters, have access to experienced, talented people who are not actively looking for employment but might consider moving to a new e-business if they perceive it to be an exciting opportunity. To add credibility to your startup e-business and satisfy investors' requirements for experienced management, you may need to hire an experienced chief executive officer or other senior management up front. In addition to soliciting leads from advisors, you may turn to one of several well-known global professional recruiters, such as Heidrick & Struggles, Korn/Ferry International, and Spencer Stuart, which focus on filling senior management positions.[11] Figures 5-4 through 5-6 illustrate the Web sites of various professional recruiters.

FIGURE 5-4 Heidrick & Struggles

FIGURE 5-5 Korn/Ferry International

FIGURE 5-6 SpencerStuart

E-CASE

Springing into Action

By the early years of the twenty-first century, Dan Isaacs and Adam Moore had been co-workers in Austin, Texas for nearly a decade—first as marketing managers at Dell Computers (now Dell, Inc.), then as partners at the T3 (The Think Tank) advertising agency, where they handled client advertising that generated millions in revenues. During these years of working together, Isaacs and Moore had often thought about starting an e-business. Their e-business idea? An interactive marketing firm that would provide their clients a menu of services such as strategic planning, consulting, financial modeling, marketing planning, advertising, and Web site development.[12] Blessed with $150,000 in seed money and years of cumulative business experience in marketing, client management, and small business Web site development, Isaacs and Moore finally decided to exploit these advantages and try to create a thriving new business.

On March 1, 2004, Isaacs and Moore got to work on a business plan that included a very tight budget. They knew that a large percentage of small businesses fail within the first two years of operation, and with this in mind, they deferred their own salaries until the end of their company's first fiscal year so they would be able to hold down costs. Armed with seed money, experience, an ambitious but pragmatic business plan, and a fierce determination to succeed, Isaacs and Moore launched their interactive marketing firm, named Springbox, just two weeks later.

continued

As part of their next step, Isaacs and Moore began tapping into the deep pool of contacts they had accumulated over the years and succeeded in picking up clients like Lance Armstrong, the Tour de France champion, and Advanced Micro Devices, a microprocessor manufacturer. As a result of their hard work, Springbox, the little advertising shop located in the heart of downtown Austin, began to turn a profit in its fourth month of operation.

Once Springbox got off the ground, Isaacs and Moore focused on finding and retaining the best creative staff available in order to cultivate the firm's growing stable of clients—and fortunately, the two principals had the confidence to hire people whose skills equaled or surpassed their own. To keep new talent invested in the company, Isaacs and Moore instituted performance-driven incentives like profit-sharing and bonuses. They also made a conscious effort to keep the work environment conducive to creativity by making it fun and energetic. So far, their hiring strategy has worked well. With its talented team of employees, Springbox is on pace to exceed its second-year financial targets in 2005 and is expected to double its staff in 2006.

QUOTES ON SUCCESS

"Finding great people is another challenge. Great people require financial reward, job stability, growth opportunity, and a great work environment. It can also be difficult convincing candidates that your new startup can provide them with those benefits."

Dan Isaacs, co-founder of Springbox

After you've identified your e-business's personnel needs, you should begin the hiring process by researching the salary and benefit expectations of prospective employees in your region. The Web is a good place to begin this research. Various government organizations, nonprofit groups, and business magazines often post free salary survey information at their Web sites. The U.S. Department of Labor, Bureau of Labor Statistics, and Salary.com are good sources of information on compensation and benefits. Figures 5-7 and 5-8 illustrate Web sites that provide information on compensation and benefits.

When it comes to attracting employees for key positions, a startup e-business often must not only offer the prospective candidates competitive salary and benefits, but also some ownership equity in the e-business. But monetary and equity compensation alone may not be enough to attract and keep talented employees. An attractive corporate culture that promotes and rewards creativity and personal growth, along with a pleasant work environment, may be just as important in attracting key employees.

After hiring your first round of key employees, you should turn to the important challenge of deciding how your customers will pay for your products and services. Nothing is more critical to your bottom line than determining what electronic payment methods your customers can use.

FIGURE 5-7   U.S. Department of Labor, Bureau of Labor Statistics

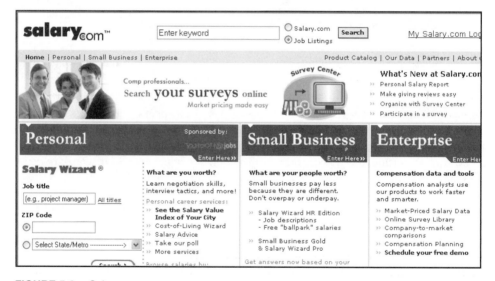

FIGURE 5-8   Salary.com

ELECTRONIC PAYMENT METHODS

You should base the selection of appropriate electronic payment methods on the kinds of products and services your e-business will offer and the type of customers who will buy those products and services. For example, a B2C e-business selling products to adults will likely accept credit, debit, or charge cards and may accept electronic checks. However, a B2C e-business that targets teenagers and young adults should remember that many potential customers in this demographic do not have access to credit, debit, or charge cards; therefore, stored value cards or electronic cash in the form of micropayments might be the best payment approach. In contrast, a C2C e-business might find it more efficient and cost-effective to use a person-to-person (P2P) electronic payment method. A B2B e-business may use a combination of electronic payment methods, such as credit cards, electronic checks, or electronic funds transfer (EFT), in addition to traditional paper checks. To help you make a selection that is appropriate for your e-business, the upcoming sections provide details on a variety of common electronic payment methods.

Credit, Debit, and Charge Card Payments

To pay for online purchases, consumers most commonly expect to use one of three types of payment cards—bank-issued credit cards, debit cards, or charge cards. While similar in nature, these three payment cards have important distinctions, which, as you will see, can have different consequences in terms of transaction liability. A **credit card** allows a consumer to pay for items using a revolving line of credit. A **debit card**, also called a check card, enables a consumer to pay for items with a debit directly from the consumer's checking account. A **charge card** is a payment card that requires a consumer to pay the balance in full each month.

Most consumers prefer to pay for their online purchases by credit or charge card in part because these payment cards offer limited liability for any unauthorized purchases that are made when the card number is stolen. The federal Truth in Lending Act limits liability for unauthorized credit and charge card charges to $50, and many card issuers waive all liability for unauthorized charges. But liability for unauthorized debit card withdrawals is set by provisions of the Electronic Funds Transfer Act and is based on how soon a consumer notifies his or her financial institution that the card has been used unlawfully. Liability for unauthorized withdrawals varies from $50 (if notification occurs within two business days), to $500 (notification within 60 days), to unlimited losses (notification after 60 days).[13]

In order to accept payment cards, your e-business must have a merchant account, payment processing software, and, for real-time processing, a link to a secure payment gateway. Additionally, you must have procedures in place to protect your customers and your e-business from fraud.

Merchant Accounts and Payment Gateways

A **merchant account** is an account at a financial institution, such as a merchant bank, into which payment card receipts are deposited. A **payment gateway** is a secure online service that submits payment card transactions to the merchant bank's card processing network and, as shown in the steps below, notifies the e-business's customer if the payment transaction is approved or declined.[14] The processing of a payment card transaction at an e-business Web site typically involves the following steps:

- *Step 1*: The customer enters and submits payment information at the e-business's Web site.
- *Step 2*: The e-business transmits the payment information to the payment gateway.
- *Step 3*: The payment gateway connects to the merchant bank's card processing network to route the transaction to the card issuer.
- *Step 4*: The card issuer approves or declines the payment transaction, sends back a notification, and, if the transaction is approved, transfers funds to the e-business's merchant account.
- *Step 5*: The payment gateway notifies the e-business and the customer of the results of the transaction.

Remarkably, this payment card transaction process, illustrated in Figure 5-9, takes just a few seconds. Examples of secure payment gateway service providers include Authorize.Net, CyberSource, ICVERIFY, and VeriSign.

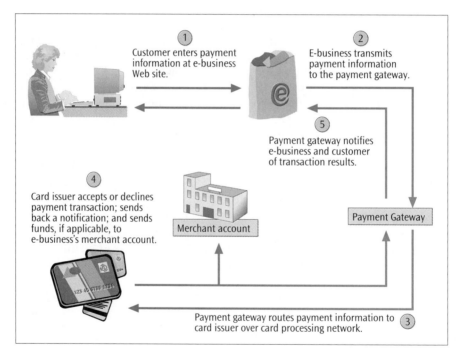

FIGURE 5-9 Payment card processing

You can apply for a merchant account for your e-business either directly at a financial institution or through a merchant account provider. A **merchant account provider** partners with financial institutions and payment gateway services, and can provide the

158

entire payment card processing package: a merchant account, payment processing software and equipment, and a connection to a payment gateway. You apply for a merchant account much in the same way you would apply for a personal credit card account, by supplying requested financial information. The kinds of information requested by a financial institution or merchant account provider may vary but will likely include:[15]

- your credit history and that of your e-business
- how long your e-business has been in operation
- the kind of products and services offered
- payment cards to be accepted
- the anticipated average size of each transaction
- the anticipated monthly volume of transactions

In establishing its fees for a given business merchant account, the financial institution or merchant account provider assesses the risk level of the business, which depends on the business owner's credit worthiness and a number of other factors. Several "high-risk" factors can also influence the fees charged for a business merchant account. For example, businesses involved in telemarketing, catalog sales, and online sales generally pay higher fees for a merchant account than brick-and-mortar businesses. These types of businesses involve a greater risk of payment card fraud because the cardholder is not present at the point-of-sale. This is called the **cardholder not present** risk factor.

Another factor is the chargeback risk. A **chargeback** occurs when a consumer refuses to pay a charge on his or her credit card account for a variety of reasons, including product returns, billing errors, and fraudulent charges. Financial institutions and merchant account providers maintain chargeback risk statistics for specific types of businesses. If an e-business falls into the "high-risk" category by offering online gambling or other non-traditional (namely, pornographic) products or services, it will likely pay higher fees for its merchant account.

Although the fees you can expect to be charged for your e-business merchant account vary, they will generally include: a one-time setup fee, a per transaction fee, a monthly minimum charge, a per-transaction authorization fee, a fee for a monthly transaction statement, and a percentage of each transaction (called a **discount rate**). Table 5-1 offers examples of the merchant account fees typically charged by a merchant bank and by various merchant account providers. Because fees vary, it is a good idea to shop around to find the best value before you set up your e-business merchant account.

TABLE 5-1 Sample merchant account fees

SAMPLE MERCHANT ACCOUNT FEES				
Fees	Merchant Bank #1	Merchant Account Provider #1	Merchant Account Provider #2	Merchant Account Provider #3
One-time set up fee	$149.00	N/A	N/A	N/A
Monthly fees				
Minimum monthly fee	$35.00	$9.95	N/A	$10.00
Payment gateway access	Included	Included	$15.95	$15.00
Monthly statement fee	Included	Included	$9.95	$10.00
Per-transaction fees				
Discount rate	2.40 to 3.25%	2.33%	2.19%	2.25%
Authorization fee	$0.40	$0.30	$0.25	$0.25

As you determine the payment methods your e-business will accept, remember that the security of personal and financial information is of utmost importance to your customers. It is vital for e-businesses accepting credit, debit, or charge card payments to provide adequate security for the card information as it is being transmitted through the payment process. The next section outlines ways to provide this security.

Secure Transmission of Payment Card Numbers

To provide secure transaction processing, your e-business must use some type of encryption and digital authentication to protect data while it's being transmitted to and from your customers.

Encryption is the translation of data into a secret code, and a **protocol** is a standard or agreed-upon format for electronically transmitting data. One of the earliest Internet security protocols developed is **Secure Sockets Layer (SSL)**, which provides encrypted server-side transactions for electronic payments and other forms of secure Internet communications. To use the SSL protocol, an e-business must contact its ISP or Web hosting company, which then places the Web page for the e-business's online order form on a secure server. The URL for the SSL-secured online order form then begins with https:// instead of just http://, indicating that the information entered in the form is transmitted using the SSL protocol. Most browsers indicate that information is being transmitted using encryption by displaying a warning message or some type of icon, such as a lock or key.[16] Figure 5-10 illustrates the process of encrypting and decrypting credit card transaction information. Figure 5-11 illustrates the secure transmission indicators you might see in a Web browser.

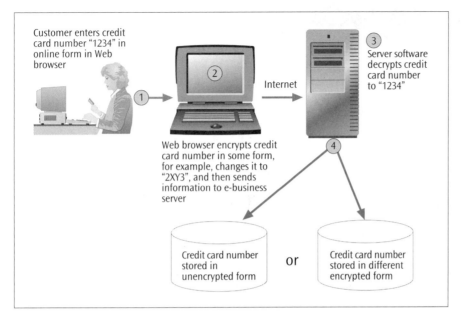

FIGURE 5-10   Encrypting and decrypting credit card information

FIGURE 5-11   Web browser secure transmission indicators

SSL transmissions rely on authentication using digital certificates. **Authentication** is the process of verifying identity, and a **digital certificate** is an electronic message attachment that verifies the sender's identity. Because the SSL protocol uses digital certificates to authenticate an e-business's identity during secure data transmission, an e-business offering secure data transactions must have a digital certificate.[17] Digital certificates are issued by a **certificate authority**, a trusted third-party organization such as VeriSign, which guarantees the identity of the sender for a fee. To get a digital certificate, an e-business's ISP or Web site hosting company generates a Certificate Signing Request (CSR), which is then submitted to a certificate authority. The certificate authority authorizes the digital certificate, and the ISP or Web hosting company then installs the digital certificate at the e-business's Web site. Figure 5-12 illustrates a digital certificate.

FIGURE 5-12 Digital certificate

Corporate and Customer Identity Theft Risks

Identity theft is the number one fraud-related complaint received by the Federal Trade
Commission (FTC).[19] Unfortunately, identity theft—called "the fastest-growing white-
collar crime in the U.S."—has become so profitable that domestic and international organized crime is now involved.[20] Identity theft involves stealing consumers' names, addresses,
Social Security numbers, and credit card numbers. Corporate identity theft, which
involves stealing and misusing corporate identity information, is also becoming a growing
problem. Corporate identity theft often targets small and medium-sized businesses where
the theft of business names, logos, addresses, and Web site information might "fly under the
radar" long enough for the criminals to use the information in some type of scam.

QUOTES ON SUCCESS

"Identity theft has emerged as one of the dominant white collar crime problems of the 21st Century."

Chris Swecker, Assistant Director of Criminal Investigative Division, Federal Bureau of Investigation

Fraudulent credit card charges cost businesses and consumers billions of dollars each year. For example, in its sixth annual fraud report, the online payment processing company CyberSource reported that Internet fraud cost e-retailers about $2.6 billion in 2004 alone.[21] To steal consumer information, criminals often use the Internet to hack into business networks and databases. In fact, in one short period of time in early 2005, the media reported more than 10 significant data thefts, including thefts of consumer information at major corporations such as Lexus-Nexus, Bank of America, CardSystems Solutions, Inc., and Ameritrade.[22] Criminals also use online "phishing" schemes, which involve sending e-mails with links to fake versions of well-known Web sites to "fish for" and catch credit card numbers and personal information from unsuspecting consumers.

Criminals profit from stealing credit card numbers and business or personal information in a number of ways: by going on a shopping spree themselves, by reselling the numbers or information to other criminals, or by setting up fake e-businesses similar to real businesses. For example, one way criminals profit from both corporate and consumer identity theft is by taking advantage of the less stringent requirements of some merchant account providers and opening bogus merchant accounts with stolen business name and address information. The criminals then make charges small enough to be easily approved using consumers' stolen credit card numbers, and they submit these charges through the bogus merchant accounts' payment gateway. By doing this, they can then clean out the credit card payment funds deposited in the merchant accounts before anyone notices the scam.[23]

As noted earlier in the chapter, consumers who pay online with credit cards (the most popular payment method) are protected against fraudulent charges. Banks that issue credit cards generally charge back fraudulent credit card charges to the e-businesses who accepted the cards. As an e-business entrepreneur, you must, therefore, consider all possible measures to reduce your e-business's exposure to these chargebacks. Some of those measures include:[24]

- sending orders by national shipping companies and requiring a signature on delivery to avoid having customers claim that the products were not received
- getting complete billing and shipping information, including a complete address and phone number, before shipping orders
- accepting orders placed from ISP-based or domain-name-based e-mail addresses only, not from free, Web-based, or e-mail forwarding addresses, so that transactions can be traced to an actual person
- calling the phone number on a questionable order to verify the order
- waiting to ship items until authorization and verification are ensured, especially for orders being shipped out of the country
- carefully checking very large orders and overnight delivery requests

- double-checking international orders where the shipping address is outside the U.S. but the charge, credit, or debit card is issued by a U.S. bank; for example, double-checking an order to be shipped to central Europe paid for by a credit card issued by a Houston, Texas bank
- using the Address Verification System and other fraud prevention tools that are available through your merchant account provider or the financial institution that hosts your merchant account

One tool that both consumers and e-businesses can use to protect against the risk of credit card fraud at the point-of-sale is disposable or **virtual credit card numbers**. Citi's Virtual Account Numbers and Discover Card's Discover Deskshop are two examples of virtual credit card number programs. A virtual credit card number is designed such that it can be used only at the Web site where it was first used; therefore, it has little value to a thief because it is not re-usable. As an additional security measure, some virtual credit card numbers have an expiration date. While intended to protect the consumer, the virtual credit card number may also protect an e-business by minimizing the risk associated with a theft of information from the e-business's databases.[25] A disposable credit card number is not, however, suitable for all online transactions—such as guaranteeing airline, hotel, or rental car reservations, or for recurring credit card charges such as automatic payment of utility bills or automatic payments for monthly pet medications.

E-CASE

A Real Problem: Fake E-Businesses

In late 2004, the owner of a small software company in New York that doesn't take credit cards was notified that more than $10,000 in credit card charges had been made from his business. During an investigation, it became apparent that the software company was one of about 50 legitimate businesses—none of which accepted credit cards—that were victims of an elaborate scam.

First, the perpetrators of the scam used the businesses' names and similar addresses to set up Web site addresses (URLs). Next, the scammers created dummy Web sites. Then the scammers applied to various merchant account providers and financial institutions, such as Global Payments, North American Bancard, First Data, Wells Fargo, and Beacon Bank, to open new merchant accounts. They used stolen business names and business addresses together with the fake Web sites to make the applications appear to be from legitimate businesses. The scammers managed to open several merchant accounts; once a merchant account was set up, the scammers were able to use stolen credit card numbers already in their possession to process charges through the merchant accounts.[26] Although the exact loss figures are still not known, it is believed that the perpetrators were able to steal several hundred thousand dollars before the business owners and merchant account providers caught on to the scam.

continued

Investigators think the scam may have involved international organized crime because it closely resembled the PharmacyCard.com scam of 2004, in which almost 90,000 consumers were charged a total of $10 million in fraudulent credit card charges before the scam was stopped. The Federal Trade Commission (FTC) traced the monies from the PharmacyCard.com scam to Nicosia, Cyprus.[27] What can a small business owner do to protect his or her business from such a scam? Be aware of the risk and search the Web on a regular basis for Web sites that might be using your business's name or other information.

Other Electronic Payment Methods

Although the preferred electronic payment methods among consumers today are credit, debit, and charge cards, e-businesses can offer their customers several other electronic payment methods, such as stored value cards, electronic checks, electronic cash (micropayments), and person-to-person (P2P) payments.

Stored Value Cards

Thousands of teens and children under the age of 12 would like to spend their allowances or part-time-job incomes shopping online for clothes, music CDs, and other popular items. Most young shoppers lack access to the credit cards accepted by e-retailers, and parents are often reluctant to allow children to use their credit cards. Additionally, there are thousands of adult consumers who do not have or want credit, debit, or charge cards. To sell products and services to these two groups of consumers, e-businesses can accept stored value cards.

A **smart card** is a small card approximately the size of a credit card that contains an electronic memory chip. Smart cards are used for a variety of purposes, such as storing medical records, personal identification data, and cash. A **stored value card** is a smart card that stores cash. Examples of stored value cards include prepaid telephone cards, government benefit cards used to purchase food, and gift cards. Some stored value cards, such as single-purpose gift cards, can be used only where they were purchased. For example, a Target store gift card can be used at Target's brick-and-mortar stores and at the Target Web site, but not at any other store or Web site. In contrast, multi-purpose stored value cards, such as Visa Cash, MasterCard prepaid cards, and American Express Travelers Cheque cards, can be used anywhere—both online and offline—that regular Visa, MasterCard, and American Express cards are accepted. The Federal Reserve Bank of New York expects that, by 2006, transactions involving all types of stored value cards will total more than $72 billion.[28] Figures 5-13 through 5-15 illustrate card issuer Web sites where stored value cards are described.

FIGURE 5-13   Visa Cash cards

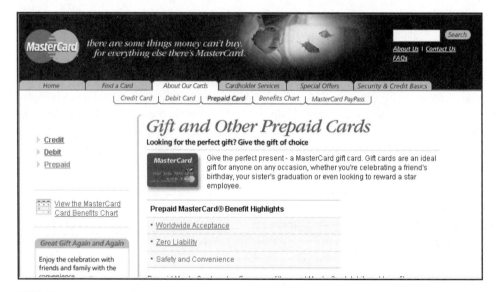

FIGURE 5-14   MasterCard prepaid cards

FIGURE 5-15 American Express Travelers Cheque cards

Electronic Checks

An **electronic check**, sometimes called an **e-check** or **ACH check**, is the electronic version of a paper check. It not only contains the same information as a paper check, but is based on the same legal framework as that of a paper check, and thus can be used in any transaction in which a traditional paper check is used. An electronic check is, however, much less expensive to process than a paper check. While the costs for processing a paper check range anywhere from 75 cents to $3, it costs only a few pennies to process an electronic check.[29]

Like paper checks, electronic checks are processed at the point of sale. To pay for a purchase at a brick-and-mortar store using an electronic check, the buyer writes out a check and gives it to a cashier. The cashier then scans the check through a MICR (magnetic ink character recognition) device that captures the account number, serial number, and financial institution routing number that appear on the check. The scanned data moves electronically to a check authorization service such as TeleCheck, which verifies that the check is drawn on an open account and that the buyer does not have a record of writing "hot" checks (checks drawn from accounts with insufficient funds). The paper check is then voided and maintained by the seller, or it may even be returned to the buyer. After verification from an authorization service, the electronic check data then goes to the ACH (Automated Clearing House) network, where the amount of the check is credited to the merchant's account and the data is forwarded to the buyer's financial institution, where the amount is debited from the buyer's account.

Similarly, to use an electronic check at an online store, buyers fill out a Web-based form at checkout, providing the merchant with their checking account number and their bank's routing number. This data is submitted to a check authorization agency and sent on through the ACH system just like an electronic check presented at a brick-and-mortar store. Figure 5-16 illustrates the electronic check process at an online store.

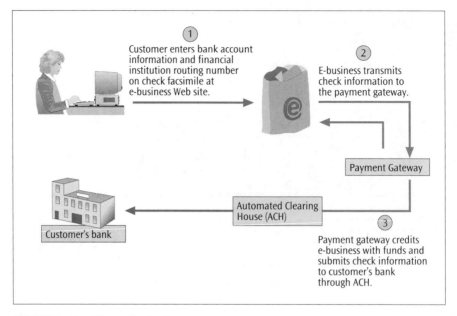

FIGURE 5-16   Electronic check process

Electronic Cash and Micropayments

Electronic cash, sometimes called **digital cash** or **e-cash**, is a payment method that allows buyers to pay for goods or services online by transmitting a unique electronic number (or other identifier that carries a specific dollar value) over the Internet. There are two primary advantages of using electronic cash instead of a credit card: lower processing costs for the seller, and no credit card authorization for the buyer. Although electronic cash is a popular online payment method in Europe and Japan, it has not been widely accepted by U.S. consumers, who still prefer to use credit cards to pay for their online purchases.

TIP

The late 1990s saw the launching of many new e-businesses, such as beenz and Flooz, that offered electronic cash consumer rewards programs or provided electronic cash for prepaid accounts targeted at children and teens. Despite high-level promotion efforts, this type of electronic cash just never took off with young consumers or their parents, and these companies did not survive the dot.com bust of 2000-2001.

E-PIONEERS

Unfortunate Timing for a Good Idea

David Chaum is a renowned cryptographer and mathematician. He began his work in cryptography in the 1980s as a graduate student in computer science at Berkeley, where he wrote his master's thesis on an electronic mail system that was so private it could be used for voting. After receiving a doctorate at Berkeley, Chaum went to the Center for Mathematics and Computer Science in Amsterdam, where his interest in privacy and the Internet led to his inventing a form of electronic money whose legitimacy could be verified while its source could not be traced. Chaum named his electronic cash "eCash," and in 1990 he founded an e-business called DigiCash in order to market eCash to banks.[30]

DigiCash's eCash product gained widespread attention in 1994 and 1995 as banks and other businesses began to look at the commerce opportunities the Internet provided, and soon DigiCash succeeded in arranging licensing agreements with several banks, including Deutsche Bank, Den Norske Bank, Bank Austria, Advance Bank of Australia, and Mark Twain Bank of St. Louis. These banks hoped to offer their customers a new service: the ability to use the funds in their bank accounts to make purchases online without disclosing critical bank account information.

Unfortunately, however, DigiCash's eCash product was ahead of its time. In 1995, commercial Web browser technology was still in its infancy, and the idea of using the Internet to purchase goods and services was so new that the term "e-business" had not yet been invented. Also unfortunate for DigiCash was the fact that as online shopping eventually began to take off, consumers turned out to be much less resistant to the idea of providing personal credit card information over the Internet than industry analysts had originally supposed.[31]

Lack of success with the eCash pilot programs, the complexity of the eCash system, and internal management problems led to DigiCash filing for bankruptcy in November 1998.[32] But the idea of providing consumers with electronic cash was still attractive to some entrepreneurs and investors. In May 1999, a group of entrepreneurs bought DigiCash's software and patents and formed a company called eCash-Technologies, Inc., known simply as eCash. But eCash was unsuccessful at reviving Chaum's electronic cash idea. In 2002, InfoSpace, Inc., a wireless and Internet technologies company, purchased eCash Technologies' intellectual property. Meanwhile, Chaum, who had no involvement with eCash Technologies or InfoSpace, Inc., continues to pursue his interest in cryptography, privacy issues, and secure election voting.

It was once thought that **micropayments**, payments for purchases ranging from a few cents to $5 in cost, would spur the growth of online shopping. Early attempts at introducing micropayment solutions involved charging consumers small amounts for access to Web content. These attempts were not successful for a number of reasons, including a general unwillingness by U.S. consumers to be "nickeled and dimed" for the Web content they expected to access for free.[33]

But the recent success of charging consumers to download music—individual songs downloaded for less than $1—may indicate a revival for the micropayment concept. Several new e-businesses, such as BitPass, Peppercoin, Paystone, and Yaga, are betting that an increasing number of online consumers will be willing to pay for all types of content, including music, games, photos, and online publishing (e-books, comic books, and short stories)—and their bets may just pay off. According to a 2004 survey by the market research firm Ipsos-Insight, micropayment purchases in the U.S. are growing at more than 300 percent a year, and an estimated 30 million U.S. consumers are somewhat likely to make a micropayment purchase.[34] TowerGroup, a financial services research and consulting company, predicts the global micropayment market will grow to more than $10 billion by 2009.[35] Figures 5-17 and 5-18 show Web sites for some of the micropayment technology and service providers in the emerging micropayment marketplace.

FIGURE 5-17   Peppercoin

FIGURE 5-18 Yaga

E-CASE

A Fortune in Nickels and Dimes?

Shortly after the dot.com bust of 2000-2001, Kurt Huang and his roommate at Stanford, Gyuchang Jun, were dismayed to find that some of their favorite Web sites had begun charging a monthly subscription fee for access to content that had previously been free. Neither wanted to pay for a monthly subscription just to be able to access a single article or report that had caught their interest. Following in the footsteps of earlier Stanford graduate students such as Larry Page and Sergey Brin (who launched Google) and David Filo and Jerry Yang (founders of Yahoo!), Huang and Jun decided to start an e-business. Their e-business idea was to offer customers a service that enabled them to quickly and easily pay for individual items of online content—using micropayments—instead of signing up for a monthly subscription.[36]

Jun developed the micropayment software, and Huang pitched their e-business idea to a number of venture capitalists, including Guy Kawasaki, managing director of Garage Technology Ventures. Kawasaki liked the e-business idea and the software; in 2002, Garage Technology Ventures provided $1.5 million in funding for the new e-business, now named BitPass. BitPass offers consumers a prepaid account from which they can make micropayments for online purchases. Opening a prepaid account with BitPass is free, and consumers can fund with it as little as $3. Once consumers have an account, they can use their BitPass funds to pay for content at Web sites that offer the BitPass payment service. At the BitPass Web site, consumers can find a directory of Web sites that accept the BitPass micropayments. BitPass makes money by charging these Web sites a percentage of each micropayment transaction.[37]

continued

The future might be bright for BitPass. Two experienced e-business and financial services managers are now at the helm as CEO and COO, and a former chairman and CEO of American Express sits on the BitPass board. In late 2004, BitPass announced that it received a second round of funding in the amount of $11.75 million from Worldview Technology Partners, Steamboat Ventures (the Walt Disney Co.), RRE Ventures, and others. Since its inception, BitPass has signed up more than 2,000 providers of digital content (Figure 5-19).[38, 39]

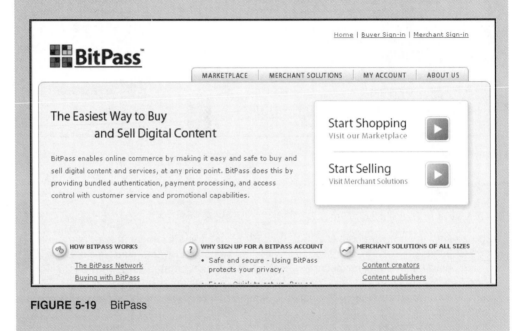

FIGURE 5-19   BitPass

QUOTES ON SUCCESS

"Three major things have changed to conspire to make this [micropayments] a big opportunity for BitPass. People are no longer scared to make a financial transaction online and are much more willing to use their credit card online. Another change is that advertising can no longer support all content models, which drives a willingness to charge [for content]. Thirdly, the technology has really stabilized. Original providers required a download and [Web browser] plug-in in order to use their service whereas our [service] does not."

Kurt Huang, co-founder of BitPass

Person-to-Person Payment Systems

P2P (or person-to-person) **payment systems** enable a person to send money to anyone anywhere around the world just as long as that other person has an e-mail address. Consumers can pay for online purchases through their personal P2P account using a credit card or debit card, or by directly transferring funds from their bank account. E-businesses can use a business P2P account to accept these payments. P2P payment systems are used extensively by buyers and sellers in the online auction marketplace. In the late 1990s, several companies jumped on the P2P payment bandwagon; most have either gone out of business (such as Yahoo! DirectPay) or changed their focus to online payment systems for B2C transactions, such as Western Union MoneyZap and ProPay.com. Left standing are Bid-Pay and the major player in the P2P payment system marketplace, PayPal, which has more than a 60 percent share of the market.[40] Figures 5-20 and 5-21 illustrate the Web sites of e-businesses offering P2P payment systems.

Determining just how your customers will pay for your products and services is an important step for an e-business startup. You must also understand the types of technologies that can help you operate your e-business, and become familiar with the vendors who provide these technologies.

FIGURE 5-20 BidPay

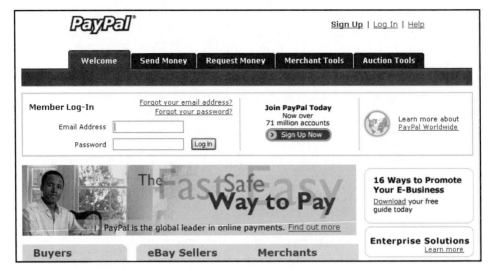

FIGURE 5-21 PayPal

E-BUSINESS TECHNOLOGIES

As an entrepreneur, you will make many technology-related decisions for your new e-business. One of the most important of these involves determining what software you will need to manage all of the various aspects of running your e-business, such as processing sales transactions and payments, monitoring inventory, accounting, and other internal recordkeeping. You must also decide how your e-business will connect to the Internet and whether you should host and maintain your Web site or outsource the Web site operations. Your technology decisions will be based on your e-business's front-end and back-end systems needs.

Front-End and Back-End Systems

Front-end systems are the aspects of a business with which the business's customers interface and over which customers can exert some control. For your e-business, front-end systems would be the Web site and other related processes that your customers use to view information and purchase products and services. **Back-end systems** are the aspects of a business that are not directly accessed by customers. For your e-business, back-end systems would include the systems that handle the accounting and budgeting, manufacturing, marketing, inventory management, distribution, order-tracking, and customer support processes.

Your e-business's front-end systems will probably require much of the same data that is already stored in its back-end systems, such as product availability and pricing. Additionally, new data being gathered by your front-end systems—for example, order and customer information—must be made available to the back-end systems to facilitate internal business processes such as accounting, billing, payment processing, and order fulfillment. Integrating front-end and back-end systems not only provides an e-business with more useful information about its own operations, but also enables the e-business to reduce its costs by allowing common data to be shared across front-end and back-end applications. Figure 5-22 illustrates the integration of an e-business's front-end and back-end systems.

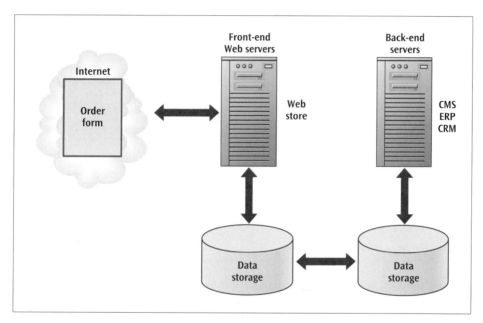

FIGURE 5-22   Front-end and back-end systems

Many e-business operations depend on the integration of front-end and back-end systems. For example, in order to authenticate and authorize a given credit card payment, the credit card information (front-end data) must be transferred either to a credit card processing provider, or processed internally using credit card authorization software (both of which are back-end business processes). Also, if a customer purchases a product at your Web site, the order information must be routed to your warehouses for fulfillment, and the product inventory records must be updated. The transaction must also be recorded in your accounting system. If you have outside suppliers, it may be necessary to transmit the order transaction to systems outside of the e-business, such as those of designated suppliers or shipping agents. Figure 5-23 illustrates an example of typical order transaction processing for an e-retailer.

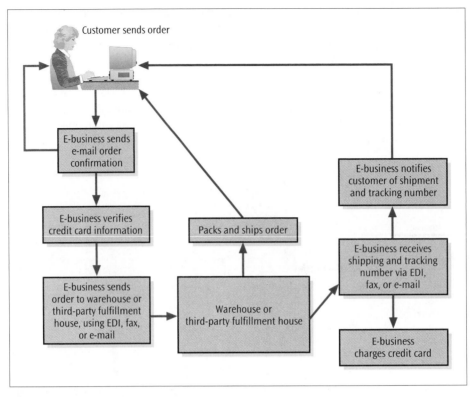

FIGURE 5-23   Example of online transaction processing

Entrepreneurs who are adding an e-business component to their existing brick-and-mortar businesses have an additional challenge when making technology decisions—how to integrate their new systems with their existing or **legacy systems**. Integration of new systems with existing systems requires the following considerations.

- *Real-time or batch processing*: Do the various systems involved require real-time updating of data—for example, checking the inventory system before allowing the customer to order an item? If there is no need for real-time data exchange, or if this updating is simply not feasible, then the new system must store the data to be updated in a queued list to be sent at a specified time. Many e-businesses do this with their credit card authorization when their systems are not capable of real-time authorization. They store all of the credit card numbers to be authorized during the day, and then authorize them in a batch at the end of the day.

- *Security*: Because e-business operations usually involve several interconnecting systems, each system must be secure. It is possible that the e-business will exchange data with an outside business such as a supplier or shipping agent. Such connections need to be secure as well, in both the old and new systems.
- *Compatibility*: Compatibility is always a consideration when integrating systems not originally designed to communicate with each other. When selecting software components for online transaction processing, it is important to make sure that these software components can share data with the other components. To facilitate the exchange of data, it may be necessary to write special data translation programs, or to purchase specialized software, often called **middleware**, for making connections between two systems.

As your e-business grows, the software technologies you need to operate and integrate your front-end and back-end systems can range from simple storefront software, which can help you create and operate a small online store, to the expensive and extensive Web content management, enterprise resource planning, and customer relationship management technologies that are used by larger or more mature e-businesses.

E-Business Software

The amount of information that your e-business must keep track of is tremendous. What's more, this information must be processed quickly. For example, with the swift pace of business today, you may need to rapidly update product and pricing changes in your online catalog, post current industry announcements, or add up-to-date syndicated news headlines, weblogs, and articles to your Web site in order to keep your Web site current and competitive.

You must also ensure that the products in your customers' shopping carts are actually in inventory. To fill customers' orders in a timely manner, you must be able to quickly route order and delivery information to a warehouse for order fulfillment. To retain customers as your e-business grows, you will need to learn more about your customers' shopping habits and buying preferences. Fortunately, there are several types of management software systems that your e-business can use to store and process its vast quantities of data, convert that data into useful information, and then use the power of the network to make the information quickly available to employees, vendors, and customers. Table 5-2 summarizes different categories of e-business management software systems.

TABLE 5-2 Management Software Systems

Software	Description	Sample Vendors
Storefront Software	Builds and maintains Web pages and underlying databases for an e-retail site	GoEcart StoreFront.net Yahoo! Small Business
Content Management System (CMS)	Manages and controls dynamic Web content from authorship to publication	Eprise OpenText Corporation Interwoven PaperThin Vignette
Enterprise Resource Planning (ERP)	Integrates all aspects of business operations including manufacturing, purchasing, sales, and accounting	Oracle PeopleSoft Oracle J. D. Edwards SAP Plumtree Software
Customer Relationship Management (CRM)	Organizes and manages interactions with customers including sales, payments, and customer support	salesforce.com Siebel Oracle PeopleSoft SAP

Storefront Software

If your e-business is an e-retailer, you will need to set up and maintain an online store. You can create online store by licensing and installing storefront software on your server. **Storefront software** creates an online store for your Web site and enables you to accept order and payment information and then process that information. Small to medium-sized e-retailers can choose from a number of pre-packaged online store options, called hosted storefront software. **Hosted storefront software** can provide everything a small e-retailer needs to build an online store: Web site hosting services, easy-to-use templates to build online store Web pages (such as product catalogs), and shopping cart software. **Shopping cart software** manages customer purchases by temporarily storing information about items selected for purchase and then handling the complete checkout process, which generally includes shipping and tax calculations, credit card authorization, and payment processing. Larger e-retailers may choose to license customizable storefront software with more complex features, such as inventory management, and make their own Web site hosting arrangements. Vendors that offer storefront software include Yahoo! Small Business, GoECart and StoreFront.net. Figures 5-24 through 5-26 illustrate the Web sites for these popular storefront software vendors.

FIGURE 5-24   Yahoo! Small Business

FIGURE 5-25   GoECart

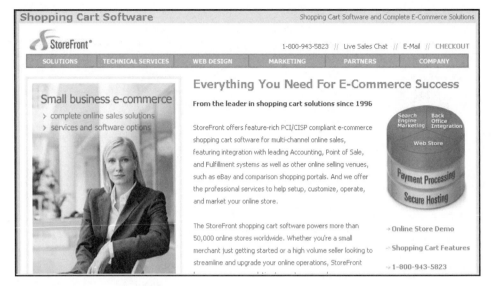

FIGURE 5-26   StoreFront.net

Content Management System (CMS)

As your e-business grows, the content on your Web site is likely to become more complex. To make your site more interesting and useful to your customers, you may choose to provide syndicated news headlines, industry updates, and other dynamic material from a variety of sources. In order to manage complex Web site content but also make certain that all of your Web pages maintain a professional and standardized appearance, you must be able to control the quality, accuracy, and timing of the content that comes in from these sources. A **content management system (CMS)** is a system that controls all the processes involved in Web content development, including authoring, reviewing, editing, and then publishing the content in a timely way.

Databases, called **content repositories**, are at the center of a content management system. Content repositories contain Web page templates and style sheets that help control the appearance of all the Web pages at a site, as well as commonly used graphics (such as logos), text documents, and syndicated content (such as news headlines). Web content can be built "on the fly" from the components stored in these content repositories. In addition, time-sensitive content, like press releases, can be stored in content repositories and then published according to a predetermined schedule. Vendors that provide content management system technologies include OpenText Corporation, Interwoven, and Vignette. Figures 5-27 through 5-29 illustrate the Web sites for these content management system vendors.

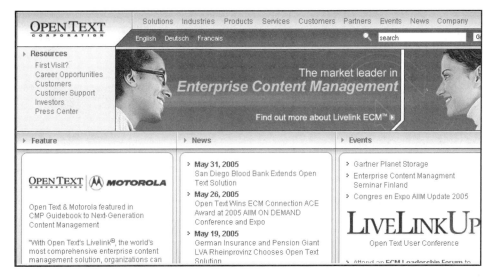

FIGURE 5-27   OpenText Corporation

FIGURE 5-28   Interwoven

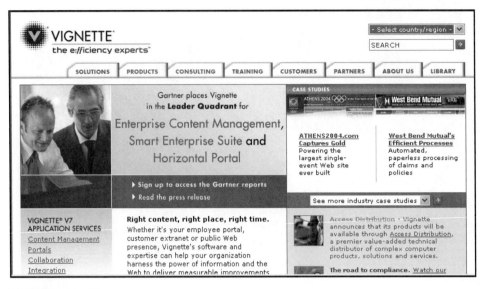

FIGURE 5-29 Vignette

Enterprise Resource Planning (ERP)

The term "enterprise" is often used in business to identify any organization that uses computers and a network to interact with employees, suppliers, and customers, but it is more often associated specifically with large businesses. Large e-businesses may use **enterprise resource planning (ERP)** systems to integrate all aspects of their business operations, including product planning, manufacturing, human resources, accounting, finance, sales, and marketing. ERP systems had their origins in the 1960s with manufacturing companies, who began to incorporate elements such as raw material planning, purchasing, shop floor management, and distribution into their manufacturing systems. In the early 1990s, other enterprise activities, such as engineering, project management, accounting, finance, and human resources, were added to ERP systems. Today, many large businesses use Web-based interfaces to connect their ERP system components. The leading ERP systems providers for both brick-and-mortar enterprises and e-businesses are Oracle and SAP.

Customer Relationship Management (CRM)

Getting and keeping customers is a top priority for any business. Another large enterprise system that, like ERP, uses Internet and Web-based technologies is a **customer relationship management (CRM)** system. CRM applications compile customer information into databases and use the information to match customer needs with products and services, to remind customers when new products or service updates are available, and to analyze customer shopping and buying behaviors. ERP systems providers such as SAP and Oracle also provide CRM systems technologies.

In addition to acquiring a basic understanding of the various technologies that can support your e-business, you must make decisions about which e-business operations can be managed in-house and which can be outsourced.

OUTSOURCING OPERATIONS

The wide array of outsourced services available to your startup e-business includes Internet access for your employees, hosting services for your Web site, software application rental, Web development services, Web strategy consulting, data file storage and backup, product warehousing and fulfillment services, and all conceivable combinations of these and other services. The major advantage of outsourcing some of your e-business's operations is reduced costs. The major disadvantage is your lack of control over all aspects of those outsourced operations. Table 5-3 shows examples of service providers to which you can outsource some of your e-business's operations.

TABLE 5-3   Service Providers

Service Provider	Description	Sample Vendors
Internet Service Providers (ISPs)	Provides Internet access, e-mail services, small business Web hosting, and wireless connectivity services	Earthlink Time-Warner Cable SBC
Web hosting companies	Provides a variety of Web hosting options, including co-location and managed hosting	Rackspace Managed Hosting Hostway Affinity Hosting.com
Application Service Providers (ASPs)	Distributes software services over the Internet	Amerivault (data backup) USInternetworking (ERP/CRM) Thomson Elite (time and billing)
Fulfillment houses	Warehouses products and picks, packages, and ships orders	Webgistix Innotrac Turnaround

Outsourcing Web Hosting to an ISP or Web Hosting Company

Internet service providers (ISPs), sometimes called Internet access providers (IAPs), are the companies that provide connections to the Internet. Because they already have the hardware, software, and Internet connection necessary to operate Web sites, many ISPs, such as Earthlink, also offer personal and business Web hosting services. In contrast, for **Web hosting companies**, such as Affinity, Hosting.com, and Rackspace Managed Hosting, managing and maintaining outsourced commercial Web site operations at their data centers is the primary focus. Outsourcing your e-business's Web site operations to one of these Web hosting service providers has several advantages over hosting your own site. In particular, a hosting service provider:

- has a staff available 24 hours a day that have the technical knowledge necessary to keep the servers and network connections running
- can provide a level of redundancy (multiple high-speed Internet connections, backup power from generators, and backup servers) that an e-business may not be able to provide for itself

Outsourcing Web site operations can be a particularly cost-effective choice for a small e-business. Instead of assuming all the costs related to Web site hardware, software, and operations by itself, an individual e-business ends up sharing these costs with the other e-businesses whose sites are also outsourced to the service provider. When evaluating Web hosting service providers, you should look for those providers that have scalable servers, high-speed connections, 24/7 technical support, server and network redundancy, secure server facilities, and tested disaster-recovery plans.[41]

Many hosting service providers offer different levels of service at different prices, so you must determine the level of service your e-business requires. Your management team should consider questions such as: How much server storage space and bandwidth does your e-business need? Can your e-business share a server with other e-businesses, or do you need a server that is dedicated to your e-business's operations? Do you have the in-house expertise to manage your Web site remotely, or do you need a managed hosting arrangement where the hosting provider's staff handles all routine server management? Determining the answers to these questions will help you decide the level of hosting service that would best serve your e-business.[42]

An e-business that prefers to control its server operations can provide its own servers and software and simply **co-locate** its servers at the hosting service providers' facilities. Co-locating servers involves renting space and Internet connectivity from a hosting service provider for an e-business's own servers. Co-locating its servers allows an e-business to maintain control over its servers while taking advantage of the high-speed transmissions and secure data center facilities offered by the service provider.

In addition to outsourcing Web site operations, you may consider outsourcing your e-business's software applications to an application service provider.

Outsourcing Web-Based Software Applications

Application service providers (ASPs) use private networks or the Internet to deliver and manage software applications and other computer services from remote data centers. ASPs offer a diverse menu of services. For example, ASPs typically provide IT infrastructure, software licenses, application support, database administration, system administration, software upgrades, and technical-problem resolution for their clients. Client companies access their software applications remotely over the Internet, and may pay either a monthly fee for these services or a per-transaction fee. Examples of ASP-provided applications that your employees could access over the Internet are time and billing, calendaring, data backup, ERP, and CRM applications. One leading ASP is USInternetworking, which partners with major companies such as Oracle PeopleSoft (human resources and financial applications), Microsoft (e-business, exchange hosting, and messaging applications), and Siebel Systems (customer information applications) to provide a variety of packaged application software.

For some startup e-businesses, the advantages of using an ASP are great, and include exploiting the ASP's superior level of IT experience and services while reducing hardware, software, and personnel costs; getting access to ERP and CRM systems too expensive to be supported internally; and allowing the e-business's employees and management team to focus on core business functions. The disadvantages of using an ASP are lack of control in terms of application performance and data, data security concerns, and the additional costs that may be incurred to customize application features.[43] Before you contract with an ASP, you should become familiar with the size and location of its data centers and its business practices, including management policies, data security procedures, customer support levels, and user training options.[44]

If your e-business is an e-retailer, effectively managing the order fulfillment process—getting customers' orders filled, shipped, and delivered in a timely manner—is critical.

Outsourcing Order Fulfillment Management

Unless your e-business ships thousands of orders each day, it may be more cost-effective for you to outsource your order fulfillment processes instead of building expensive in-house warehousing, order picking, and order distribution systems. A **fulfillment house**, sometimes called a third-party logistics provider or 3PL, is an independent company that provides order fulfillment services: warehousing your products and picking, packaging, shipping, and tracking your customers' orders. By outsourcing your order fulfillment management to a fulfillment house, you may reduce costs by avoiding the expense of maintaining warehouse space (including related costs such as warehouse heating, cooling, maintenance, and insurance) and the cost of hiring employees to oversee order processing. Additionally, partnering with a fulfillment house may enable your e-business to offer its customers various other benefits, including increased customer service and support after the sale; additional marketing options, such as promotional gifts and gift wrapping; and access to better shipping rates. Webgistix and Turnaround, as shown in Figures 5-30 and 5-31, are examples of fulfillment houses.

FIGURE 5-30   Webgistix

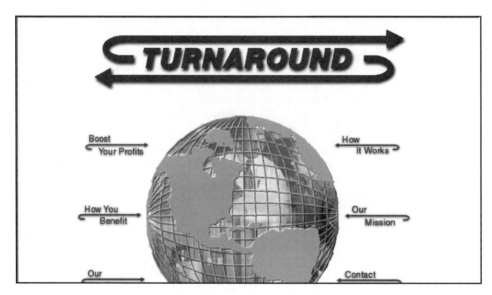

FIGURE 5-31   Turnaround

Outsourcing critical functions such as order fulfillment means that you lose a measure of control over the processes that lead directly to customer satisfaction. In order to establish an effective e-business partnership with a fulfillment house, you should consider several factors. For example, a fulfillment house should:

- maintain adequate inventory levels to satisfy anticipated demand
- maintain a high order-fill rate (the percentage of orders that get filled on the same day the order is placed)
- have a reasonable cutoff time for orders to be filled the same day
- use a variety of well-known shipping companies
- support real-time communication to your e-business regarding the status of inventories and orders, instead of communicating these details via simple batch processing

The questions you should ask a fulfillment house before partnering with it include: Are there any restrictions on the types of items that can be warehoused and shipped? Are there required minimum or maximum quantities of each order? Are fees are charged for each service—shelf space, picking, packaging, shipping—and how much are the fees? What types of management reporting on inventory and shipments is available?[45]

TIP

Some e-businesses choose to maintain a virtual inventory—that is, a record of products that do not belong to the e-business itself, but rather to its partners, third-party manufacturers, or distributors, who own, warehouse, and ship the products. This arrangement allows the e-business to avoid tying up its cash in inventory. When an e-business maintains a virtual inventory, its partners control the fulfillment process, but the e-business is still responsible to its customers for the effectiveness and timeliness of the order fulfillment processes.

QUOTES ON SUCCESS

Everyone had been out there believing these crazy predictions about incredible [e-business] growth and changing-the-world technology and it didn't happen, and people got scared and left. These two companies [eBay and PayPal] are two of the only ones whose predictions were right. Their technologies are changing the way people do things."

John Draft, financial analyst with D.A. Davidson & Co.

. . .MONEY FROM E-MAIL

Confinity officially launched its person-to-person electronic payment service, now called PayPal, in December 1999, by having celebrity spokesperson James Doohan, a.k.a. "Scotty" on the original Star Trek™ television program, use his PDA to beam money to selected e-mail users.[46, 47] Confinity's faith in the network effect and viral marketing was well-placed—customers began signing up for its service in droves. Customers loved the Confinity electronic payment service, but they weren't using it for wireless devices. Instead they were using their personal computers and PayPal accounts to send and receive payments for their online auction purchases and sales. By January 2000, PayPal had more than 10,000 new customers, about 60 percent of whom were eBay buyers and sellers. In March 2000, PayPal, which had quickly become a leading payment system for eBay customers, merged with X.com, an online banking service; and by May 2000, PayPal had 1.5 million customers.[48] In 2000-2001, the dot.com bust was taking its toll on PayPal's competitors, including eBay's own in-house person-to-person payment service, BillPoint. But PayPal just keep rolling along, riding successfully on its first-mover advantage. Within a year, the e-business had 6.5 million customers and total investor financing of more than $200 million. Strengthened by its dominant position in the market, PayPal began charging its high-volume customers transaction fees.[49]

It wasn't all rosy, of course. Some customers complained that PayPal made it too difficult to get a check for the balance of their account; international criminals scammed PayPal accounts, creating fraud losses that the company had to bear; and regulators tried to find ways to keep PayPal from operating in certain states.[50] But by this point PayPal had become an unstoppable juggernaut.

In February 2002, PayPal began to be profitable, with more than 10 million customers and a customer base that was growing at about 18,000 new customers a day. Fueled by its success, PayPal launched a successful IPO on February 15, and by the end of the day, the e-business's stock market value was $1.2 billion.[51] But the best was yet to come. Within five short months after launching its IPO, eBay jettisoned its in-house BillPoint service and bought PayPal for about $1.5 billion.[52]

Today, eBay's award-winning PayPal person-to-person payment system has more than 70 million customers in 45 countries around the world.[53] Was the eBay buyout a good deal for the co-founders, Thiel and Levchin? The young entrepreneurs received millions in eBay stock and moved on to new opportunities. Thiel is now president of Clarium Capital Management LLC, a private investment fund, and Levchin is involved with a new e-business venture named Yelp!, an online person-to-person referral service.

Chapter Summary

- To help resolve the startup and operational challenges of starting a new e-business, one of the first things an entrepreneur should do is establish a relationship with an attorney experienced with business startups (preferably e-business startups).

- A copyright is a legal protection for the author of an original work such as a piece of writing, drama, music, art, or some other type of intellectual property. A trademark or service mark is a distinctive symbol, word, or phrase that identifies a business and its products and services.

- An e-business should post a liability disclaimer on its Web site to protect itself from lawsuits; it should also post a statement of its privacy policy regarding the handling and securing of data collected at its Web site.

- The laws affecting e-business taxation continue to change and should be monitored.

- An e-business should create a facilities plan before searching for commercial office space, and it should use the services of a commercial real estate broker to locate this space and negotiate its lease.

- Startup e-businesses often hire the employees they need as they obtain the funding required to cover additional salaries; first hires will likely be at the senior management level—namely, a chief technology officer and a marketing professional.

- Electronic payment methods include payment cards (credit, debit, and charge), stored value cards, electronic checks, electronic cash, micropayments, and P2P payment systems.

- To accept credit, debit, and charge cards, your e-business must have a merchant account and access to a payment gateway.

- You can apply for a merchant account directly with a financial institution or through a merchant account provider.

- The two major risks associated with accepting credit cards online are the cardholder not present risk and chargebacks. When the cardholder is not present, the risk of credit card fraud is greater. Chargebacks occur when a consumer refuses to pay for a credit charge because of fraud, billing errors, or product returns.

- Credit card numbers should be transmitted using the Secure Sockets Layer (SSL) encryption protocol.

- To manage the various aspects of running your e-business, you will require technologies that integrate your e-business's front-end systems (business activities with which customers interface and over which they can exert some control) with its back-end systems (business activities that are not directly accessed by customers). Technologies for an e-business include storefront software, content management systems, enterprise resource planning, and customer relationship management software.

- E-businesses may elect to outsource some of its operations to ISPs, Web hosting companies, ASPs, and fulfillment houses.

Checklist

Operating Your E-Business:

❏ Review copyright, trademark, service mark, content liability, and customer information privacy statements, as well as the most current e-business taxation requirements with your attorney and accountant.

❏ Create a facilities plan and budget before looking for new commercial office space.

❏ Research the compensation and benefits expectations in your region before hiring new employees.

❏ Evaluate the different electronic payment methods available, and determine which will best suit your e-business based on the kinds of products and services you offer and the type of customers who will buy those products and services.

❏ Evaluate and select storefront software, if applicable to your e-business.

❏ Evaluate and select the technologies and technology vendors applicable to your e-business operations.

❏ Consider the pros and cons of outsourcing some of your e-business operations to ISPs, Web hosting companies, application service providers, and fulfillment houses.

Key Terms

ACH check	e-check
application service providers (ASPs)	electronic cash
authentication	electronic check
back-end systems	electronic wallet
burn rate	encryption
cardholder not present	enterprise resource planning (ERP)
certificate authority	front-end systems
charge card	fulfillment house
chargeback	hosted storefront software
co-locate	legacy systems
content management system (CMS)	merchant account
content repositories	merchant account provider
copyright	micropayments
credit card	middleware
customer relationship management (CRM)	P2P payment systems
debit card	payment gateway
digital cash	protocol
digital certificate	Secure Sockets Layer (SSL)
digital wallet	service marks
discount rate	shopping cart software
e-cash	smart card

stored value card

storefront software

trademark

virtual credit card numbers

Web hosting companies

Review Questions

True/False Questions

1. An application service provider partners with financial institutions and payment gateways to provide B2C payment processing packages. True or False?

2. The discount rate is a percentage of each payment transaction processed through a merchant account. True or False?

3. Hosted storefront software can provide everything a small e-business needs to operate an online store. True or False?

4. Web hosting companies provide a level of redundancy in Web site operations that an e-business may not be able to provide for itself. True or False?

5. Monetary compensation is the only thing that matters to prospective employees. True or False?

Multiple Choice Questions

1. One of the first things an entrepreneur should do is:
   a. hire an ASP.
   b. create an ERP system.
   c. file a state sales tax return.
   d. establish a relationship with an experienced attorney.

2. A fulfillment house:
   a. focuses on hosting commercial Web sites.
   b. is a non-profit organization that focuses on privacy issues.
   c. delivers and manages software applications for multiple customers from remote locations over the Internet.
   d. provides warehousing plus order picking, packaging, and shipping services.

3. Which of the following is important when evaluating a Web hosting service?
   a. 24/7 technical support
   b. scalable servers
   c. tested disaster-recovery plans
   d. All are important factors.

4. The e-business risk associated with returned products, billing errors, and fraudulent charges is a:

   a. merchant account.

   b. payment gateway.

   c. micropayment.

   d. chargeback.

5. A tool used to protect consumers and e-businesses from credit card fraud at the point of sale is:

   a. a debit card.

   b. a stored value card.

   c. a virtual credit card number.

   d. a micropayment.

Exercises

1. Using links on the student online companion to this text, review the privacy policy statements available at the Web sites of three e-businesses discussed in this chapter, then write down the answers to the following questions: What types customer information do the three e-business's retain? How do they use the information? Do the e-businesses' privacy policies make you more likely or less likely to conduct transactions at their sites, and why? Use your research to discuss the importance of Web site privacy statements with your classmates.

2. Using links on the student online companion to this text, review the Web sites for e-businesses that offer micropayment solutions, such as BitPass. Then write down the answers to the following questions: As an e-business entrepreneur, would you offer micropayments at your Web site? As a consumer, would you use micropayments? Explain the reasons for your answers. Use your research to discuss the viability of micropayments with your classmates.

3. Using online search tools and other relevant resources, locate and review the current salary range in your area for the following positions: chief information officer (CIO), accountant, marketing director, Web designer, Webmaster, database programmer, and executive assistant. Create a table that compares the salary range for each position. Include the source of your data.

4. Using links on the student online companion to this text, online search tools, and other relevant resources, locate and review the Web sites of three professional recruiting firms in your area and one global executive recruiting firm. Then write a brief paragraph about each firm, describing its recruiting business scope.

5. Using online search tools and other relevant resources, research the current status of e-business taxation, including collection and payment of sales taxes. If possible, speak with an attorney or accountant who is knowledgeable about tax issues. Present your research to a group of classmates.

Case Projects

1. Your new e-business has outgrown its temporary office, which is located in a spare bedroom in your house. You need to hire an assistant and a technical support person, and there is no room for two more employees in the home office. Additionally, you are ready to set up meetings with prospective clients and need a more professional atmosphere in which to conduct those meetings. Create a detailed facilities plan for the commercial office space you need, including a budget, a description of the desired area in which you want to locate your office, and a list of the things your office space will need (such as divided work areas, wiring requirements, room for office equipment, parking, security, and so on). Then research the available commercial office space in that area. If possible, arrange a meeting with a commercial real estate broker to discuss your options.

2. You are creating a startup e-business following the B2C model and need to decide whether to use an ISP or a Web hosting company to host your Web site. Use links on the student online companion to this text, online search tools, or other resources to locate and review the options provided by two ISPs and two Web hosting companies. Create a table comparing the cost and services of each of the four companies, make a decision on which company you want to host your Web site, and use the table to discuss your analysis and final choice with a group of classmates.

3. You and your friend Liz are planning to start an e-business that sells equipment and uniforms for women's sports and provides auctions for women's sports memorabilia. Based on research and personal experience, you believe your e-business idea targets three age groups: young adults who are 19–30 years old; teenagers, 13–18 years old; and children, 10–12 years old. You and Liz want your e-business to offer payment methods that are appropriate for sales to each of these age groups and to auction participants. Create an outline detailing the different electronic payment systems you plan to offer, including the reasons for selecting each payment system. Use the outline to discuss your choices with a group of classmates.

Team Project

You and two classmates are starting a B2C e-business. Working together, define the e-business, its products and services, and the type of customer(s) who will buy those products and services. Next, identify the electronic payment methods that are most appropriate for your e-business. Then, using links on the student online companion to this text, online search tools, or other relevant sources, perform the following tasks:

- Review merchant account providers and choose one for your e-business, if applicable.
- Review micropayment providers and choose one for your e-business, if applicable.
- Review storefront software providers and choose the software for your e-business. You may choose a package that includes access to a merchant account and payment gateway or just the software.
- If Web hosting is not part of your storefront software solution, review ISPs and Web hosting companies and choose one for your e-business.
- Review fulfillment houses that provide e-business order fulfillment services and choose one for your e-business, if applicable.

Using Microsoft PowerPoint or other presentation tools, create a presentation of 10 to 15 slides defining your e-business idea and its potential customers. Then identify the electronic payment method(s) you selected and the reasons for your choice(s). If you selected a method that could be vulnerable to fraud, list the ways in which your e-business will protect against potential fraud. Identify the storefront software solution you have chosen and the reasons for your choice. Identify the hosting solution your selected and the reasons for your choice.

Give your presentation to a group of classmates who have been selected by your instructor to critique your e-business idea, your payment selections, fraud-prevention procedures, storefront software, and Web hosting solution.

For Further Study

Here are some resources that might help you in further investigating the topics covered in this chapter.

Student Online Companion

Check out the *Creating a Winning E-Business, Second Edition* student online companion Web site for links to the sites discussed in this chapter and to other useful Web sites.

Articles and Books

"Hiring Tips from Top Headhunters." Dunhill Professional Search. www.dunhillhouston.com/headhunter_tips.htm. 2005.

"Hiring Tips." Monster Hiring Center. hiring.inc.com/tips.html. 2005.

Brouillard, Sarah. "Consider Geography, Building Type When Choosing Specialist." Upsizemag.com. www.upsizemag.com/article.asp?issueID=5&articleID=47. October 2003.

Cantor, Sheryl. "Creating an Online Store." PC Mag.com. www.pcmag.com/article2/0,1759,1463690,00.asp. February 17, 2004.

Jackson, Eric. *The PayPal Wars: Battles with eBay, the Media, the Mafia, and the Rest of Planet Earth.* Los Angeles: World Ahead Publishing, Inc. 2004.

Johnson, Jennifer. "Choosing a Web Hosting Company." PowerHomeBiz.com. www.powerhomebiz.com/vol48/webhost.htm. 2005.

Levchin, Max. "PayPal Slide Show." www.levchin.com/paypal-slideshow/index.html. 2005.

End Notes

[1] Levchin, Max. "PayPal Slide Show." www.levchin.com/paypal-slideshow/index.html. 2005.

[2] Plotkin, Hal. "Beam Me Up Some Cash." *Silicon Valley Correspondent.* www.halplotkin.com/cnbcs029.htm. September 8, 1999.

[3] Lillington, Karlin. "PayPal Puts Dough in Your Palm." *Wired News.* wired-vig.wired.com/news/technology/0,1282,20958,00.html. July 27, 1999.

[4] "Copyright Basics." U.S. Copyright Office. www.copyright.gov/circs/circ1.html. 2005.

[5] "Digital Information." Copyright Website. www.benedict.com/Digital/Digital.aspx. 2005.

[6] "About Truste." Truste. www.truste.com/about/index.php. 2005.

[7] McCullagh, Declan. "States Yearn to Collect Online Sales Taxes." C/Net News.com. netscape. com.com/States+yearn+to+collect+online+sales+taxes/2100-1028_3-5672198.html?tag= st.rn. 2005.

[8] Pedroza, Gisela M. "Hiring Telecommuters." Entrepreneur.com. www.entrepreneur.com/article/ 0,4621,295960,00.html. January 2002.

[9] Green, Jeff. "Employees By the Round: Who's the Key Hire? Who Should You Put Off Until Round Three?" *Business 2.0*, 168. March 2000.

[10] Ibid.

[11] Williams, Geoff. "Now Hiring: There Comes a Time When Every Start-up Entrepreneur Just Can't Do it Alone Anymore. Is it Time to Find Your First Employee?" Entrepreneur.com. www.entrepreneur.com/article/0,4621,303820,00.html. November 2002.

[12] Cummings, Danny. "Interview With Co-CEOs Dan Isaacs and Adam Moore of Springbox." *Hoovers*. 2005.

[13] "Consumer Handbook to Credit Protection Laws." The Federal Reserve Board. www.federalreserve.gov/pubs/consumerhdbk/electronic.htm. 2005.

[14] "Authorize.Net—Frequently Asked Questions." Authorize.Net. www.authorize.net/resources/ faqs/. 2005.

[15] Morris, Charlie. "Accepting Credit Cards: Getting a Merchant Account." Web Developer's Virtual Library. wdvl.com/Internet/Commerce/MerchantAccounts/. April 29, 1999.

[16] "Secure Sockets Layer." Whatis.com, searchSecurity.com. searchsecurity.techtarget.com/ sDefinition/0,290660,sid14_gci343029,00.html. 2005.

[17] Ourshop.com. "Straightforward Explanations of SSL and HTTPS." www.ourshop.com. 2005.

[18] "Digital Wallet." ecommerceguide.com. e-comm.webopedia.com/TERM/D/digital_ wallet.html. 2005.

[19] Konrad, Rachel. "Burned by ChoicePoint Breach, Potential ID Theft Victims Face a Lifetime of Vigilance." *InformationWeek*. informationweek.com/story/showArticle.jhtml? articleID=60403319. February 24, 2005.

[20] Krass, Peter. "The New Face of Identity Theft." CFO.com. www.cfo.com/article.cfm/3737810/1/ c_3759578?f=insidecfo. March 15, 2005.

[21] "Online Credit Card Fraud." *CyberSource Online Fraud Report for 2005*. www.cyveillance.com/ web/online_risks/credit_card_fraud.htm. 2005.

[22] Brissett, Jane. "Customers' Data Stolen from DSW Shoe Warehouse Computers." *Duluth News Tribune*. www.duluthsuperior.com/mld/duluthsuperior/11636633.htm. May 13, 2005.

[23] Sullivan, Bob. "Feds Probe Mysterious Credit Card Charges." MSNBC.com. www.msnbc.msn. com/id/7150531/. March 10, 2005.

[24] "10 Ways to Beat Credit Card Crooks." ECHO Credit Card Processing. www.echo-inc.com/loss_ prevention.html. 2005.

[25] Lazarony, Lucy. "Perishable Credit Card Numbers Take the Fear Out of Web Shopping." Bankrate.com. www.bankrate.com/brm/news/cc/20021011a.asp. March 23, 2004.

[26] Sullivan, Bob. "Fake Companies, Real Money." MSNBC.com. www.msnbc.msn.com/id/ 6175738/. October 7, 2004.

27 Sullivan, Bob. "Feds Probe Mysterious Credit Card Charges." MSNBC.com. www.msnbc.msn. com/id/7150531/. March 10, 2005.

28 "Stored Value Cards: An Alternative for the Unbanked?" Federal Reserve Bank of New York. www.newyorkfed.org/regional/stored_value_cards.html. July 2004.

29 Samaad, Michelle. "Echecks and Electronic Check Presentment: New Technologies Aim to Wire Your Checkbook." Bankrate.com. www.bankrate.com/brm/news/chk/19981124.asp. January 11, 2001.

30 Pitta, Julie. "David Chaum." *Forbes*. 159(14), 320(2). July 7, 1997.

31 Stock, Helen. "DigiCash Idea Finds New Life in More Flexible eCash." *American Banker*, 165(67). April 6, 2000.

32 Power, Carol and Kutler, Jeffrey. "Bankrupt DigiCash to Seek Financing, New Allies." *American Banker*, 163(216). November 10, 1998.

33 Millard, Elizabeth. "The Death of Micropayments?" *E-Commerce Times*. www.ecommercetimes.com/story/32566.html. January 12, 2004.

34 "Micropayments are Multiplying." *eMarketer* as reported by *WebMetro*. www.webmetro.com/ news1detail1.asp?id=1106. October 7, 2004.

35 Fost, Dan. "Tiny Bills = Big Deal, New Firms Line Up to Enable Micropayments for Net Buys." *San Francisco Chronicle*. www.sfgate.com/cgi-bin/article.cgi?file=/chronicle/archive/2004/09/ 08/BUGDH8L53E1.DTL. September 8, 2004.

36 Ibid.

37 Kaiser, Nathan. "nPost.com Interview with Kurt Huang." nPost.com. www.npost.com/interview. jsp?intID=INT00081. March 22, 2004.

38 Fost, Dan. "Tiny Bills = Big Deal, New Firms Line Up to Enable Micropayments for Net Buys." *San Francisco Chronicle*. www.sfgate.com/cgi-bin/article.cgi?file=/chronicle/archive/2004/09/ 08/BUGDH8L53E1.DTL. September 8, 2004.

39 "About BitPass." BitPass. corp.bitpass.com/aboutus/. 2005.

40 Kuchinskas, Susan. "Yahoo Says So Long to PayDirect." ecommerce-guide.com. www.ecommerce-guide.com/news/news/article.php/3427721. October 27, 2004.

41 Stuart, Anne. "The Perfect Host." Inc.com. www.inc.com/articles/2004/03/theperfecthost.html. March 2004.

42 Harper, Roy. "Evaluate Your Web Host Options." WorkZ. www.workz.com/content/view_content. html?section_id=507&content_id=5772. February 4, 2005.

43 Wainewright, Ivan. "An Introduction to Application Service Providers (ASPs): The Pros and Cons." Techsoup. www.techsoup.org/howto/articlepage.cfm?articleid=59&topicid=2. May 1, 2000.

44 Wheatley, Malcolm. "Cover Your ASP." CSO Online.com. www.csoonline.com/read/010104/asp. html. January, 2004.

45 Schiff, Jennifer. "Outsourcing Your Fulfillment: What You Need to Know." Small Business Computing.com. May 26, 2005. www.smallbusinesscomputing.com/emarketing/article.php/ 3507976.

[46] *PRNewswire*. "PayPal.com and Star Trek's 'Scotty' Put the Power to Beam Money in the Palm of Your Hand." www.paypal.x.com/html/pr-121799.html. December 17, 1999.

[47] Weitzman, Jennifer. "Star Trek Promise Fulfilled: Wireless Cash Transfer." *American Banker,* 164(235). December 9, 1999.

[48] "Interview: Peter Thiel, PayPal's Co-founder and CEO." *eFinance Insider.* www.efinanceinsider. com/paypalinterview31501.htm. March 15, 2001.

[49] Ibid.

[50] Anderson, William. "The Genius and Struggle of PayPal." Mises Institute review of *The PayPal Wars: Battles with eBay, the Media, and the Rest of Planet Earth* by Eric M. Jackson. Los Angeles, CA: World Ahead Publishing. www.mises.org/fullstory.aspx?Id=1710. 2005.

[51] Glasner, Joanna. "PayPal: IPO Omen or Anomaly?" *Wired News.* www.wired.com/news/ business/0,1367,50461,00.html. February 15, 2002.

[52] Goldfarb, Jeffrey. "PayPal and Its Owners are Partying Like It's 1999." *Reuters Company News* as reported by Clearstone Venture Partners. www.clearstone.com/news/news_PAY070902. html. July 8, 2002.

[53] "About Us." PayPal. www.paypal.com. 2005.

CHAPTER 6

MARKETING YOUR E-BUSINESS

LEARNING OBJECTIVES

In this chapter, you will learn to:

- Describe the marketing mix
- Explain the importance of branding
- Describe primary and secondary market research
- Create a marketing plan
- Identify marketing tools

THE SAGA OF A BRAND . . .

In 1994, nationally recognized wine expert Peter Granoff and his brother-in-law Robert Olson, an engineer knowledgeable about computer software, were enjoying a lovely dinner when the discussion turned to interactive software, the Internet—and wine.[1] Why couldn't people buy wine and related products from an online store? Surely consumers would flock to a Web site that provided in-depth information about the wines and the wineries that made them. What if the online store featured only the best wines from the best vintners, to ensure that customers would be satisfied with every wine purchase? And what if it used sophisticated interactive software tools to make buying wine and related products as easy and convenient as possible? Was an online store that catered to wine aficionados really a viable e-business idea? To find out, Granoff and Olson ran their idea by several wineries. Encouraged by the positive responses they received, Granoff and Olson became e-business pioneers by launching Virtual Vineyards in early 1995. Virtual Vineyards hit the mark with wine buyers, and by 1996 sales were about $1 million a year.[2] But the saga of Virtual Vineyards was just beginning.

MARKETING MIX

Marketing is the process of developing the mutually satisfying relationships between your e-business and your customers that result in sales and profits. Specifically, marketing involves developing the strategies (plans) and selecting the tactics (actions) necessary to build these relationships. Classic marketing strategies are based on the Four Ps model, which is composed of product, place, promotion, and price (Figure 6-1). The Four Ps, also called the **marketing mix**, is a concept that was originally developed more than 40 years ago by consultant and marketing educator E. Jerome McCarthy.[3] The Four Ps marketing mix model identifies those elements of your marketing strategies and tactics over which you have control:[4]

- *Product*: all the elements of each individual product or service that is to be sold, including branding, product features, packaging, installation, service, and warranties
- *Place*: activities pertaining to how the product or service is distributed to the customer, including distribution channels and intermediaries, warehousing, order fulfillment, and shipping
- *Promotion*: activities pertaining to how information about products and services is communicated to customers, including advertising, public relations, promotion, and customer education
- *Price*: the cost of the product or service to the customer

FIGURE 6-1 The Four Ps marketing mix model

While the Four Ps marketing mix model looks at marketing from an "inside-out" perspective (from your e-business to your customer), another way to look at the marketing mix is from an "outside-in" perspective (from your customer to your e-business). In this case, the marketing mix is referred to as the Four Cs model (Figure 6-2); and it involves the following elements, all specific to the customer's point of view:[5]

- *Customer Needs and Wants*: product variables are redefined in terms of what the customer needs and wants
- *Convenience*: product distribution variables are broadened to include all the elements involved in obtaining and using a product or service
- *Communication*: promotion variables are broadened to include all communications with customers designed to increase sales
- *Cost to the Customer*: price variables change to be the total cost to the customer of consuming a product or service

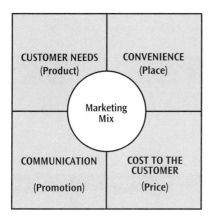

FIGURE 6-2   Comparison of the Four Cs marketing mix model and the Four Ps marketing mix model

TIP

Since the marketing mix concept was first developed, educators and marketing professionals continue to define various extended marketing mix models that include additional variables such as people, process, physical evidence, public image, online scope, Web site operations, Web site content, online communities, and networking technologies.[6, 7, 8]

When used together, the Four Ps, the Four Cs, and other extended marketing mix models can provide you with a broad view of the marketing variables you will need to define as you develop your e-business's marketing strategies. An important first step in developing those marketing strategies is branding your e-business and your Web site.

BUILDING YOUR BRAND

A brand incorporates all aspects of your customers' perceptions about and experiences with your business. Traditionally, the effectiveness of business branding is measured in terms of marketplace awareness—name and product or service recognition. For an e-business, building an effective brand often leads to more than just a spike in market-place awareness—it leads directly to actions taken by consumers at your e-business Web site.

What Is a Brand?

What exactly is a brand? According to the American Marketing Association, a **brand** is a combination of name, logo, and design that identifies a business's products or services and differentiates the products and services from those of competitors.[9] It is the difference between a soda and a Coke, a tissue and Kleenex, drinking water and Evian, running shoes and Nikes, and online search and Google. A brand comprises the subjective experience that consumers have with a product or service, as well as the assumptions they make about it and the trust they have in it satisfying their needs. In many respects, a brand is much like an identity or personality—it is the "face" a business presents to the world.[10, 11]

TIP

Trademark is the legal term for brand.

Having a trusted brand that is recognizable to your customers and to strategic business partners can drive sales and increase your e-business's chances of achieving success. Building a successful brand for your e-business involves understanding how your customers perceive your products and services, selecting an appropriate brand name and e-business' logo, and registering a domain name (URL) that ties closely to your products and services in the minds of consumers.

Building Your E-Business's Brand

To build your e-business's brand, you must first define how you want your customers to perceive the products or services offered by your e-business. This perception is the core identity or essence of your e-business's brand. Building this perception and a successful brand involves a lot of homework. For example, you must:[12, 13]

- understand the core elements of your e-business and how these elements differ from those of your competitors
- identify how your e-business's products or services solve customers' problems or fulfill their specific needs
- determine how to convince potential customers that your e-business is the best source for the products or services you offer
- consider methods you can use to build customer loyalty
- select the words, phrases, and images that put the best public "face" on your e-business

Next, you name your brand by choosing your e-business's name, logo and symbols, and domain name.

Naming Your Brand

After doing your homework, the next step in branding your e-business is selecting a name and logo image that you want potential customers to identify with your specific products or services. Brand names should be kept simple so that they are easy to remember, spell, and understand. Traditionally, many successful brand names have been proper names; for example, Coca-Cola, Microsoft, Ford, Kodak, and Sony all resonate with customers who identify these proper names with specific products. Early and successful e-businesses such as Yahoo!, Amazon.com, eBay, and priceline.com are also branded with proper names. In addition to being short, easy to remember, and easy to spell, Internet brand names should also have "snap"—meaning they should evoke a memorable emotional response and bring to mind thoughts about specific products and services. For example, consider Hotmail as a name for an e-mail provider, or EarthLink as a name for an ISP. Both names are snappy, and they embody a clear, intuitive connection to the products or services these respective e-businesses offer.

Personalizing a brand by naming an e-business after a well-known individual can help make the brand memorable. While not many people are likely to remember the generically-named PC Limited and its products and services, most people will immediately recognize this e-business by the name it uses today, Dell (Figure 6-3). Switching the brand name from PC Limited to the proper name Dell (based on the name of the founder, Michael Dell) in 1987 enhanced the publicity potential of the e-business.[14] A consequence of this is that all the attention and publicity Michael Dell receives can directly benefit (or tarnish) the Dell brand. Another example of an e-business that built its brand around a personality (a fictional one, in this case) is the original Ask Jeeves online search tool (Figure 6-4). The Ask Jeeves brand was personified by Jeeves the butler, a figure who embodied impeccable personal service and suggested that the e-business, by association, had a similarly attentive personality. The character of Jeeves was the distinctive hook that resonated with users and stayed in their minds. You should be aware, however, that branding can be a dynamic and evolving process. In 2005, for example, the company that owns the Ask Jeeves search tool announced plans to rename the search tool and revamp its features. In addition, as part of a new branding effort designed to emphasize the search tool's enhanced features, the company announced plans to retire the portly butler.

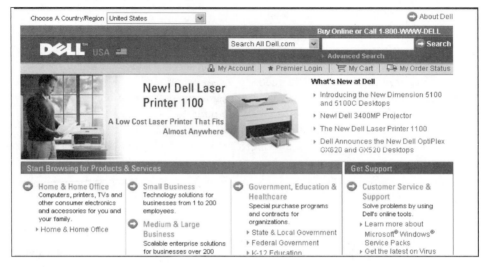

FIGURE 6-3 Dell

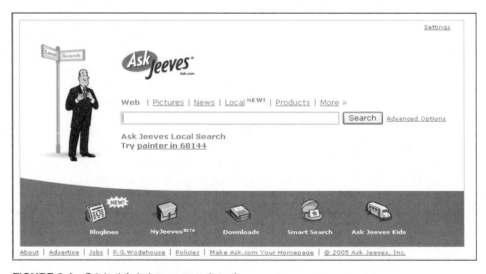

FIGURE 6-4 Original Ask Jeeves search tool

TIP

Once you develop a brand, you should be vigilant about protecting it from abuse or misuse by posting a copyright notice on your Web site, by listing rules for how your logo and company name should be used, and by monitoring the Web for instances of unauthorized use of your Web site logo, art, and text.[15]

A **domain name** or **Uniform Resource Locator (URL)** is the address associated with a Web page that a Web browser uses to locate the page. (You'll learn more about the structure of domain names later in the chapter.) In the early days of e-business, before single-word domain names (URLs) were gobbled up by entrepreneurs, names based on common words plus the domain—such as Business.com and Office.com—were popular. One disadvantage of these types of generic names is that customers may not be able to make a direct connection to the specific products and services offered at the respective e-businesses. For example, can you tell what types of products or services are available at Business.com or Office.com? Without a trip to the respective Web sites, it is difficult to know. (Business.com is actually a B2B directory, and Office.com is a small business portal offering tips, articles, and links to other e-businesses.)

Another disadvantage of names based on common words is that although such names are short and easy to spell, they lack snap and are not especially memorable. In general, there is some doubt among professional marketers about whether generically named e-businesses can build a successful brand over the long term. Some professional marketers are also critical of the common practice of choosing an e-business name that includes a specific reference to the Internet and Web, such as Cyberchefs. Names tied directly to the Internet and the Web are thought to be limiting in the long term, perhaps because they may inadvertently bear some of the negative connotation that has lingered after the failure of so many e-businesses during the dot.com bust of 2000-2001. If you have an e-business whose name includes prefixes such as "cyber," "net," or "tech," or domain name suffixes such as ".com" or ".biz," you should consider whether these prefixes and suffixes might limit the usefulness of your e-business's name over time.[16]

In choosing a brand name, it is wise to indulge in a little "creative brainstorming." For example, you could create a list of words or phrases that describe your e-business. Then play with the list to create different combinations of words, eliminating those words or word combinations that do not work well. Next, you could invite family, friends, and advisors to critique the short list of words or word combinations that remain.[17] Alternatively, some companies look to branding professionals such as The Namestormers and NameLab to help them create a memorable brand name for their business (Figures 6-5 and 6-6).

QUOTES ON SUCCESS

"Branding is the art and cornerstone of marketing."

Philip Kotler, author, consultant, and marketing educator

After selecting a brand name, be certain to check the U.S. Patent and Trademark Office's searchable trademark database (TESS) to determine if this name is available. After this, your next step is to check to see what domain names are available.

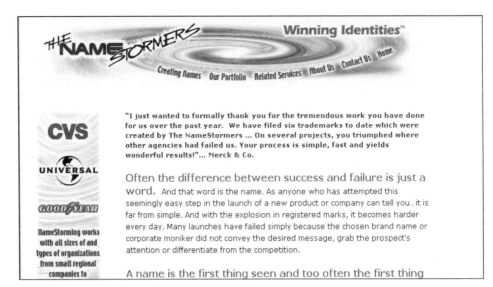

FIGURE 6-5 The Namestormers

FIGURE 6-6 NameLab

Domain Names

As you learned earlier in the chapter, domain names (URLs) are Web page addresses recognized by a Web browser. For example, the domain name "www.rackspace.com" is the address of Rackspace Managed Hosting's Web site's home page. The "www" portion is called a subdomain and points to a specific part of Rackspace's Internet services. By convention, most companies use the "www" subdomain to point to their Web site. The "rackspace" portion of the domain name identifies the organization or entity associated with the domain name. The "com" portion is called the top-level domain (TLD) and identifies the general category in which the domain name is registered. For example, the top-level domain ".com" indicates the commercial domain. Originally, there were seven TLDs designating the major kinds of organizations that published Web pages. Recently, in an attempt to satisfy the exploding demand for domain names, the seven original TLDs have been expanded into sixteen TLDs, shown in Table 6-1. As of this writing, there are other TLDs that have been proposed and are being considered.

TABLE 6-1 Top-Level Domains (TLDs)

Original Domain Categories	Domain	Additional Domain Categories	Domain
Commercial	.com	Aviation	.aero
Educational	.edu	Businesses	.biz
Government	.gov	Cooperatives	.coop
Military	.mil	All uses	.info
Networking	.net	Museums	.museum
Non-profit	.org	Individuals	.name
International	.int	Professionals	.pro
		Travel	.travel
		Human Resources	.jobs

Domain name registration is managed by the Internet Corporation for Assigned Names and Numbers (ICANN), a non-profit organization that operates under the auspices of the U.S. Department of Commerce. ICANN actually contracts the domain name registration process to approved private companies (called **accredited registrars**) such as Network Solutions, GoDaddy.com, and Register.com. Figures 6-7 through 6-9 illustrate the Web sites of these accredited registrars.

You can use search tools at an accredited registrar's Web site to search a database of already assigned domain names to determine if your domain name choice is available or if someone has already registered it. If the domain name you desire is available, you can register it for a fee ranging from about $7-$35 per year.

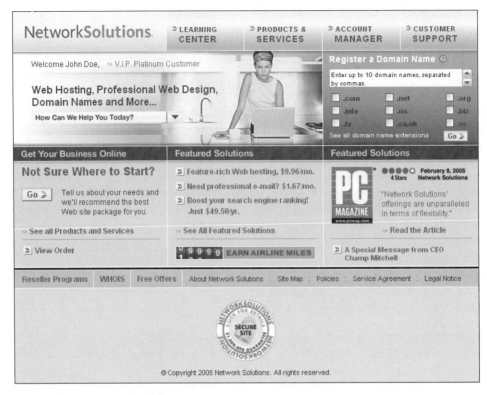

FIGURE 6-7 Network Solutions

FIGURE 6-8 GoDaddy.com

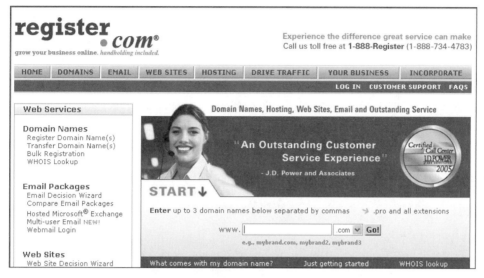

FIGURE 6-9 Register.com

As you have seen, some e-businesses—Amazon.com and Register.com, for example—chose to build their brand by making sure that their e-business's name is the same as their domain name. But an e-business's name and domain name do not have to be the same. Consider the domain name for Barnes & Noble (online bookstore), which is bn.com, and for Ernst & Young (major accounting firm), which is ey.com. Note how these brief domain names are easy to remember and easy to type in a Web browser; yet they still make a good connection to the business name. When a domain name is closely associated with the e-business name, products, or services, it becomes another tool for branding your e-business.

Be aware that today the availability of short, meaningful domain names in the .com domain is very limited. In fact, most English words have already been registered as domain names in the .com domain. You may choose to use another business-oriented domain, such as .biz, to register your domain name. Alternatively, to register in the more familiar .com domain, you might need to invent a fun new word for a domain name, such as "google"—a wordplay on the mathematical term "googol," which describes a very large number (10^{100}).[18]

Another alternative is to buy the domain name you want from someone who has already registered it. A quick Web search using the keywords "domain name resellers" will provide a list of many e-businesses who operate as domain name resellers. Reselling domain names has become big business. According to the *Ecommerce Times*, the volume of domain name reseller transactions tripled in 2004, and the average domain name resale price was just over $9,000.[19] Two types of companies or individuals register multiple domain names for resale: legitimate domain name brokers and cybersquatters. Domain name brokers register generic domain names that they think will be easy to resell. Cybersquatters register domain names that are often a company name or acronym with the intent of selling the domain name to its rightful owner. Cybersquatting became illegal in 1999 with the passage of the U.S. Anticybersquatting Consumer Protection Act.[20]

TIP

To have a little fun while checking the availability of domain names, check out the nameboy and e-gineer Domainator Web sites. These sites allow you to enter words and phrases to view a list of available domain names related to the words or phrases you enter.

A domain name is not a trademark. In order to function as a trademark, the domain name must serve as a product identifier, and not merely as a Web address. If the domain name acts separately as a product identifier, the domain name must be registered with the U.S. Patent and Trademark Office as a trademark or service mark. If registered trademark owners want to use their trademarks as their domain names, they must file for domain name registration with a registrar accredited by ICANN.

Brand names, domain names, and trademarks have become interchangeable in consumers' minds. Since domain names have economic value, many e-businesses are registering their domain name as a trademark. Thus, when choosing your domain name, be sure to check whether the name you choose infringes on another e-business's registered trademark. The e-business NameProtect offers a free search tool that checks for both registered trademarks and registered domain names. You should also check county and state business name databases to determine the availability of the name, as you may also be required to register your name with the country and state in which your e-business resides.

The ultimate goal of building your brand name is to develop marketing strategies that target and draw customers. Before you can define your marketing strategies, however, you must understand the marketplace in which your e-business operates. Market research helps you develop that understanding.

MARKET RESEARCH

Market research involves collecting and analyzing the data you need to make informed decisions about how to go about selling your business's products or services in a specific marketplace. Market research is used to identify needs, trends, customer preferences and opinions, and even your competitors' stake in a marketplace. The two types of market research are primary research and secondary research.

Primary research involves physically collecting marketplace, consumer, and competitor data, organizing and manipulating it, and then analyzing and publishing the results. Primary research can involve both **quantitative research** (collecting data that can be analyzed using statistical methods) and **qualitative research** (collecting data that requires informed interpretation and cannot be analyzed using statistical methods). For e-businesses, quantitative and qualitative research usually focuses on data collected from online customer surveys, focus groups, Web server logs, customer transaction databases, and electronic tracking tools such as cookies. **Cookies** are small text files that are placed on the hard drive of a viewer while that viewer (that is, potential customer) is visiting a Web site.[21]

Gathering and then analyzing the data necessary for primary research can be time-consuming and expensive for a startup e-business. The alternative is the most widely used and least expensive type of research, **secondary research**, which involves collecting data through secondary sources, such as market research companies, who collect data, analyze the data, and then sell research reports based on their analyses. Companies such as Forrester, eMarketer, and NPD Group, shown in Figures 6-10 through 6-12, are examples of market research companies.

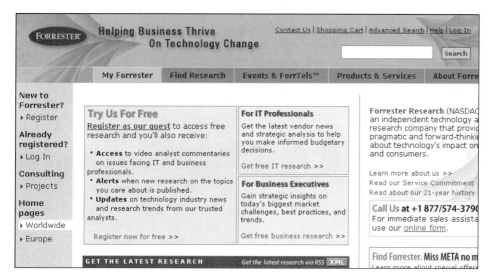

FIGURE 6-10 Forrester

FIGURE 6-11 eMarketer

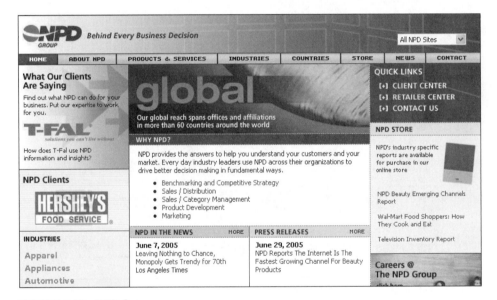

FIGURE 6-12 NPD Group

Other sources of secondary research data for an e-business include industry white papers, magazine articles, U.S. and foreign government databases, domestic and international trade associations, and articles published in professional journals.[22] The Web is a great source of both free and paid secondary research material. Six Organizations that provide secondary research that might be useful to an e-business entrepreneur include STAT-USA Internet (U.S. government statistics), DismalScientist (economic trends), AdAge.com (advertising trends and news), Bitpipe (IT industry data and news), and The Direct Marketing Association (direct marketing industry data and news). Figures 6-13 through 6-15 depict the Web sites of some of these organizations.

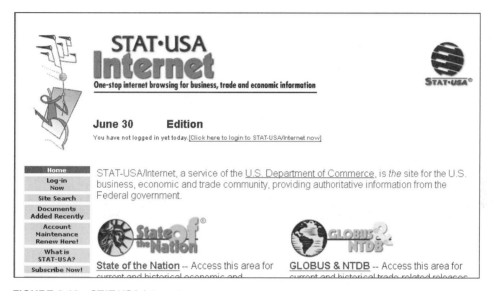

FIGURE 6-13 STAT-USA Internet

Your branding decisions, market research, and analysis should come together in your marketing plan, a document that shows how your e-business will manage its marketing mix, enhance its brand, drive sales, and generate profits.

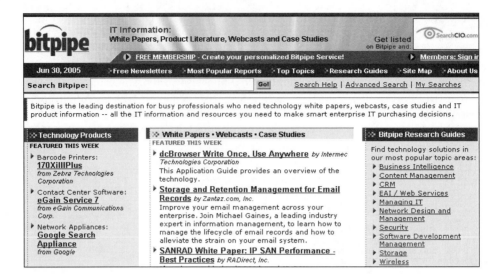

FIGURE 6-14　Bitpipe

FIGURE 6-15　The Direct Marketing Association

CREATING A MARKETING PLAN

In Chapter 3, you were introduced to business planning and the various elements of a formal business plan. As you may recall, one of these elements was a marketplace analysis summary, which provides general information about the industry within which your e-business operates and a brief overview of your e-business's targeted customers and competitors. A **marketing plan** provides the details behind this marketplace analysis summary. It specifies how your e-business will manage the marketing mix in order to generate sales and profits. A sufficiently detailed marketing plan helps you:[23]

- define the market in which your e-business will operate
- zero in on your customers and competitors
- determine effective ways to acquire and keep your customers
- plan ways your e-business can deal with change

Like overall business planning, marketing planning is dynamic. Your original marketing plan will change over time as you add new products or services and modify your strategies and tactics to adapt to changes in the marketplace.

Marketing Plan Elements

Your marketing plan, just like your business plan, may contain many different elements. Plan elements may be organized and formatted in many different ways, depending on your e-business's needs. For example, the elements of a marketing plan for a startup e-business may be different in scope and number from the elements in a plan that was created by an existing e-business in order to introduce a new product or service. With that said, most marketing plans, at a minimum, contain the following elements:[24, 25]

- *Executive Summary*: summarizes the overall marketing plan
- *Situational Analysis*: explains what you know about the marketplace
- *Objectives, Strategies, and Tactics*: describes your overall marketing mission, identifies what you want to accomplish and how you will do it, explains how and when your marketing tactics will be implemented, and identifies who inside or outside your e-business will be responsible for the implementation
- *Budget and Performance Measures*: details the estimated costs associated with accomplishing your objectives and how the results are measured

In creating a marketing plan, you might find it useful to consult examples of marketing plan outlines and sample marketing plans. Fortunately, you can find many online resources that will help you do this, including samples of marketing plan outlines and complete marketing plans available from the U.S. Small Business Administration (SBA), marketing services companies that publish sample plans and tips for writing plans, and businesses that sell business and marketing planning software, such as those you learned about in Chapter 3.

Executive Summary

The Executive Summary portion of your marketing plan is similar to the Executive Summary of your overall business plan. In no more than three pages, it briefly summarizes the features of your marketing plan. Also like your business plan Executive Summary, your marketing plan Executive Summary should be written last, after you create your other plan elements.

Situational Analysis

The Situational Analysis portion of your marketing plan is based on your market research and analysis. This section describes key aspects of the overall marketplace, such as market size, market segments, your target market, your competition, and the individual products or services your e-business offers. Being thorough in your preparation of this section is more important than meeting a specific page length.

Market size can be defined in a number of ways: for example, the number of potential consumers, consumer purchasing power, projected sales volume, and so forth. Estimates of the market size for the market in which your e-business will operate should be gleaned from reliable sources and cited appropriately. The description of market size should also explain how the market is growing or shrinking and why.

A **market segment** is a subgroup of the overall market. Markets can be segmented by a number of factors, including geographic region, demographic characteristics of consumers (such as age, gender, education, family size, and income), psychographic characteristics of consumers/regions (such as buying behavior and price sensitivity), and so forth. Understanding how the market in which your e-business operates is segmented will help you identify your target market.

Your e-business's **target market(s)** is the pool of potential customers to which you will direct your marketing efforts. A description of your target market(s) should be both qualitative and quantitative and should include geographic, demographic, and psychographic profiles of potential customers. Additionally, you should describe exactly how your products and services meet the needs and wants of these potential customers. An analysis of the political, economic, social, and technological factors that influence your potential customers' buying decisions—sometimes called a PEST analysis—is often included.[26] These qualitative descriptions of your target market(s) should be supported by data from reliable sources, such as the total number of potential customers in the market or the average income of potential customers.

QUOTES ON SUCCESS

"Determining the right target audience is probably the most important part of your marketing efforts, because it doesn't matter what you're saying if you're not saying it to the right people."

Lee Ann Obringer, author of "How Marketing Plans Work" and marketing communications consultant and designer

The competitive analysis section of your Situational Analysis must be a comprehensive summary of your top competitors: who they are, what their market share is, what their strengths and weakness are, and how they manage the elements of their marketing mix (product, price, distribution, and promotion). To get first-hand knowledge of your competitors, you should examine their Web sites, pour over their advertising and promotional materials, compare their pricing structures, and try out their competing products and services.[27]

Lastly, your Situational Analysis should include a product and services overview section that identifies each product or service your e-business offers. For each product or service, you should include the following:

- a general description
- an explanation of how the product or service is used
- a list of special features that differentiate your product or service from competitors' similar products or services
- a description of online and offline distribution channels (how the product or service will reach customers)
- a list of the benefits customers will enjoy by using the product or service (why customers should buy the product or service)

T I P

A Situational Analysis is also called a SWOT analysis, as it involves uncovering the strengths, weaknesses, opportunities, and threats associated with various elements in the marketplace.

Objectives, Strategies, and Tactics

Objectives are specific, attainable, measurable, realistic, and time-specific goals. Your overall business plan, as discussed in Chapter 3, includes sales objectives for a given planning period. These are the anticipated sales in dollars (and perhaps units) for each month or year of the plan, which will eventually be measured against actual sales results. You may choose to reiterate your sales objectives in the Objectives, Strategies, and Tactics section of your marketing plan. Marketing objectives are distinct from sales objectives in that they define the marketing activity that will help your e-business meet its sales objectives. For example, a sales objective might be to have $5 million in annual sales within the next three years. A complementary marketing objective might be to increase the number of return customers by 25 percent in year 1, 50 percent in year 2, and 75 percent in year 3.

A **strategy** is a plan of action—what you are going to do to accomplish a goal. **Marketing strategies** describe how you will manage the marketing mix in order to sell more products and services to current customers, acquire new customers, retain current customers, or develop new products and services.[28] In the Objectives, Strategies, and Tactics portion of your marketing plan, you should include marketing strategies that describe the specific actions or **tactics** your e-business will employ, and it should list the marketing tools (such as traditional advertising, PR, or online advertising) that you will use to meet your marketing objectives. Returning to the previous example, a marketing strategy to increase the number of return customers might include the following tactics: making online coupons or discounts available for return customers' purchases, sending e-mail to existing customers (who agree to receive e-mail) that contains links to new product pages, automatically enrolling return customers in an online contest each time they make a purchase, and so forth. Including a schedule of who inside or outside your e-business is responsible for implementing each tactic and a timetable for the various implementations is also useful.

Budget and Performance Measures

Your **marketing budget** will be structured similarly to the expenses portion of the projected income statement included in your overall business plan. The budget should reflect the details of what it will cost to implement your marketing plan by itemizing how much you are going to spend, on what, and when. You may choose to organize the budget by objective, by product, or by time period.

The final element of a marketing plan is a description of **performance measures**—that is, the ways in which you will measure the success of your marketing strategies and tactics. One basic performance measure is to track sales results and the ratio of sales to marketing expenses. Other important measures involve keeping track of the number of direct mail and permission-based e-mail responses, the number of visitors to your Web site, the number of Web site visitors who actually make a purchase, the number of people who click through to your Web site from other Web sites, and so on. (In Chapter 9, you will learn how to measure activity at your Web site and how to interpret and make use of these measurements.)

After drawing up a marketing plan and developing various marketing strategies, you will need to learn about some of the marketing tactics and tools you can use to implement your marketing strategies.

MARKETING TOOLS

One of the fallacies of the marketing approach of many early e-businesses was the attitude that "If we build it, they will come." Although this approach may work in the brick-and-mortar world, where hundreds of people may drive by every day and see a new business, it doesn't suit the e-business world, where a new Web site can be like a brick-and-mortar store built on a dead-end street with few people driving by. Drive-by Web "viewers," those who are simply doing undirected browsing, may, in fact, never find a specific e-business. Even when visitors set out to search for Web sites that are closely attuned to their interests, they may never arrive at the Web site of a specific e-business that best fits their needs. This is because the huge volume of available Web sites may make it very difficult for any specific e-business's Web site to appear near the top of a search results list. Given this situation, unless you make a special effort to let potential customers know how to find your e-business's Web site, the site is likely to have few visitors. One way to increase traffic to your site is to make certain that all major search tools have information about your site in their indexes.

Search Tool Submissions

A majority of people find the products or services they are looking for online by searching for them as opposed to going directly to a Web site they already know about. In fact, a 2004 report by the Pew Internet & American Life Project indicates that more than 80 percent of Americans who are online use search tools (both keyword searches and directory links), and that on any given day, more than half of Internet users are using search tools.[29]

Online **search tools** include search engines, meta search engines, and directories. They use a variety of methods to gather Web page information and build indexes of Web pages. **Search engines** are search tools that use software programs (known as **spiders**, **bots,** or **crawlers**) that automatically update their indexes of Web pages by examining existing pages and moving from page to page across the Web via hyperlinks. Because of the growing volume of Web pages, any changes made to a Web site, such as pages added or deleted, may take some time to be indexed by search engine spiders. Examples of popular search engines include Google, Teoma, and MSN Search, as shown in Figures 6-16 through 6-18.

FIGURE 6-16 Google

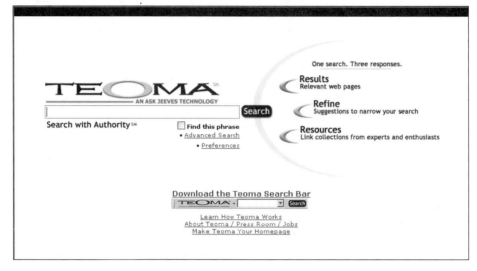

FIGURE 6-17 Teoma

FIGURE 6-18 MSN Search

Another type of search tool is a meta search engine. A **meta search engine** allows users to perform a keyword search on multiple search tool indexes and then combines the results of that search action into one report. Meta search engines include metacrawler, Dogpile, KartOO, and Mamma, as shown in Figures 6-19 through 6-22.

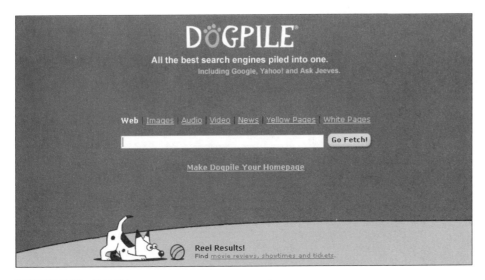

FIGURE 6-19 metacrawler

FIGURE 6-20 Dogpile

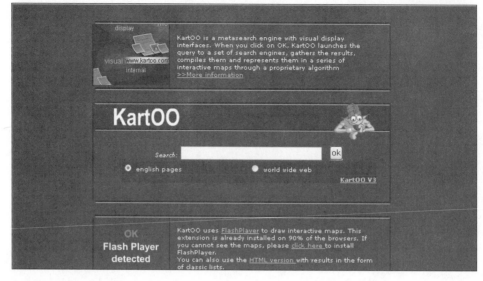

FIGURE 6-21 KartOO

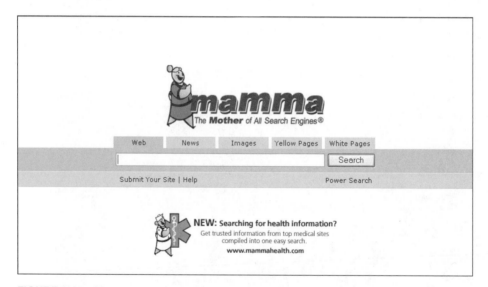

FIGURE 6-22 Mamma

A **directory** is a search tool that generates its Web page index with information submitted by people using forms available at the directory's Web site. The information is then categorized in a way that's similar to how the old library card catalog system worked. To find information using a directory, a visitor clicks multiple links to "drill down" from category to subcategory until the desired Web page is found. The first popular directory was Yahoo! Other directories include LookSmart Directory and the Open Directory Project; Web sites for these directories are shown in Figures 6-23 through 6-25.

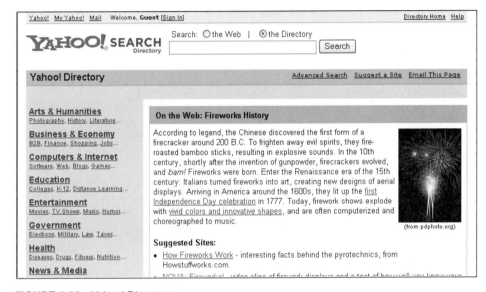

FIGURE 6-23 Yahoo! Directory

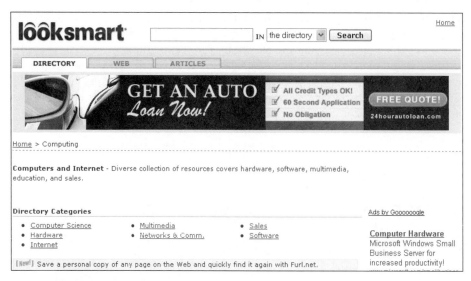

FIGURE 6-24 LookSmart Directory

FIGURE 6-25 Open Directory Project

TIP

For more details on how individual search tools work, how they make money, and other interesting search tool information, use a link on the student online companion to this text to check out the information at the SearchEngineWatch Web site.

Today, most search tools combine the features of both a search engine (keyword search) and a directory (category links). For example, Google and Yahoo! have both a keyword search index and a directory of category links, and many search tools actually rely on the indexes of other search tools to deliver search results. For example, the search results for the AltaVista and All the Web search tools are provided by the Yahoo! Search index. So submitting your e-business's Web page information to one or two search tools such as Yahoo! Search means your pages will likely appear on the search results lists of many other search tools.

Search tools that accept direct submissions will have a link (likely on their home page) to a page describing how your e-business can submit its Web page information plus a Web-based form in which you enter your information. The information you may be required to submit can include a short description of your e-business, primary keywords or phrases that potential customers might enter when using the search tool to locate Web sites similar to yours, and general information about your e-business.

Some search tools that accept submissions allow you to do so for free but with no guarantee that your pages will be indexed; some operate a **paid-inclusion program** in which an e-business is guaranteed accelerated inclusion of its Web pages in a search tool's index for a fee. Because individual search tools operate differently, it is a good idea to check out the submission and paid-inclusion programs of the search tool in which you are interested.

Getting your Web pages indexed by a spider or submitting your Web page information through a free or paid-inclusion program does not guarantee that your pages will appear at or near the top of a search results list. Since search tools are such an important means of attracting customers, you should try to design your e-business's Web pages so that you optimize their chances of appearing prominently in a search results list.

Search Engine Optimization (SEO)

Search engines that use spiders to find Web pages build their indexes, create search results lists, and rank pages in their search results list through a variety of methods. For example, some search engines get their indexing information from page content and page titles (the text that appears in the Web browser title bar). Some use the information in meta tags (used in Web page HTML coding) to find keywords for indexing and for descriptions of pages in their search results lists. Some evaluate the number of relevant incoming links to a specific page to establish a page ranking for their search results lists. Some accept payment in exchange for prominently listing Web sites on a search results page that contain specific keywords. Most search engines use a combination of these and other methods to create and report search results.

The basic way a search engine works is that it uses the keywords a user enters together with a ranking algorithm to first identify which pages to include in a search results list and then to rank those pages (generally called hits) in some order of relevancy. Different search engines have different ranking algorithms. Given the sheer volume of Web pages that are available today, a search results list of hits can be very long. Depending on the search strategies used by the search engine and the keywords or phrases entered by the user, a search results list can consist of hundreds or perhaps thousands of pages—most of which are of questionable relevance. Users generally expect to find the most relevant Web page links near the top of a search results hit list, certainly within the top 10 or 15 Web pages listed. Thus, as an e-business entrepreneur, your goal is to get the searcher's attention by making sure your Web pages appear at or near the top of a search results list.

Search engine optimization (SEO) is the art of building your Web pages and Web site links to maximize the chances of your pages getting positioned at or near the top of as many search results lists as possible. Short of paying to become a prominent search result (which may not be a viable option for a cash-strapped startup e-business), there is currently no guaranteed method of ensuring that your e-business's Web pages will be listed at or near the top of a search results list for every one of your relevant keywords. There are, however, a number of steps that you can take to increase your Web pages' chances of being well-positioned in a search results list. The first step involves making sure you write effective descriptions of your products and services, and then pick as many highly relevant keywords as possible when you submit your Web pages to a search tool. Other search engine optimization techniques include:

- writing clear and on-topic Web page content that a search engine can easily index
- using descriptive Web page titles
- avoiding frames and dynamic content
- using text navigation links among pages at your Web site
- using meta tags
- arranging to feature relevant inbound links from other Web sites, as these are used by some search tools to determine a Web page's ranking or relevance

Page Content and Page Title

Some search tools assess the number of times the keywords being searched appear on a Web page and their location on the page to determine the page's relevance and thus its rank in a search results list. You should therefore carefully write the content on each of your Web pages to focus on a specific topic and include the keywords and phrases a user might enter in a search tool to find pages on that topic.[30]

The Web page title is the text that appears in the title bar of the Web browser when you visit a Web page. Some search engines use the Web page title to index and report Web pages. Other search engines look for keywords in a Web page title in order to rank the page in a search results list.[31] Additionally, some search tools use the Web page title to identify the page in the search results list. You should make sure that the title of each of your Web pages accurately reflects the page's content.

Web pages are created using the Hypertext Markup Language (HTML), a markup language that identifies page components so that the page can be displayed in a Web browser. A Web page title is created using the following HTML tags or codes: <title> and </title>. An example of the HTML tags that specifies the title of a Web page is <title>Rackspace Managed Hosting</title>. Figure 6-26 illustrates how this title appears in the Internet Explorer Web browser title bar. Figure 6-27 illustrates how this title appears in a Teoma search results list that was generated by entering the keyword "rackspace."

Frames, Dynamic Content, and Navigation Links

Web page creators use HTML to divide a browser window into viewing areas, called frames, in which different Web pages at the same site can be viewed all at once. Many Web page designers avoid frames, however, in part because some search engine spiders are not able

Rackspace Managed Hosting - Microsoft Internet Explorer

File   Edit   View   Favorites   Tools   Help

FIGURE 6-26   Title text in a Web browser title bar

Rackspace Managed Hosting
Rackspace Managed Hosting provides **managed** solutions
providing dedicated servers running on Linux and Microsoft.
www.rackspace.com/ | Cached
[Related Pages]

FIGURE 6-27   Title text in a search results list

to "see" the Web pages that appear in the frames.[32] Search engine spiders may also have trouble with pages that are loaded with images and dynamic "flash" content. When designing a Web page, keep in mind that the simpler the content, the easier it will be for a search engine spider to "see" and index the page. To index multiple pages at your Web site, a search engine spider follows the hyperlinks that connect these various pages. To enable the search engine spider to easily locate and follow your hyperlinks from page to page, make certain that any Web pages that contain image-based hyperlinks also have corresponding text-based hyperlinks that the search engine can read.[33]

TIP

To see your Web page the way a search engine spider "sees" it, use the link on the student online companion Web page to the Web site for Search Engine World (not to be confused with Search-EngineWatch) and try out the Sim Spider tool.

Meta Tags

HTML **meta tags** are used to add information to a Web page that only a Web browser can see. A few search engines get additional information—such as a page description and relevant keywords—from the text inserted between HTML meta tags. Meta tags are placed at the top of a Web page between the heading HTML tags <head> and </head>, and they use the following format:[34]

<meta name="description" content="Description goes here.">
<meta name="keywords" content="Keywords go here, separated by commas.">

Figure 6-28 illustrates the meta tags on the Rackspace Managed Hosting home page.

TIP

You can see the HTML codes of a Web page by using the Windows Notepad utility. In the Internet Explorer browser, visit a Web page whose code you want to see, click View on the menu bar, and then click Source to open Notepad. In Firefox, click View on the menu bar and then click Page Source to see the HTML codes of the Web page in a text window.

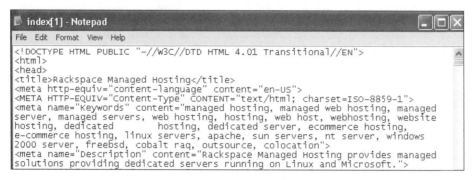

FIGURE 6-28 Meta tags

E-CASE

Success in Doing What You Love

E-business entrepreneurship comes in many flavors—selling all types of products, providing a variety of services, and making all kinds of information available to targeted markets. For Danny Sullivan, a California newspaper researcher and reporter, the e-business itch came in the mid-1990s when he first experienced the Web. Sullivan realized immediately that he wanted to be part of the Internet and Web revolution, but how? The newspapers Sullivan worked for had not yet discovered the power of the new communication medium. So Sullivan left newspaper publishing and joined a Web development company that not only built Web sites, but also used all types of online marketing tools to promote those sites.[35]

One day, when a client asked a question about which keywords would best position his Web site in a search results list—and no one at the company could give him the answer—Sullivan's experience in newspaper research kicked in. He began researching and learning everything he could about search engines and directories and how they worked. He then used his research to publish search engine optimization tips and other search engine information online in the *Webmasters Guide to Search Engines.* The *Guide* became very popular with Web development, design, and marketing professionals. Before long, Sullivan left the Web development company and set up his own e-business site, SearchEngineWatch, which was devoted to all things related to search engines and directories.[36]

Much of the general search engine information at SearchEngineWatch, such as a directory of search engines and search engine rankings, was free to site visitors; but there was also lots of in-depth information that Web page designers, e-business marketers, and other professionals could access for a subscription fee. The Search-EngineWatch site continued to grow in popularity, providing up-to-date search engine news and search engine optimization information to its viewers. In late 1997, Sullivan harvested his new e-business by selling it to the Web publishing powerhouse Jupitermedia. Sullivan stayed on as editor, continuing to do what he loved—researching and writing about the search industry.[37]

continued

Today, Sullivan is recognized as a leading authority on search engines and search engine marketing. He continues to edit SearchEngineWatch and also manages a search industry blog, SearchEngineWatch Forums. Together, the Web sites for SearchEngineWatch and SearchEngineWatch Forums generate more than 200,000 page views (requests to load a page by a browser) each day.[38] Sullivan also hosts search industry conferences.

Link Popularity

Link popularity is a measure of the number of relevant, high-quality, inbound links there are to your Web pages. Many search engines now look at a page's link popularity to establish keyword relevancy and to rank the page in a search results list. You can find a number of online tools to check a page's link popularity, such as the Marketleap Link Popularity Check, the Webmaster Toolkit Link Popularity Checker, and LinkPopularity.com, shown in Figures 6-29 through 6-31.

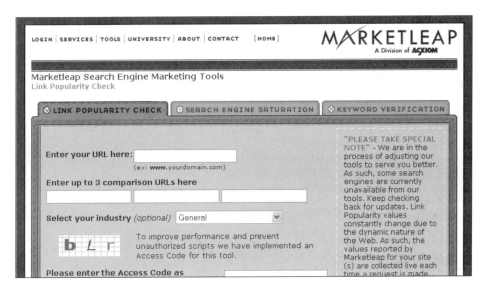

FIGURE 6-29 Marketleap Link Popularity Check

At each of these pages, you can enter the URL for your Web page to see how many other sites are linking to it. Later in this chapter, you learn about acquiring high-quality inbound links for your Web pages.

Given the complexity involved in ensuring that your e-business's Web pages appear among the top search results hits for specific keywords, you may at some point need to seek assistance from search engine optimization professionals (called SEOs) to optimize your pages. Professional SEOs can offer a variety of tools to evaluate how effective your e-business's Web page content, page titles, internal links, meta tag content, and link popularity are at getting your pages listed near the top of a search results list.

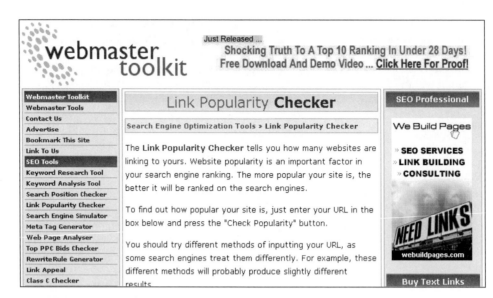

FIGURE 6-30 Webmaster Toolkit Link Popularity Checker

FIGURE 6-31 LinkPopularity.com

Because search engine optimization is an ongoing process, you should periodically conduct some online searches using various relevant keywords to review how your e-business's Web pages appear in the search results lists. Tracking your Web page's ranking on search results lists will also help you assess how well your Web page search engine optimization techniques are working. Monitoring search results lists can provide other useful information as well, such as the presence of a new competitor and how well your competitors' Web pages are optimized.

In addition to submitting information to search tools and optimizing Web pages for search engine spiders, you should also consider using traditional marketing tools such as public relations and advertising to get your e-business and its Web site noticed.

Public Relations

Public relations (PR) is the process of establishing and maintaining a company's public image. When it works well, a public relations effort is one of the more cost-effective marketing tools a business has at its disposal. Public relations activities for an e-business can include developing strong and positive media, community, consumer, and governmental relations; informing the general public about the e-business's policies, procedures, activities, and accomplishments; and, if necessary, managing crisis communications.

Part of the art of a public relations effort is writing effective press releases and sending those releases to the appropriate destinations at just the right time. A **press release** is a short announcement of a newsworthy item that is sent to members of the press. A timely and effectively written press release can attract the attention of writers and editors, and if it's published, it can be a boon to your e-business. Sending the release to the right members of the press also makes a big difference. Another important consideration is to make sure that the news media's audience matches that of the e-business's target market(s). For example, a news radio station may have audience demographics similar to that of an e-business offering online stock trading, while a rock 'n' roll radio station may not. While it's clear in this example on which station an e-business should focus its PR efforts, not all PR choices may be so clear. It is a good idea to work with a public relations professional who can handle this aspect of your marketing effort.

Advertising

As an e-business entrepreneur, you may need to use both online and traditional advertising for two reasons: to let the public know about your e-business (i.e., to build your branding name) and to drive traffic to your Web site (to generate direct sales). Online advertising includes banner, sidebar, pop-up, pop-under, floating, and streaming media ads; permission-based e-mail advertising and newsletters; featured or sponsored placement at other Web sites; and paid placement on the search results lists. Traditional advertising includes radio, television ads, and print media advertising.

TIP

According to a report by PriceWaterhouseCoopers, online advertising ad space is expected to grow to $32 billion by 2009.[39]

Banner, Sidebar, Pop-Up, Pop-Under, Floating, and Streaming Media Ads

Banner ads are rectangular images, or banners, that appear fixed in place on a Web page and that link to the advertiser's Web site. The typical size for a banner ad is 468 pixels wide by 60 pixels high. A **sidebar ad**, also called a skyscraper ad, is vertical rather than horizontal and is typically 600 pixels high by 120 pixels wide. **Pop-up ads** and **pop-under ads** open in their own window on top of a Web page or underneath it on the desktop, respectively. Pop-up and pop-under ads that appear in their own window, and thus must be must manually closed, are unpopular with many viewers. In fact, most Web browsers now contain a pop-up and pop-under ad blocking feature that can be turned on or off, or customized to allow ads only from specific Web sites.

Floating ads, also called **Shoshkele ads** (pioneered by United Virtualities and named Shoshkele after the nickname of the company founder's daughter[40]), appear to float across a Web page for a few seconds, then settle in a specific area of the page or automatically close. Unicast was the first company to use streaming media to present ads with audio and video in their own window. Floating ads and streaming media ads belong to a category of ads known as **rich media**, because they contain interactive elements, audio, and video.[41] You can see examples of floating ads and streaming media ads at the United Virtualities and Unicast Web sites. Figure 6-32 illustrates the United Virtualities Web site.

FIGURE 6-32   United Virtualities

Online ads are generally priced according to the number of **impressions**, or the number of times the ad is viewed. This pricing is expressed in cost per thousand (CPM) impressions. The CPM for an online ad can range from a few pennies, to a few dollars, to much more, depending on the popularity of the site on which the ad appears, the demographics of the site's users, the demand for ad space, and the type of ad. Some online ads are priced on a **pay-per-click** basis; you pay only when a viewer clicks through to your Web site using the ad. The least expensive (and least effective) online ads are banner ads. Because of their size, sidebar ads can be more effective and are therefore more expensive than banner ads. Pop-up, pop-under, floating, and streaming media ads are more effective than banner or sidebar ads, and thus are the most expensive types of online ads.[42]

Some online ads can be targeted to the specific demographics of the registered users at a Web site. For example, an online ad could be targeted to Web site viewers living in a specific zip code with a median income greater than $75,000, or it could be targeted to an even more specific or narrow audience by being designed to appear only on selected pages at a given Web site. Targeted ads are usually more expensive than those that are not targeted, but the higher cost of targeted ads is often worth it, because effectively targeted ads can generate high click-through rates. The **click-through rate** is the percentage of viewers who click an online ad to view the advertised Web site. Click-through rates for targeted ads may be many times higher than the click-through rates for non-targeted ads.

233

> ## TIP
>
> **Click fraud**—creating fraudulent pay-per-click data with phony clicks—is becoming a major concern for online advertisers that follow the pay-per-click revenue model. Different types of click fraud have been reported, including businesses that repeatedly click competitors' ads to run up advertising expenses and e-businesses that bill advertisers at their sites for click-throughs that were not really made. Some e-businesses that provide software or services to help prevent click fraud include PPCTrax.com, ClickFacts.com, and WhosclickingWho.com.[43, 44]

Opt-In E-Mail Advertising and Newsletters

Permission-based e-mail advertising, also called **opt-in e-mail advertising**, is a feature of **permission marketing**, the process of getting consumers to voluntarily learn more about a company and its products and services. Opt-in email advertising involves sending e-mail advertising messages or e-mail newsletters to potential customers who have agreed to receive or "opt-in" to the messages. It is important to remember that opt-in e-mail message or newsletter recipients *choose* to receive these messages. Opt-in e-mail advertising is not unsolicited junk mail, commonly called **spam**.

Opt-in e-mail advertising can be effective because the target audience has chosen to receive the e-mail advertising and, therefore, consists of a group of consumers who are more predisposed to purchase a business's products or services. In its fifth annual report on consumer e-mail advertising, DoubleClick reports that in 2004, 32 percent of opt-in advertising recipients clicked through to the Web site being advertised in the e-mail message and made an immediate purchase; a slightly smaller percent clicked through to get more information and then returned later to make a purchase; and 12 percent clicked through to view more information but made their purchases offline.[45] According to the DoubleClick report, many consumers prefer opt-in e-mail advertising to other forms of direct sales techniques such as telemarketing, in-person direct sales, or direct mail.[46]

Some e-businesses who use opt-in e-mail advertising build their own databases of e-mail addresses by encouraging Web site visitors to complete and submit an online form that will enable them to get additional product information or to subscribe to an industry newsletter. Because of the negative consumer attitudes about spam, an e-business that offers opt-in e-mail sign-up at its Web site should be certain of each sign-up's validity. The best way to do that is to set up a **double opt-in** process, in which a confirmation e-mail message is sent to the consumer who signs up. The consumer's e-mail address is not added to the e-business's database until he or she responds to this confirmation e-mail message. Requiring the consumer to take to two separate actions—the original sign up and responding to the confirmation—also helps to ensure that the consumer is a willing recipient of the opt-in e-mail advertising and should result in higher response rates.

A number of e-businesses, such as Return Path and Yesmail, maintain and rent lists of opt-in e-mail subscribers. These lists can be useful for an e-business that does not want to build its own opt-in database. Generally, an e-business that rents a particular list submits its e-mail advertising message to the list owner, who then sends the opt-in e-mail message to all of the members of the list. The e-business never receives a copy of the list members' e-mail addresses.

An advantage of an opt-in e-mail advertising program is that the time required to execute it is very short, just a matter of days. This means that you can execute this portion of your e-business's advertising strategy quickly, and analyze the results in real time as they come in. This speed allows you to refine your e-business's opt-in e-mail program and then try it again, all in the same time frame it might take just to get an online or traditional ad published.

A disadvantage of opt-in e-mail advertising is the "false positive" response to opt-in e-mail messages that is produced by spam blockers that mistakenly treat legitimate opt-in e-mail as spam. Around 80 percent of all e-mail messages are spam.[47] Consequently, many ISPs, business networks, and individual consumers now employ spam filtering software to delete suspected spam messages or to direct suspected spam into a special e-mail folder where it can be reviewed and, if necessary, deleted. This spam blocking process often removes legitimate opt-in e-mail messages by mistake. A study by Pivotal Veracity, an e-mail advertising optimization services provider, indicates up to 20 percent of opt-in e-mail messages fail to reach destination mailboxes because of spam blockers. More than half of the businesses who participated in the study reported "false positive" effects on their opt-in e-mail advertising because spam blockers had deleted or redirected their messages.[48]

Search Tool and Portal Advertising

Many e-businesses ensure that their Web sites appear on search results lists for specific keywords by participating in various types of advertising programs sponsored directly by search tools or by e-businesses who represent a network of search tools and portal sites. One such type of advertising program is a search tool pay-per-click program in which an e-business pays for placement in a search results list. To participate in this type of pay-per-click program, you first identify search keywords relevant to your e-business. Then you create a brief ad for each keyword, consisting of a title, description, and URL. You submit the ad and a bid—the amount you are willing to pay each time a viewer clicks through to your Web site—to have the ad presented in a search results list. Most bids range from a few pennies to a few dollars. The higher the bid, the higher the placement of your e-business's ad in the search results list. The advertising provider then distributes your e-business ad across its network of search tools and other Web sites. Today, most search tools are careful to identify these listings as paid or "sponsored listings" to avoid confusion with their non-paid search results.

Search tool and portal advertising providers include Yahoo! Search Marketing, MIVA (formerly FindWhat.com), Enhance Interactive (formerly ah-ha.com), goClick.com, Google Adsense, and Kanoodle.com. Figures 6-33 through 6-35 illustrate the Web sites of three e-businesses that offer search tool and portal advertising programs.

FIGURE 6-33   Yahoo! Search Marketing

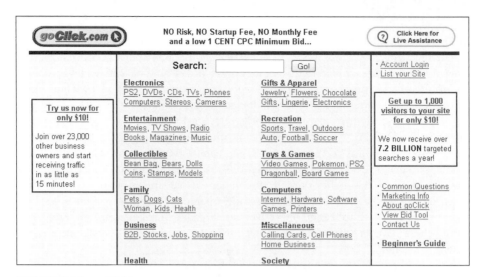

FIGURE 6-34 goClick.com

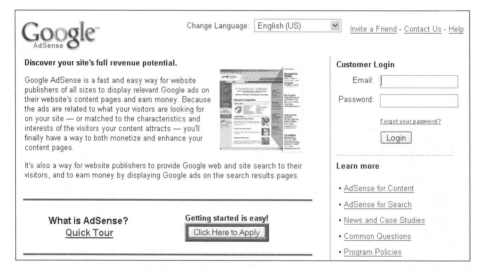

FIGURE 6-35 Google Adsense

E-PIONEERS

Bidding for Placement

In the late 1990s, one search engine—GoTo.com—changed the search industry forever by allowing e-businesses to bid for placement in its search results lists. An e-business could select keywords relevant to its products and services and then submit a bid for each keyword to GoTo.com. Each time a GoTo.com user entered one of the keywords, the e-business's title, description, and URL appeared in the search results list. How high in the list the information appeared depended on the size of the bid. The information from the highest bidder was placed at the top of the search results list, the information from next highest bidder was placed next, and so forth, until all bidders' information was listed. Below these bidders would appear the actual search results list, which listed the Web pages of businesses that did not participate in the bidding process. It wasn't long before other search tools began developing their own pay-for-placement programs.

But soon there was a tremendous controversy raging over these programs. While pay-for-placement search tool options were a boon to e-businesses who wanted to be the first in line to get a search tool user's attention, search results containing paid placements at the top of the search results list were not necessarily a good thing for search tool users. For one, users frequently found that the top listed results were not necessarily the most relevant results for their needs. Plus, some search tools obscured the fact that their top placements were paid placements, and eventually questions began to arise about the editorial purity of the search results. In July 2001, the consumer watchdog group Commercial Alert filed a deceptive advertising complaint with the Federal Trade Commission (FTC) charging that several major search tools were misleading consumers by disguising ads as legitimate search results hits.[49] In June 2002, the FTC responded to this complaint by issuing a letter to major search tools requesting that they make a clear distinction between paid and nonpaid search results hits.[50] Search tools began cleaning up their act, and today most search tool pay-for-placement advertising is clearly marked as sponsored or paid listings.

How did GoTo.com fare throughout all this controversy? Very well, indeed. Originally incubated by the commercial business incubator Idealab, GoTo.com went public in June 1999; in 2001, it changed its name to Overture Services, Inc. and broadened its advertising offerings; and in 2003, it was purchased by Yahoo!, one of its most important strategic partners, for a reported $1.6 billion![51] In March, 2005, Overture Services was re-branded as Yahoo! Search Marketing.

Traditional Advertising

As part of your overall advertising campaign, you should consider using traditional advertising for your e-business. Traditional advertising includes print media, radio, television, outdoor advertisements, and direct mail. Online advertising has not replaced traditional media. In fact, the two complement each other. Costs for traditional media are often measured in CPM, as they are for online advertising. Prices for advertising in traditional media vary widely, but are easily compared using the CPM measure.

The two goals of your e-business advertising program should be to acquire new customers as inexpensively as possible and to build your brand. If your e-business does not have the in-house expertise to design and manage an advertising campaign, consider hiring an advertising agency. The professional marketers at an advertising agency should have the experience and expertise to determine what does and does not work well for e-businesses similar to yours, and can help you design your entire advertising campaign, including helping you choose both the online and traditional advertising you should pursue. An advertising agency can also help ensure that all your e-business's advertising efforts are integrated, communicate the same message, and work toward building your brand.

In addition to public relations and advertising, you can use a number of other, cost-effective promotion methods to let the public know about your e-business.

Other Promotion Tools

Along with the formal marketing tools presented so far, there are many other tools that you can use to promote your e-business. One basic way is to make sure that your e-business Web site's URL and e-mail address appear on all of your e-business's printed materials, including business cards, letterhead stationery and envelopes, and brochures. In addition, any building signs, company uniforms, delivery vans, shipping boxes, Yellow Page listings, and trade organization directories you use—basically, any vehicles, packaging, or venues that may be seen by potential customers—should display your e-business's Web site URL. Other promotion tools include link exchanges, newsgroup and Web-based forum postings, word of mouth, weblogs, RSS feeds, affiliate programs, and Web rings and awards.

Link Exchanges

As you learned earlier in this chapter, link popularity is a very important component of search engine optimization. Acquiring relevant, high-quality inbound links is, therefore, important for your e-business. Many e-businesses cultivate a variety of partnerships in order to have complementary Web sites with which to exchange links. For example, an online pet store might exchange links with a veterinary clinic's Web site, or a travel tour site might exchange links with sites in its touring areas, such as restaurants, hotels, bed-and-breakfast accommodations, and so forth.

The first step to developing **link exchanges** is to identify Web sites that complement your e-business. Once you've identified a few such Web sites, you should send an e-mail message to each that describes the complementary elements between your two sites and propose a link exchange. Many e-businesses or organizations are happy to add additional links to their Web sites and to have others link to their Web site; however, some e-businesses will not exchange links, or are very selective about the Web sites with which they exchange links. Occasionally, an e-business might be willing to provide a link to your Web site while not requiring you to establish a reciprocal link, because it's simply interested in providing additional information to its own customers. Another way to encourage link exchanges is to offer the complementary Web site some free content, such as an article of interest to the site's visitors, in exchange for a link to your site. E-businesses such as LinkPartners.com can help in selecting high-quality reciprocal links (Figure 6-36).

FIGURE 6-36 LinkPartners.com

Beware of exchanging links through a **link farm**, a group of sites with unrelated content that link to each other solely for the purpose of building link popularity. Most search engines have editorial policies against ranking pages with a link farm connection and have very sophisticated methods of sniffing out link farm connections. Note that link farming is sometimes called **link stuffing**.[52]

QUOTES ON SUCCESS

"I think it's unfortunate that people today seem to think about links only in terms of what Google might like, rather than asking themselves what's the intrinsic value behind getting a link. Think a link will send you an audience you want in and of itself? Then ask for or pay for that link. Think your visitors would want to find a link for a particular resource on your site? Then install that link. These are the key criteria of linking, to me."

Danny Sullivan, founder and editor of SearchEngineWatch

TIP

Here's a quick way to search for Web pages that link to your Web pages using the Google search engine. Just type the word "link" and a colon. Then type the URL of your Web page—for example, link:rackspace.com—and click the Google Search button to see a search results list of the Web pages that link to yours.

Newsgroups and Web-Based Forums

A **newsgroup** is a topic-specific electronic "bulletin board" accessed over the Internet. Newsgroup participants submit messages and respond to others' messages using a Web browser or e-mail software plug-in called a newsreader. **Web-based forums** are discussion groups that allow participants to read and post messages through a Web site and Web browser. Newsgroups and Web-based forums in which topics related to your e-business are discussed can be a great source of information and can also serve as an indirect venue for marketing your e-business. Participating in topic-related newsgroups and Web-based forums by getting involved in the group and answering questions posted by other participants allows you to build relationships. In this way you can establish your expertise in the topic, and indirectly market yourself and your e-business to the other participants. A side benefit of being involved in a newsgroup or Web-based forum is that you quickly become aware of negative postings about your e-business's products or services, and are able to check out any postings about your competition.

But remember that your participation only indirectly benefits your e-business—never post blatant advertisements to the group. These groups are forums for information, not for advertising. After you identify a newsgroup or Web-based forum to join, it is a good idea to simply observe the message postings for a while and learn what is acceptable and what is not acceptable to the group before you begin posting your own messages.

> **TIP**
>
> Certain rules should be followed when responding to or posting messages to newsgroups and Web-based forums. These rules are often referred to as "netiquette" (short for network etiquette). Keep messages short, to the point, and relevant to the topic. For more information, just search the Web using the keyword "netiquette."

Old-Fashioned Word of Mouth

When customers tell their friends about your e-business, and those friends tell their friends, and so on, an exponential explosion can occur in the number of visitors who view your Web site and purchase your products or services. In Chapter 2, you learned about the power of the network effect and viral marketing. At the heart of viral marketing is old-fashioned **word of mouth**—one customer telling others of his or her experience. Today, a message or recommendation can travel through "electronic word of mouth" over the Internet from person to person around the world in seconds.

Remember Hotmail, the e-mail provider whose early success was a direct result of the network effect and viral marketing? Amazon.com offers another great example of the power of the network and word of mouth viral marketing. The first Amazon.com test Web site (called a beta site) was created in late spring, 1995. Not long afterward, Jeff Bezos and his team made the site available to 300 family members and friends who were asked to test the site's features.[53] The following month, the Amazon.com site was made available to the public, and to promote the site, Bezos and his team simply asked the 300 original testers to tell others about the site. The result was phenomenal—in the first 30 days that the new site was available to the public, Amazon.com sold books to new customers in every U.S. state and in more than 40 countries![54]

New-Fashioned Word of Mouth—Business Blogs

Weblogs, or **blogs**, are online diaries and have been around for quite a while. It wasn't until the 2004 U.S. presidential election, however, that "blog" became a household word. During the election, thousands of people started "blogging"—posting their thoughts on blogs hosted by news sites that were covering the campaign and election process, or on Web sites that supported one of the political candidates. By the end of 2004, 27 percent of Internet users—32 million people—were reading blogs, and 8 million people had created one![55]

Blogs are fast becoming an influential communication medium in the business world. For example, many businesses now host CEO blogs that help put a "human face" on a business. Business blogs have become the new medium for word of mouth marketing and can be useful to your marketing efforts in several ways. For one, publishing your own blog can help you reach out to your e-business's target audience and build brand loyalty. In addition, tracking the blogs of others can provide you with a wealth of market research because they can serve as venues where your happy and not-so-happy customers discuss your business. In general, businesses consider any product- or service-related feedback that comes from the marketplace—whether it's good or bad—to be valuable; in the case of blogs, you can not only tap into potentially constructive primary research, but you do so at very little cost.

Companies as diverse as Microsoft, Stonyfield Farm, GM, and Google are using blogs in a variety ways: as a tool that fosters internal employee collaboration; enables employees and the public to discuss new technologies; and helps the business create a permanent historical record of company activities.[56] Some e-businesses are even buying ad space on popular blogs. A line can be crossed here, however, and many marketing professionals warn e-businesses not to publish a fake blog or "flog," a business blog hosted by a fictitious person and designed specifically to tout a new product. Many bloggers are highly critical of any attempt to "spam" the blogging world, sometimes called the blogosphere, with flogs.[57, 58]

One additional warning: Remember to never underestimate the power of the network effect and viral marketing. Old-fashioned or new-fashioned word of mouth referrals from satisfied customers can be a major source of new customers and revenues. But beware the "dark side" of the network effect and viral marketing—bad word of mouth from customers who have had negative experiences with your products or services. Although a satisfied customer may or may not tell others about his or her experience with your e-business, an unhappy customer is highly likely to do so.

RSS

Another way to promote your e-business is to offer RSS syndication of some of your Web site's content to consumers. **RSS (Really Simple Syndication)** is an XML technology used to publish blogs and to syndicate other types of content (such as headline news stories) from Web sites. For a consumer, subscribing to an RSS feed (syndication) is as simple as clicking the small red or blue RSS or XML button at an RSS-enabled Web site and then subscribing to the RSS feed. Figure 6-37 illustrates the XML buttons used to subscribe to various RSS feeds at the RSS in Government Web site.

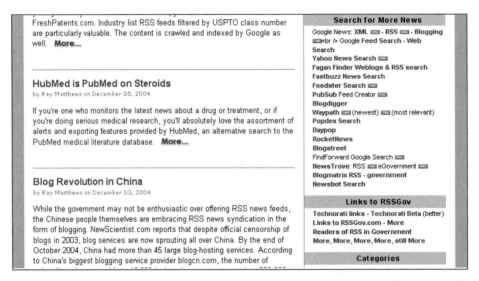

FIGURE 6-37   RSS in Government

The RSS syndicated content to which the user has subscribed then becomes automatically available to the viewer through his or her Web browser. The current versions of the most popular Web browsers, such as Internet Explorer 7 and Firefox, feature a built-in RSS reader that viewers can use to read RSS content. Viewers can also download free RSS reader plug-ins for older browsers that do not support RSS, such as the Pluck download for older versions of the Internet Explorer browser.

Podcasting is the RSS of audio files, such as music or radio programs, and it provides yet another way for e-businesses to reach their targeted audiences. Consumers can subscribe to a podcasting feed from a Web site in the same way that they subscribe to any RSS feed.

Affiliate Programs

One marketing tool that many e-businesses have used effectively is an affiliate program. An e-business **affiliate program** is an arrangement by which an e-business pays a flat referral fee or a commission for all sales sent to it from another Web site. To operate an affiliate program, an e-business must identify and draw affiliates and then connect to each affiliate Web site via a customized link. When a visitor at an affiliate site clicks through one of these links and makes a purchase at the e-business's site, the affiliate is given credit for the click through. In addition to the sale, the benefit to e-businesses is that another Web site has an incentive to take on the marketing of the e-business's products and services. A pioneer in e-business affiliate programs is Amazon.com. Affiliate programs are discussed in more detail in Chapter 7.

Web Rings and Awards

Sometimes a group of similar e-businesses will create a "chain" of links among their Web sites that allows a visitor to link through to all of the sites in the chain. This circular chain of links is called a **Web ring**. Participating in a Web ring can increase the number of visitors to your e-business Web site, but it will also increase the number of visitors who leave the site to follow the Web ring.

Another way to draw potential customers to an e-business Web site is to win an award, such as a "cool site of the day" designation or Web design awards. Winning an award can also help give a fledgling e-business credibility.

QUOTES ON SUCCESS

"Winning an award demonstrates your Internet investment is paying off. It's an opportunity to promote your accomplishment to the media, clients, prospects and employees."

William Rice, president of the Web Marketing Association

. . .THE SAGA OF A BRAND

Selling alcoholic beverages across state boundaries involves negotiating a myriad of laws and regulations. Virtual Vineyards moved quickly to establish strategic partnerships with alcoholic beverage importers, distributors, and wholesalers as well as a law firm that specialized in alcoholic-beverage law. As a first mover in the industry, one of Virtual Vineyards' goals was to position itself as a leader in developing an e-business model that complied with state-by-state regulations related to the selling of alcoholic beverages as well as becoming a lobbyist for changes in the law to make selling wine across state lines easier.[59]

continued

Despite experiencing some ups and downs that are common to startup e-businesses, Virtual Vineyards was still standing in late 1999 when it merged with rival online wine merchant Wine.com. The merged company retained the Wine.com name and URL, and became the "new" Wine.com. Not long afterward, the "new" Wine.com had raised about $100 million from VCs and others and was on its way to becoming the number one wine portal.[60, 61] The San Francisco advertising agency kirshenbaum bond + partners was hired to design a successful brand awareness campaign for Wine.com, and millions were spent on advertising and promotion.[62, 63]

By the end of 1999, it looked as though the sky was the limit for e-businesses in general, and Wine.com in particular seemed poised for success. But it didn't turn out that way. Wine.com found that customer acquisition was slower than expected, distribution costs were higher than expected, and navigating the arcane state-level alcoholic-beverage laws and regulations designed to protect in-state businesses at the expense of out-of-state businesses was almost overwhelming.[64, 65] By late 2000, Wine.com was hemorrhaging cash, and the company merged again—this time with startup rival Wineshopper.com. The merged company retained the by now well-branded name, Wine.com. Unfortunately, however, the merger didn't save either of the two e-businesses. The economic downturn of 2000-2001 coupled with the dot.com bust was dealing death blows to many e-businesses who were already in trouble—and Wine.com was one such victim. By the spring of 2001, a Portland, Oregon-based rival, eVineyard, was able to purchase the Wine.com name, URL, customer list, and Web site content for less than $10 million.[66]

In July 2001, eVineyard relaunched its Web site as "Wine.com by eVineyard" using the www.wine.com URL. Announcing the relaunch, eVineyard Vice President Brett Lauter stated, "There was a lot of brand recognition and awareness for the Wine.com name and logo." eVineyard CEO Larry Gerhard also noted, "In adopting the Wine.com look and feel, we are also capitalizing on the more than $100 million that the original Wine.com investors spend on advertising and branding."[67]

Despite these high hopes, "Wine.com by eVineyard" fared little better financially than its previous incarnations. In 2002, eVineyard restructured its operations, sought new funding, changed its name back to—you guessed it, Wine.com—and moved its operations to San Francisco. By the fall of 2004, the latest incarnation of Wine.com remained the premier wine e-retailer, and thus managed to draw more than $30 million in funding—$20 million of which was from a venture capital firm Baker Capital of New York. In the funding press release, Joseph Saviano, general partner of Baker Capital, explained, "What we found most interesting and promising about Wine.com is the company's tremendous brand, uniquely positioned as the category leader in wine."[68]

Another very important event in the saga of the Wine.com brand took place in May, 2005 when the U.S. Supreme Court ruled that individual states can no longer restrict alcohol sales from out-of-state businesses to protect in-state businesses—now the restrictions, if there are any, must be applied to both in-state and out-of-state sales. It is expected that this ruling will make state laws less restraining and, in turn, make it easier for e-businesses like Wine.com to operate.[69]

Remember Peter Granoff and Robert Olson, the entrepreneurs who co-founded Virtual Vineyards, and started the saga of the Wine.com brand? As of this writing, both remain e-business entrepreneurs. Olson left Wine.com in 1999 and now serves as Vice President of Engineering for PostX, a technology startup. Granoff left Wine.com in 2001 and co-founded a new business, Ferry Plaza Wine Merchant and Wine Bar in San Francisco and the Ferry Plaza Wine Merchant Web site and online store.

Chapter Summary

- Marketing is the process of developing the mutually satisfying relationships between your e-business and your customers that result in sales and profits.

- The classic marketing mix model consists of the Four Ps: product, place, promotion, and price. Other marketing mix models include the Four Cs model (customer needs and wants, convenience, communication, cost to the customer) and a number of extended marketing mix models that add additional variables.

- A brand is a combination of a name, logo, and design that identifies a business's products and services. A brand name should be something that attracts attention but is easy to remember, spell, and understand.

- A domain name, or URL, is a Web page's address on a Web server. A well-chosen domain name can be an effective tool in building an e-business's brand.

- Market research involves collecting and analyzing the data needed to create informed decisions about selling products or services in a specific marketplace. The two types of market research are primary and secondary research.

- A marketing plan is a document that provides the details of the marketplace analysis summary portion of your business plan.

- Marketing plans vary, but the document should include the following elements: an Executive Summary; Situational Analysis; Objectives, Strategies, and Tactics; and Budget and Performance Measures.

- The Executive Summary is similar to a general business plan Executive Summary and gives an overview of a marketing plan in no more than three pages.

- The Situational Analysis portion of your marketing plan describes what you know now about the marketplace.

- The Objectives, Strategies, and Tactics portion of a marketing plan itemizes your attainable, measurable, realistic, and time-specific goals; your strategies for achieving those goals; and the specific tools you will use.

- The Budget and Performance Measures portion of your marketing plan reflects the costs of your marketing strategies and contains a section that describes how you will measure the success or failure of your marketing strategies.

- The marketing tools you can use to promote your e-business include traditional tools (such as radio, TV, and print ads) and public relations efforts.

- Online marketing tools you can use include search tool submissions, search engine optimization of your Web pages, online ads, opt-in e-mail and newsletter advertising, search tool and portal advertising, link exchanges, newsgroup and Web-based forums, business blogs, RSS feeds, affiliate programs, and Web rings and awards.

Checklist

Promoting Your E-Business:

- ❑ Have you created a marketing plan?
- ❑ Does your marketing plan describe your marketing goals and list the strategies you will follow and the specific tactics you will employ to meet those goals?
- ❑ Have you submitted your Web site URL to all the major search tools?
- ❑ Have you added relevant meta tag keywords to the HTML coding in your Web pages?
- ❑ Are your Web page titles brief and descriptive?
- ❑ Have you arranged a link exchange with complementary Web sites?
- ❑ Do you have a PR professional on staff, or have you hired a PR professional to prepare your press releases and other important public communications?
- ❑ Have you considered publishing a blog?
- ❑ Do you participate in relevant industry newsgroups or Web-based forums?
- ❑ Have you checked out the possibility of participating in an industry Web ring?
- ❑ Are you using banner, sidebar, pop-up, pop-under, floating, or streaming media ads? If so, have you measured their effectiveness lately?
- ❑ Are you using opt-in e-mail advertising?
- ❑ Do you regularly review search results hit lists using keywords relevant to your e-business?

Key Terms

accredited registrars	link stuffing
affiliate program	market research
banner ads	market segment
blogs	market size
bots	marketing
brand	marketing budget
click fraud	marketing mix
click-through rate	marketing plan
cookies	marketing strategies
crawlers	meta search engine
directory	meta tags
domain name	newsgroup
double opt-in	objectives
floating ads	opt-in e-mail advertising
impressions	paid-inclusion program
link exchanges	pay-per-click
link farm	performance measures
link popularity	permission marketing

podcasting

pop-under ads

pop-up ads

press release

primary research

public relations (PR)

qualitative research

quantitative research

rich media

RSS (Really Simple Syndication)

search engine optimization (SEO)

search engines

search tools

secondary research

Shoshkele ads

sidebar ad

spam

spiders

strategy

streaming media ad

tactics

target market(s)

Uniform Resource Locator (URL)

Web ring

Web-based forums

word of mouth

Review Questions

True/False Questions

1. A USENET newsgroup or Web-based forum is a perfect place for an e-business to place advertisements for its products and services. True or False?

2. The marketing mix identifies the elements of your marketing strategies over which you have control. True or False?

3. A trademark and a brand are the same thing. True or False?

4. Qualitative market research involves using statistical methods to analyze data. True or False?

5. It is a good idea for an e-business to periodically review search engine results, using the keywords already submitted to search engines and meta tag keywords, to see how high its URL is returned in a hit list. True or False?

Multiple Choice Questions

1. Which of the following depends on user-submitted information to compile its index?
   a. search engine
   b. directory
   c. blog
   d. affiliate program

2. Which of the following best describes marketing objectives?
   a. attainable
   b. measurable
   c. time-specific
   d. all of the above

3. Shoshkele ads are also called:

   a. banner ads.

   b. floating ads.

   c. pop-up ads.

   d. streaming media ads.

4. The art of building Web pages and Web site links that are easy for a search spider to index is called:

   a. search engine blogging.

   b. search engine linking.

   c. search engine advertising.

   d. None of the above.

5. Which of the following marketing tools can help boost your site's link popularity as measured by some search engines?

   a. opt-in e-mail

   b. blogging

   c. link exchanges

   d. streaming media ads

Exercises

1. Using a link located on this text's student online companion, visit the Web page that contains the Search Engine World's Sim Spider tool. Enter the URLs of five Web pages of your choice in the tool to see how each page looks to a search engine spider. Print the pages and use them to discuss search engine optimization with a group of classmates.

2. Using links located on this text's student online companion, visit a Web site that offers a tool to test link popularity. Enter the URLs for at least five different Web pages of your choice in the tool to test each page's link popularity. Write down the URL of each tested page and the number of pages that link to it. Use your information to discuss the importance of link popularity with a group of classmates.

3. Use online search tools to search the Web for articles on the effectiveness of rich media online ads. Then, using links located on this text's student online companion, visit the United Virtualities and Unicast Web sites and review samples of streaming media and Shoshkele rich media ads. Write a brief description of both types of rich media ads. Then discuss with a group of classmates the advantages and disadvantages of using rich media ads in an e-business advertising program.

4. Using the Web browser commands noted in this chapter, review the <title> and <meta> tags on the home page of two B2C Web sites that sell similar products. Create a list of the <title> tag text, <meta> tag keywords, and <meta> tag descriptions each page uses. Then use the <meta> tag keywords to search for the pages in at least four search tools. Compare the ranking of the pages in each search engine. Note whether or not each search engine uses the <title> tag text in the search results list.

5. Using links located on this text's student online companion or online search tools, locate three business blogs. Note the name, company, and work title of the person publishing the blog. Read several blog postings to understand the blog's content and contributors. Then write a summary describing each blog. In your summary, discuss the effectiveness of the blogs as a marketing tool for the respective business and list the overall advantages and disadvantages of blogs to an e-business.

Case Projects

1. You want to promote your C2C auction Web site by participating in a pay-per-click search tool advertising program. Identify at least 10 search keywords or key phrases that are relevant to your e-business. (You may assume any other facts about your e-business necessary to identify the keywords or key phrases.) Create a brief description for each keyword or key phrase you plan to submit.

2. Your hobby is collecting military memorabilia, including GI Joe dolls and accessories. Not long ago, you attended a local collectors' show, and the experience convinced you that an online store that sells military memorabilia could be successful. You are working on your business and marketing plans and are ready to consider the different ways you can promote your online store. Create a list of five tools you can use to promote your Web site. Then describe how you will use each tool and the e-business promotion benefits you expect to garner.

3. Your new B2C e-business Web site is not attracting the number of visitors you would like, and you decide that exchanging links with complementary Web sites will help. First, create a name and brief description of your e-business. (You may assume any facts about your e-business not already stated.) Then, using online search tools or other relevant sources, identify five real-world Web sites that are complementary to your e-business. Briefly explain how exchanging links with each site could improve your Web site's profile in the marketplace.

Team Project

You and three classmates are working on the marketing plan portion of the e-business plan for your new B2B e-business. Working together, create a name and description of the e-business and determine the major products or services your e-business will sell. Use links located on this text's student online companion to identify an available URL for your e-business. Choose a name and URL that you think will be effective in building your e-business's brand.

Use online search tools or other relevant resources to prepare some preliminary market research that identifies the characteristics of your target market and your top competitors. Then create an outline for your marketing plan that includes marketing objectives, strategies, and tactics. (You may assume facts about sales objectives and other related data not explicitly stated.)

Using Microsoft PowerPoint or another presentation tool, create a 10-15 slide presentation that describes your business name, URL, and marketing plan outline. Deliver your presentation to a group of classmates who have been selected by your instructor to critique your business name, URL, and marketing plan outline.

For Further Study

Here are some resources that might help you in further investigating the topics covered in this chapter.

Student Online Companion

Check out the *Creating a Winning E-Business, Second Edition* student online companion Web site for links to the sites discussed in this chapter and to other useful Web sites.

Articles and Books

Wikipedia. "E-Mail Marketing." en.wikipedia.org/wiki/E-mail_marketing. June 22, 2005.

Synergy Network, Inc. "Marketing Metrics and Strategic Navigation." synergynet.com/artman/publish/marketing_resources.shtml. 2005.

Baker, Stephen and Green, Heather. "Blogs Will Change Your Business." *BusinessWeek*. www.businessweek.com/magazine/content/05_18/b3931001_mz001.htm. May 2, 2005.

Brabender, Todd. "The Increasing Power of Publicity." *Mplans*. www.mplans.com/dpm/article.cfm/155. June 15, 2004.

Bradender, Todd. "I Can't Afford a Publicity/Public Relations Campaign – CAN I?" *MPlans*. www.mplans.com/dpm/article.cfm/135. May 18, 2004.

Callan, David. "How to Get Reciprocal Links." *AKAMARKETING.COM*. www.akamarketing.com/how-to-get-reciprocal-links.html. 2004.

Godin, Seth. *Permission Marketing: Turning Strangers Into Friends and Friends Into Customers*. New York: Simon & Schuster. 1999.

Franz, Catherine. "35 Quick Tips for Writing a Press Release." *Concept Marketing Group, Inc.* www.marketingsource.com/articles/view%201654. 2005.

Hamm, Steve. "Ads Gone Wild." *BusinessWeek*. www.businessweek.com/the_thread/techbeat/archives/2005/02/ads_gone_wild.html. February 4, 2005.

Hartsock, Nettie. "Mookie Tenembaum: The Power of Honest and Rich Media." *FindArticles*. www.findarticles.com/p/articles/mi_zd4149/is_200411/ai_n9476624. November, 2004.

Hewitt, Hugh. *Blog: Understanding the Information Reformation That's Changing Your World*. Nashville: Nelson Business. 2005.

Jantsch, John. "How To Write a Killer Press Release." *Concept Marketing Group, Inc.* www.marketingsource.com/articles/view/1900. 2005.

Kotler, Philip and Armstrong, Gary. *Principles of Marketing (11th Ed.)*. New York: Prentice Hall. 2005.

Leduc, Bob. "Increase Your Profits by Coordinating Online and Traditional Offline Marketing." *Concept Marketing Group, Inc.* www.marketingsource.com/articles/view/172. 2005.

Naples, Mark. "Will Floating Ads Sink Rich Media?" *iMedia*. imediaconnection.com/content/5151.asp. February 28, 2005.

Orbinger, Lee Ann. "How Marketing Plans Work." *How Stuff Works*. money.howstuffworks.com/marketing-plan.htm. 2005.

Ries, Al and Ries, Laura. *The 22 Immutable Laws of Branding*. New York: HarperBusiness. 2002.

Seda, Catherine. "Sweet Reward: Web Awards Give Your Site a Huge Boost." *Entrepreneur* as reported by *FindArticles*. findarticles.com/p/articles/mi_m0DTI. January 2004.

Tenembaum, Mookie. "Will 2005 Be the Year of Rich Media?" *MediaPost*. www.mediapost.com/PrintFriend.cfm?articleId=284679. January 3, 2005.

Whaley, Charles. "Would You Like Some URLs With That Web Ring?" *Computing Canada* as reported by *FindArticles.* findarticles.com/p/articles/mi_m0CGC/is_13_28/ai_88127425. June 21, 2002.

Wharton School of the University of Pennsylvania. "Darn Those Pop-Up Ads! They're Maddening, But Do They Work?" http://knowledge.wharton.upenn.edu/index.cfm?fa=viewfeature&id=828. August 13, 2003.

End Notes

[1] Hapgood, Fred. "What Makes Virtual Vineyards Rule?" *Inc.com.* pf.inc.com/magazine/19960615/1966.html. June 1996.

[2] Ibid.

[3] Wikipedia. "Marketing Mix." en.wikipedia.org/wiki/Marketing_mix. 2005.

[4] Wilson, Ralph F. "The 4 Ps of Marketing as Part of Your Internet Marketing Plan." *Web Marketing Today.* www.wilsonweb.com/wmt5/plan-4p.htm. May 1, 2000.

[5] McCarthy, E. Jerome and Kotler, Philip. "Four P's/Four C's Models – Marketing Mix." *ManyWorlds: The Knowledge Network of Business Thought Leadership.* www.manyworlds.com/index2.aspx?from=/authorCOs.aspx&firstname=E.%20Jerome&lastname=McCarthy. 2005.

[6] Booms, G. H. and Bitner, M. J. "Marketing Strategies and Organizational Structures for Service Firms." As reported by Donnelly, J. H. and George, W. R. *Marketing of Services*, pp. 47-51. American Marketing Association. 1981.

[7] Constantinides, Efthymios. "The 4S Web-Marketing Mix Model." *E-Commerce Research and Applications*. Elsevier Science as reported by 12Manage. www.12manage.com/methods_constantinides_4s_web_marketing_mix.html. July 2002.

[8] Barsch, Paul A. "The Demise of the 4 Ps Has Been Greatly Exaggerated." MarketingProfs.com. www.marketingprofs.com/5/barsch3.asp. June 7, 2005.

[9] *Dictionary of Marketing Terms*, American Marketing Association. www.marketingpower.com/mg-dictionary-view329.php. 2005.

[10] AllBusiness. "What is a Brand?" www.allbusiness.com/articles/QuestionsAnswers/416-2057-1640.html. 2005.

[11] Synergy Network. "What is a Brand?" *Marketing Resources*. synergynet.com/artman/publish/marketing_resources/printer_brand.shtml. May 1, 2002.

[12] Allen, Debbie. "Developing Your Brand." *The Sideroad*. www.sideroad.com/Branding/developing-your-brand.html. 2005.

[13] Schmieder, Karl. "Choosing a Company Name." *The Sideroad*. www.sideroad.com/Branding/company_names.html. 2005.

[14] Wikipedia. "Michael Dell." en.wikipedia.org/wiki/Michael_Dell. July 2005.

[15] Robinson, Fabian. "Developing a Brand: Want to Generate Loyalty Online and Off? Create a Killer Brand." *Black Enterprise* as reported in *FindArticles*. findarticles.com/p/articles/mi_m1365/is_10_32/ai_85010835. May 2002.

[16] Ries, Al and Ries, Laura. *The 11 Immutable Laws of Internet Branding*. New York: HarperCollins Publishers. 2000.

[17] Schmieder, Karl. "Choosing a Company Name." *The Sideroad*. www.sideroad.com/Branding/company_names.html. 2005.

[18] *Google: About Us*. "Google Corporate Information: Corporate History." www.google.com/intl/en/corporate/history.html. 2005.

[19] Germain, Jack M. "Domain Name Business Booming in Post-Dot-Com Era." *Ecommerce Times*. www.ecommercetimes.com/story/42657.html. May 9, 2005.

[20] Gigalaw.com. "Anticybersquatting Consumer Protection Act." www.gigalaw.com/library/anticybersquattingact-1999-11-29-p1.html. 1999.

[21] KnowThis.com. "Principles of Marketing – Marketing Research." *Marketing Virtual Library*. www.knowthis.com/tutorials/marketing/marketing_research.htm. 2005.

[22] Ibid.

[23] Mednick, Barbara K. "10 Steps to an Effective Marketing Plan." *Concept Marketing Group, Inc.* www.marketingsource.com/articles/view/1950. 2005.

[24] KnowThis.com. "KnowThis Tutorial: How to Write a Marketing Plan." www.knowthis.com/tutorials/marketing/marketingplan1.htm. 2005.

[25] U.S. Small Business Administration. "The Marketing Your Business for Success Workbook." www.sba.gov/gopher/Business-Development/Business-Initiatives-Education-Training/Marketing-Plan/. 2001.

[26] Wikipedia. "PEST Analysis." en.wikipedia.org/wiki/PEST_analysis. 2005.

[27] U.S. Small Business Administration. "Competitive Analysis." www.sba.gov/starting_business/marketing/analysis.html. 2005.

[28] SBA Online Women's Business Center. "Marketing Plan Components, A Quick Review." www.onlinewbc.gov/docs/market/mk_plan_quick.html. 2005.

[29] Pew Internet & American Life Project. "Data Memo on Search Engines." www.pewinternet.org/PPF/r/132/report_display.asp. August 12, 2004.

[30] Kessler, Cathy. "Insider Secrets of Writing for Search Engines." *Concept Marketing Group, Inc.* www.marketingsource.com/articles/view/1340. 2005.

[31] Dunn, Ross. "A 10 Minute Search Engine Optimization." Stepforth. news.stepforth.com/2003-news/ten-minute-optimization.shtml. April 16, 2003.

[32] Sullivan, Danny. "Search Engine Placement Tips." Search Engine Watch. searchenginewatch.com/webmasters/article.php/2168021. October 14, 2002.

[33] Skog, Ronny. "Busy People's Guide to Top Search Engine Position Methods." *Concept Marketing Group Inc.* www.marketingsource.com/articles/view/1263. 2005.

[34] Sullivan, Danny. "How to Use HTML Meta Tags." Search Engine Watch. searchenginewatch.com/webmasters/print.php/34751_2167931. December 5, 2002.

[35] Da Vanzo, Peter. "Ten Questions with: Danny Sullivan." Searchengineblog.com. www.searchengineblog.com/interviews/interview_danny_sullivan.htm. 2000.

[36] Ibid.

[37] Ibid.

[38] Meckler, Alan. "JupiterResearch Blogs Reap Business." *Internet Media Commentary Weblog*. weblogs.jupitermedia.com/Meckler/archives/2005_01.html. January 26, 2005.

[39] Newcomb, Kevin. "Global Advertising on Upswing." *ClickZ Network*. www.clickz.com/news/article.php/3514771. June 22, 2005.

[40] Wegert, Tessa. "Shoshkele: What Is It, and What Can It Do for You?" *ClickZ Network*. www.clickz.com/experts/media/media_buy/article.php/1479401. October 10, 2002.

[41] searchCRM.com Definitions. "Floating Ad." searchcrm.techtarget.com/sDefinition/0,,sid11_gci1008935,00.html. 2005.

[42] Brain, Marshall. "How Web Advertising Works."*How Stuff Works*. money.howstuffworks.com/web-advertising.htm. 2005.

[43] Penenberg, Adam L. "Click Fraud: Problem and Paranoia." *Wired News*. www.wired.com/news/culture/0,1284,66845,00.html. March 10, 2005.

[44] Koprowski, Gene J. "Catching 'Click Fraud' Online." *EcommerceTimes*. www.ecommercetimes.com/story/YGmch51x5Nt9k7/Catching-Click-Fraud-Online.xhtml. June 18, 2005.

[45] DoubleClick. "Double-Click's 2004 Consumer Email Study." www.emailgarage.com/html/k2_det_presentations.asp?id=57&refid=0&parid=12. October 2004.

[46] Ibid.

[47] Pivotal Veracity LLC. "False Positives: A First-Hand View of What Happens When 100 Top-Tier Enterprises, Non-profits, and Governmental Agencies Try to Communicate Via Email with Their Opt-in Customers." www.pivotalveracity.com. May 2005.

[48] Ibid.

[49] *Commercial Alert*. "Search Engines." www.commercialalert.org/index.php/category_id/1/subcategory_id/24/article_id/113. 2005.

[50] "Federal Trade Commission's Letter to Search Engine Companies About Paid Placement Search Engine Ads." As reported by *KeytLaw*. www.keytlaw.com/FTC/Rules/seplacementltr.htm. June 27, 2002.

[51] Pruitt, Scarlet. "Yahoo Buys Overture." *IDG News Service* as reported by *PCWorld*. www.pcworld.com/news/article/0,aid,111563,00.asp. July 14, 2003.

[52] Thomason, Larisa. "Promotion Tip: Link Farms Grow Spam." *Webmaster Tips*. www.netmechanic.com/news/vol5/promo_no7.htm. 2002.

[53] Quittner, Joshua. "The Background and Influences that Made Bezos the Multi-Billion-Dollar Champion of E-Tailing." *Time Magazine*. www.time.com/time/poy/bezos6.html. December 27, 1999.

[54] Ibid.

[55] Ranie, Lee. "Data Memo: The State of Blogging." *Pew Internet & American Life Project*. www.pewinternet.org/pdfs/PIP_blogging_data.pdf. January 2005.

[56] Kirkpatrick, David and Roth, Daniel. "10 Tech Trends: Why There's No Escaping the Blog." *Fortune*. www.fortune.com/fortune/subs/print/0,15935,1011763,00.html. December 27, 2004.

[57] Baker, Stephen and Green, Heather. "Online Extra: Six Tips for Corporate Bloggers." *Business Week*. www. businessweek.com/print/magazine/content/05_18/b3931007_mz001.htm?chan=mz&. May 2, 2005.

[58] *BusinessWeek*. "Blogging: A Primer." www.businessweek.com/magazine/content/05_18/b3931003. May 2, 2005.

253

[59] Bersch, Carren Louise. "I Bought It Through the GRAPEVINE – wine.com." *Success* as reported by *FindArticles*. www.findarticles.com/p/articles/mi_m3514/is_4_47/ai_64833647. September 2000.

[60] Perdue, Lewis. "Wine.com Turns Up e-Vintage Heat." *internetnews.com*. www.internetnews.com/bus-news/article.php/238341. November 12, 1999.

[61] Akin, Tim. "E-Commerce Pioneer Shares Challenges of Online Wine Sales." *Distinguished Speaker Series*, University of California, Davis. www.gsm.ucdavis.edu/innovator/winter2001/Granoff.pdf. 2000.

[62] kirshenbaum bond + partners San Francisco. "Wine.com." www.kbpwest.com/index.php/54. 2005.

[63] Junnarkar, Sandeep. "Virtual Vineyards Harvests Venture Cash." *CNET News.com*. news.com.com/2100-1017-227344.html?legacy=cnet&st.ne.180.gif.1. June 18, 1999.

[64] Totty, Michael and Grimes, Ann. "If at First You Don't Succeed..." *The Wall Street Journal*. www.domainmart.com/news/WSJ_ecommerce-stories.htm. February 11, 2002.

[65] Biskupic, Joan. "Wineries That Sell Vino Via the Internet Stand to Gain." *USA Today*. www.usatoday.com/money/industries/food/2005-05-16-wine-shipments_x.htm. May 16, 2005.

[66] Wolverton, Troy. "eVineyard Prunes Assets from Wine.com." *CNET News.com*. news.com.com/eVineyard+prunes+assets+from+Wine.com/2100-1017_3-256677.html. April 27, 2001.

[67] Goldfield, Robert. "eVineyard Converts Web Site to Wine.com." *Portland Business Journal*. www.bizjournals.com/portland/stories/2001/07/09/daily43.html. July 13, 2001.

[68] Wolverton, Troy. "Wine.com Harvests Funds, Chairman." *CNET News.com*. news.com.com/Wine.com+harvests+funds%2C+chairman/2100-1017_3-944232.html. July 16, 2002.

[69] Biskupic, Joan. "Wineries That Sell Vino Via the Internet Stand to Gain." *USA Today*. www.usatoday.com/money/industries/food/2005-05-16-wine-shipments_x.htm. May 16, 2005.

CHAPTER 7

TAKING ADVANTAGE OF AFFILIATE MARKETING

LEARNING OBJECTIVES

In this chapter, you will learn to:

- Define affiliate marketing
- Explain how affiliate programs function as a marketing tool
- Discuss how affiliate programs can serve as a revenue source
- List the elements of an affiliate agreement
- Describe affiliate tracking systems and affiliate management networks
- Identify affiliate marketing risks and challenges

SEASONAL PROFITS . . .

In the mid-1990s, two young Waukesha, Wisconsin entrepreneurs, Jalem Getz and Jon Majdoch, jumped into the seasonal retail arena by selling ties in shopping mall kiosks during the Christmas season. By the late 1990s, Getz and Majdoch had successfully expanded their seasonal retail business into Getz Majdoch Inc., an operator of a chain of seven seasonal Halloween Express stores plus two lamp and home accessories shops in the Milwaukee, Wisconsin and northern Illinois area.[1, 2]

Like many other entrepreneurs in the late 1990s, Getz and Majdoch were excited about the possibilities of doing business on the Web. They wanted to use the Internet and the Web to expand their seasonal costume business into the national and international marketplace. In August 1999, just in time for the Halloween season, Getz and Majdoch launched a Web store named BuyCostumes.com. Amazingly, despite the fact that they had little time to promote the site, the new online store garnered just about as much in Halloween sales as one of their existing brick-and-mortar stores.[3]

Encouraged by the first Halloween season's sales at BuyCostumes.com, Getz and Majdoch quickly began looking for financing in order to expand and market their BuyCostumes.com e-business. Playing matchmaker, one of their bankers arranged an introduction to Jeffrey Rusinow, a top retail executive at the Fortune 500 Kohl's store chain. Rusinow, who was getting ready to take an early retirement from Kohl's, was on the lookout for a new investment and a new challenge.[4] Was this a match made in retail Heaven?

ONLINE AFFILIATE MARKETING PROGRAMS

Affiliate marketing is a revenue-sharing approach to marketing that involves paying other e-businesses to promote your e-business Web site and the products and services you offer. In Chapter 6, you learned that you can use traditional advertising, public relations, search engine optimization, online advertising, and other online marketing tools—such as *operating* an affiliate program—to help draw customers to your e-business's Web site. Now consider the flip side of this marketing coin. Suppose your e-business publishes a Web site that sells products or services to a niche market, or publishes a Web site that provides free Web page content (content that informs or entertains, such as a newsletter or blog). You may be able to increase revenues from your Web site or generate revenues through your newsletter or blog by *participating* in another e-business's affiliate marketing program.

An affiliate marketing program is sometimes called an affiliate program, but also may be referred to as a **pay-for-performance program** or an **associate program**. An affiliate program is a marketing tool for the e-business that operates it, called the **merchant** or **advertiser**, and a source of revenue for the e-business that participates in it, called an **affiliate** or **associate** or **publisher**.[5] A successful affiliate Web site typically offers viewers information, entertainment, products, or services that complement the products and services offered at the merchant site. Affiliate sites are linked to the merchant site via customized links; when a visitor at an affiliate site clicks through to the merchant's site via these links, the affiliate typically earns a fee or commission. Three basic types of affiliate programs are: [6, 7]

- *pay-per-click* or *cost-per-click affiliate programs*: The merchant pays the affiliate a set fee each time a visitor clicks through to the merchant's site, whether or not the visitor takes any action at the site.
- *pay-per-lead* or *cost-per-lead affiliate programs*: The merchant pays the affiliate a set fee for each visitor who clicks through and takes an action at the merchant's site, such as completing an online survey, registering at the site, or opting-in to receive e-mail.
- *pay-per-sale* or *cost-per-sale affiliate programs*: The merchant pays the affiliate a percentage of the sale when a visitor clicks through to the merchant's site and makes a purchase. Some pay-per-sale programs have a residual component that enables the affiliate to continue to earn commissions on subsequent sales to the same visitor.

Figure 7-1 illustrates the pay-per-sale affiliate program process.

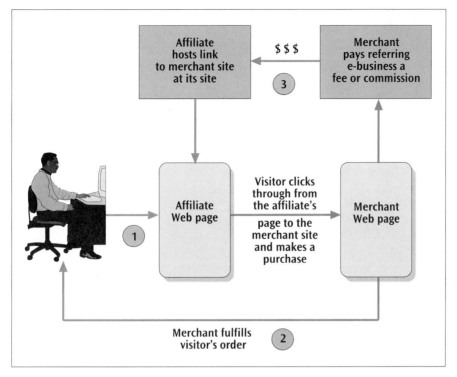

FIGURE 7-1 The pay-per-sale affiliate program process

TIP

Another type of affiliate program is a multi-tier program in which in addition to earning pay-per-click, pay-per-lead, or pay-per-sale fees for consumer click-throughs, affiliates earn referral fees for putting the merchant in contact with other affiliates.

E-business affiliate programs first appeared in the mid-1990s. While there is some debate about whether it was the first on the scene, there is no argument that the most high-profile and successful affiliate program to appear during this time was the Amazon.com Associates program.[8, 9] Launched in July, 1996—a scant 12 months after Amazon.com opened its virtual doors for business—the Amazon.com Associates program began enrolling dozens of affiliate Web sites per day. Today, the Amazon.com Associates program boasts thousands of members around the world.[10] Many other well-known merchants, such as Dell, Best Buy, and Barnes&Noble, also operate affiliate programs that serve as effective tools for marketing their products and services. Some merchants even operate multiple affiliate programs. For example, Overstock.com operates a Shopping Affiliates program for B2C online stores and an Auction Affiliates program for C2C auction sites. Figures 7-2 through 7-4 illustrate the affiliate program pages of several online merchants.

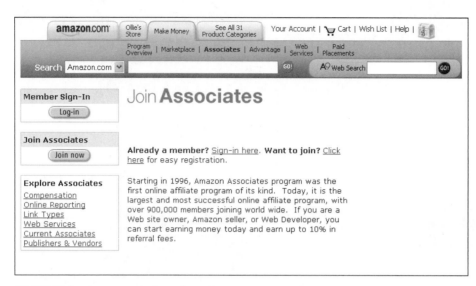

FIGURE 7-2 Amazon.com Associates Program

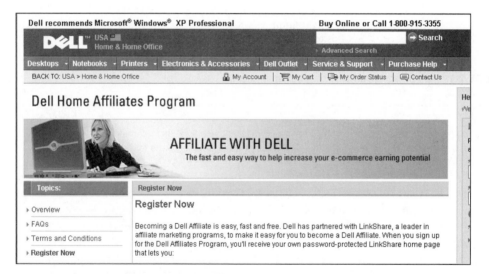

FIGURE 7-3 Dell Home Affiliates Program

FIGURE 7-4   Barnes&Noble AffiliateNetwork

E-CASE IN PROGRESS

Rackspace Managed Hosting

Rackspace Managed Hosting uses four different types of referral programs to attract potential customers: Referral Partners, Solution Partners, Development Partners, and the Rackspace Affiliate Program.

- *Referral Partners*: This program comprises anyone who refers a customer to Rackspace and receives a commission or referral fee. Referral partners are generally existing customers or customers' employees.
- *Solutions Partners*: This program includes media agencies, Web developers, consultants, application service providers, and other professionals that refer their clients to Rackspace and receive commissions or referral fees. To help promote the services it offers, Rackspace may provide training and information about the Rackspace sales philosophy to the sales staffs of its Solution Partners. Solution Partners may earn a one-time fees or recurring commissions for a referral.
- *Development Partners*: These are professional companies, such as accounting or consulting firms, that refer their clients to Rackspace but do not accept commissions or fees for themselves in order to maintain their objectivity. Development Partners may sometimes pass commissions or fees on to their clients as discounts.

continued

Taking Advantage of Affiliate Marketing

- *Rackspace Affiliate Program*: This program allows other businesses to place links to the Rackspace Managed Hosting Web site at their own Web sites and earn up to 50 percent of the first month's Web hosting fee whenever a visitor to their site clicks through to the Rackspace site and purchases Web hosting services.[11, 12, 13, 14]

A referral from a Referral Partner can be a one-time event. The referrals from Solution Partners and Development Partners, however, are recurring, and thus Rackspace maintains close, ongoing relationships with both its Solution and Development Partners. Lastly, Rackspace uses its affiliate program as an extension of its overall advertising program.[15]

Affiliate Programs as a Marketing Tool

E-business affiliate programs are electronic versions of the traditional sales referral programs used by many different types of brick-and-mortar businesses. Affiliate programs leverage other Web sites to drive potential customers to a merchant's Web site where, it is hoped, they will purchase the merchant's products and services. The two major advantages an affiliate program offers to an e-business merchant are (1) the merchant is able to measure marketing efforts by tying those efforts directly to a lead or a sale, and (2) the merchant pays only for results. In contrast, an e-business that uses traditional advertising must not only pay up front for the print, radio, or TV ads it uses, but once these ads have circulated, it may be difficult for the business to gauge how effective they were—that is, to tie the ads directly to actions taken by customers. If your e-business uses an affiliate program to market its products and services, you can track the effectiveness of the program directly and pay each affiliate only when your program gets results in the form of a customer click-through, lead, or purchase.

A survey by Shop.org and Forrester Research of 150 e-retailers found that 50 percent of the e-retailers surveyed used affiliate marketing programs. Moreover, Forrester Research estimates that the affiliate market accounts for roughly 15 to 20 percent of the estimated $72 billion online market.[16] Companies as varied as Apple Computer (iTunes), Avon (cosmetics), Oakley (sunglasses), USA Today (newspaper publishing), and eBay (online auctions) market their businesses through affiliate programs. In fact, using affiliates is Avon's fastest growing customer acquisition method.[17]

QUOTES ON SUCCESS

"Developing over 1 million affiliate sites is one of Amazon's best investments, and it is totally measurable."

Jeff Bezos, founder of Amazon.com

One approach to operating a successful affiliate marketing program is to acquire a huge and diverse affiliate membership. For example, the Amazon.com Associates program has over 1 million members, as diverse as RVPart.com (parts for recreational vehicles), Dilbert.com (cartoons and entertainment), Books for Managers (business book reviews and top ten lists), and HarperCollins.com (book publishing). What are the complementary relationships with Amazon.com that makes these four different businesses good candidates for the Amazon.com Associates program?

- RVPart.com sells plumbing, lighting, heating, towing, and other parts and accessories for recreational vehicles (RVs) and motor homes. In addition, RVPart.com uses the Amazon.com Associates program graphic and text links to promote travel books of interest to "RVers." RVPart.com customers who might not go directly to the Amazon.com site to search for a travel book—such as a popular and up-to-date campground directory—can click through from the RVPart.com site to the Amazon.com site and purchase one.

- Dilbert.com is the official site for Scott Adams' popular Dilbert comic strip, which is syndicated in 2,000 newspapers and 65 countries around the world.[18] Visitors at Dilbert.com can view cartoons, play games, send electronic greeting cards, and buy merchandise featuring Dilbert and other characters from the comic strip. Adams has also written 22 books featuring Dilbert and other characters from the comic strip. Visitors who come to the Dilbert.com Web site looking for entertainment can click through to Amazon.com to purchase one of Adams' books.

- Books for Managers compiles information on the latest business books and publications and makes it available at a central location—the Books for Managers' Web site. The information includes lists of the current top-selling business-related books provided by *BusinessWeek*, *The New York Times*, the *Wall Street Journal*, *USA Today*, and other publications. The Books for Managers e-business doesn't sell business books and publications—instead, it offers free information about them and generates its revenues by participating in several different affiliate programs. A busy business professional can visit the Books for Managers site, review its lists of top-selling business books, and then click through to Amazon.com to buy a book, or click through to a newspaper or magazine site to subscribe to a publication (both of these customer actions generate revenue for Books for Managers).

- HarperCollins uses its Web site to promote its authors and their books. However, like many book publishers, HarperCollins sells its books through bookstores and not directly to the public. A visitor at the HarperCollins site who reads about a book and wants to purchase a copy must do so through a brick-and-mortar or online book store. Membership in the Amazon.com Associates program allows HarperCollins to sell its books indirectly by offering visitors a link to the Amazon.com site.

In each of these four examples, everyone wins—the merchant (Amazon.com) cost-effectively markets its online bookstore to a diverse marketplace; the affiliates (RVPart.com, Dilbert.com, Books for Managers, and HarperCollins) earn a commission on each click-through sale; and a visitor to the affiliate site can quickly find books specifically geared toward his or her interests.

A second approach a merchant can take to operating a successful affiliate program is to focus on acquiring a smaller number of highly effective affiliates that have a high volume of Web site traffic and offer Web page content, products, and services that are directly related to the merchant's products and services. For example, an e-business that sells designer luggage might target high volume sites that sell travel services for its niche affiliate marketing program.

E-PIONEERS

Growing a Business

In the late 1970s, Jim McCann was an administrator at St. John's Home for Boys in Rockaway, New York, when he learned from a friend about a small Manhattan flower shop named Flora Plenty that was for sale. An entrepreneur at heart, McCann explored the idea of running a floral shop by working at Flora Plenty on weekends. It wasn't long before he found that he enjoyed being a florist, and decided to purchase the shop. Unable to give up his "day job" right away, however, McCann hired someone to run the shop for him.[19]

By 1986, McCann had grown his floral business from one shop to a chain of 14 shops around the Manhattan area—and he realized he was ready to give up his "day job" to run his floral business full time. It was about this time that McCann, always on the look out for new opportunities, heard about a failing floral business in Texas that had been attempting to use the new 1-800 toll-free telephone number system to sell its flowers and gifts. In 1986, many people were still unfamiliar with using the 1-800 toll-free numbers; but McCann understood how the new telephone technology could be a boost to the floral business and took a chance. He bought the failing company and its phone number, 1-800-FLOWERS. Then McCann renamed his floral chain 1-800-FLOWERS and began the long process of educating consumers, via traditional advertising media, about the convenience of ordering flowers anytime during the day or night using the 1-800-FLOWERS toll-free number. McCann and his management team also worked on building the 1-800-FLOWERS brand and increasing customer loyalty by focusing on customer satisfaction during each sale and by offering special benefits to repeat customers.[20]

As you might expect, McCann became intrigued, a few years later, with the possibilities of using another new technology—the Internet—to create an additional sales channel for his company's flowers and gifts. Convinced that the Internet was going to change retailing forever, McCann created an online presence for 1-800-FLOWERS on CompuServe's Electronic Mall in 1992, long before many consumers knew about the Internet. By 1995, 1-800-FLOWERS was operating its own e-business Web site and using online tools to market its flowers and gifts.[21, 22] Over the next several years, McCann and his management team continued to grow the business by opening new stores (both company-owned and franchised), adding new products (gifts, gourmet food, candies), and developing a strong nationally-recognized and successful affiliate program.

continued

How successful have Jim McCann and his team been in mixing sales channels (brick-and-mortar stores, call centers, and the Web) and marketing tools (such as traditional advertising and affiliate programs) to build the 1-800-FLOWERS brand? In 1999, 1-800-FLOWERS became 1-800-FLOWERS.COM and launched a successful IPO (NASDAQ:FLWS); by 2004, the company had annual sales of $604 million, net profits of $41 million, and 2,500 employees.[23] Today, 1-800-FLOWERS.COM has more than 12,000 affiliates building their own businesses by earning up to 10 percent commission on the flowers, plants, gift baskets, food, collectibles, and other items their Web site visitors purchase by clicking through to 1-800-FLOWERS.COM (Figure 7-5).[24] As of this writing, McCann and his team are expanding their internal network of company-owned and franchised stores, called BloomNet, to create a Web-based "florist transmission service" to compete head on with the granddaddy of floral wire services, FTD.[25]

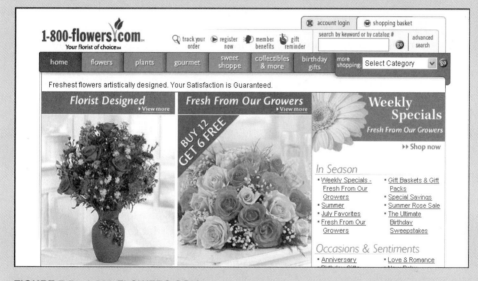

FIGURE 7-5 1-800-FLOWERS.COM

QUOTES ON SUCCESS

"There are tremendous benefits for establishing an affiliate relationship with a nationally recognized brand like 1-800-FLOWERS.COM; from our industry leading sales conversion to being recognized as the Top Online Gift Retailer and #1 Florist (by *Internet Retailer Magazine*). We view our affiliates as an external sales force; hand selected and trained to learn the best strategies to sell the freshest flowers in the business."

Renee Soulliard, Vice President of Enterprise-Wide Interactive Marketing & Web Marketing for 1-800-FLOWERS.COM

An e-business merchant can make sure that its affiliate program stands out from those of its competitors and that it more effectively generates traffic, leads, and revenues by following a few simple guidelines when dealing with affiliate partners.

- Treat affiliates like important long-term business partners and cultivate long-term relationships with them by offering products and services, customer support, and other customer benefits to the customers they refer in order to foster repeat business for them and for yourself.
- Respond to affiliates' inquiries within 24 hours. The speed of a merchant's response to an inquiry is a good indicator of the importance the merchant places on its affiliates. Failing to respond quickly to affiliates' e-mail messages may cause your affiliates to look for more responsive affiliate programs.
- Share important statistics, such as your e-business's average affiliate conversion rate (click-throughs converted to sales), sales figures, and earnings per click-through, with potential affiliates to give them an idea of the revenue they can earn. Other statistics that affiliates might find useful are the monthly earnings of top affiliates and the products or services that have the highest conversion rate (ratio of visitors to purchasers).

In a typical affiliate program, a small percentage of affiliates generate the majority of the traffic and sales for a merchant.[26] Sometimes called **super affiliates**, the members of this small group are often offered extra benefits, such as custom reporting, prioritized responses to their inquiries, and additional marketing support, to ensure that they remain active participants in the program.

Affiliate Programs as a Revenue Source

While affiliate programs serve as a marketing tool for some e-businesses, they represent an income opportunity for other types of organizations—particularly for e-businesses that sell products and services to a niche market, for Web sites that offer information or entertainment content but do not sell products or services, and for newsletter and blog publishers. Affiliate programs also offer an income opportunity for entrepreneurs who don't want to worry about the logistics of selling and shipping products, managing services sales, and providing customer support. In fact, some entrepreneurs use affiliate programs as their sole Web site revenue source. An example of this type of e-business is the Books for Managers site you learned about earlier in the chapter.

Participating in an affiliate program and then getting the most out of your participation involves:

- selecting the appropriate merchant and affiliate program for your e-business
- understanding the terms of the affiliate agreement
- adding affiliate program links to your Web pages
- building traffic at your Web site

Selecting an Affiliate Program

Your first step in selecting an affiliate program is to look for programs that are a good fit for your e-business. Remember that visitors at your site generate your affiliate income; it is critical, therefore, that the affiliate programs in which your e-business participates offer products and services of interest to your customers or to people who are likely to visit your site.[27] For example, suppose your e-business sells needlework patterns and supplies. Participating in the Amazon.com Associates program or the Barnes&Noble AffiliateNetwork would be a good choice because there is a clear overlap between your Web site's content and these merchant's products—namely, books on needlework. Participating in the PartsAmerica.com Affiliate Program, however, would probably not be a great match up for your e-business because your site visitors and customers are not likely to be thinking about purchasing auto parts when they visit your needlework site.

In addition to selecting affiliate programs that are a good fit with your e-business, you should learn as much as you can about the programs in which you are interested and the merchants who operate them. How long has the merchant been in business and what is the merchant's reputation in the marketplace? Does the program offer a fair and competitive fee or commission structure? Are fees and commissions paid on time? Finding answers to these questions can help you evaluate the effectiveness of individual affiliate programs.

You must also consider the fact that visitors who click through from your Web site to a merchant site may evaluate *your* e-business based on their experiences at the *merchant* site. Given this, it is a good idea to review the merchant's site to look for quality products and services, competitive pricing, ease of shopping and buying, customer service and product guarantees, clearly stated product return and privacy policies, and other elements that lead to satisfied customers.[28]

Once you have selected an affiliate program, the next step is to apply for membership.

Understanding the Affiliate Agreement

The online application form for membership in an affiliate program typically asks for account information such payee name, address, phone number, contact name, and tax ID number. You will also likely be asked to provide a description or profile of your Web site, its name and URL, the number of people in your organization; relate how you plan to generate referrals from your site; and indicate which of the merchant's products or services best fit the visitors to your Web site.

You will also be required to acknowledge and agree to the terms and conditions of the program's affiliate agreement. The **affiliate agreement**, created by the merchant and agreed to by the affiliate, defines all aspects of the affiliate program and typically includes:

- types of Web sites the merchant will accept into the affiliate program
- types of links allowed and guidelines regarding their use
- schedule of referral fees and commissions
- payment terms for referral fees and commissions
- terms on usage of the merchant's name, logos, and Web site content
- technical specifications that your Web site must meet, if any
- restrictions on types of content that may appear on affiliate sites
- requirements for compliance with all government laws, ordinances, rules, and regulations

- limit of the merchant's liability and other legal disclaimers
- membership termination requirements
- statement regarding the methods to be used to resolve disputes between the merchant and the affiliate

Referral fees and commissions can be structured in many different ways. For example, most affiliate agreements specify the length of time from the initial click-through to the completed purchase during which the affiliate will get credit for the purchase—this could range from 10 days to 30 days. The affiliate agreement may also specify whether or not the affiliate gets paid residual referral fees or commissions for a referred customer's ongoing purchases, such as when the customer signs up for a service with a recurring monthly service fee.

The affiliate agreement also spells out how often and when the merchant pays the affiliate, any minimum payout requirements, and the payment method. Some merchants pay their affiliates on a monthly or quarterly basis; sometimes the amount of the referral fees or commissions earned determines the payment period. The merchant usually makes payments by check, direct deposit, or credit towards purchases made at the merchant's site.

Many merchants post their affiliate agreements and the answers to frequently asked questions about them on their Web sites.

TIP

As with any legal agreement, it is a good idea to have your attorney review the terms of an affiliate agreement before you agree to participate in the program.

Adding Affiliate Program Links

After you submit an application to join an affiliate program, the merchant reviews your application for approval. If your application is accepted, the merchant then provides you with the following:

- a unique affiliate identification code or ID that is used to track your referral traffic
- instructions on how to properly link to the merchant site
- access to the merchant's databases of special affiliate banners, logos, product images, text and graphic links, search text boxes, and Web page content that can be downloaded and used on your site

Some affiliate programs, such as that of 1-800-FLOWERS.COM, offer dynamic linking, which means that when the merchant modifies graphic links or content in its databases, the changes automatically appear in the links or content on affiliates' sites.

Building Traffic at Your Site

Your success in generating revenue through affiliate program participation will obviously depend on your ability to attract visitors to your Web site and then to direct that traffic to your merchant's site. To generate interest in your Web site and draw visitors, you may use many of the marketing tools described in Chapter 6. Before you do this, however, you should carefully review any affiliate agreements you've signed to determine what restrictions, if any, are placed on the types of marketing tools you may use to attract site visitors. For example, to avoid spam, an affiliate agreement may restrict the use of e-mail advertising that contains the merchant's name, logo, or a link to the merchant's site. Press releases announcing your participation in the affiliate program may also be prohibited.[29, 30]

Whether you operate an affiliate program or participate in one, it is important for you to become familiar with the multiple methods that are used to track click-throughs from affiliate sites to merchant sites.

AFFILIATE TRACKING SYSTEMS

An **affiliate tracking system** allows a merchant to control how it credits its affiliates for click-throughs, whether per click, lead, or sale. Tracking systems can also monitor the window of time in which an affiliate gets credit for an action at the merchant site; whether sub-affiliates (multi-tier affiliates) are allowed to participate; and whether the affiliate gets credit for additional purchases made by the visitor after the first purchase is made. An affiliate tracking system also records and stores affiliate information, and provides commission or fee reports. Figure 7-6 illustrates the affiliate tracking process.

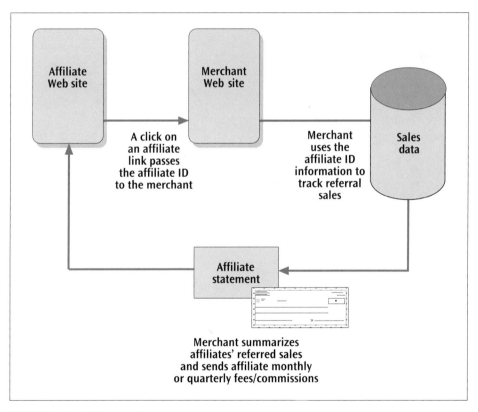

FIGURE 7-6 Affiliate tracking process

While some merchants, such as Amazon.com, manage their own programs, handling all of the details of an affiliate program can be expensive and time-consuming. Thousands of merchants, such as Best Buy and Rackspace Managed Hosting, contract the management of their affiliate programs to a special type of e-business known as an affiliate management network.

Affiliate Management Networks

An **affiliate management network** is a third-party entity that recruits affiliates, manages the registration process, tracks and properly credits all of the referral fees and commissions, and arranges for payment. In return for these services, the affiliate management network collects from the merchant a percentage of each referral transaction's fee or commission—perhaps as much as 30 percent.[31] Figure 7-7 shows the relationship between the merchants and affiliates who participate in an affiliate management network.

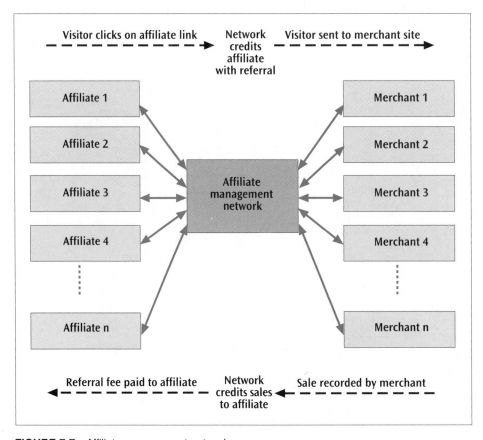

FIGURE 7-7   Affiliate management network

Commission Junction and LinkShare are two well-established affiliate management networks that manage the affiliate programs for clients such as Citibank, autobytel.com, Dollar Rent-A-Car, Rackspace Managed Hosting, Discover Card, 1-800-FLOWERS.COM, Avon, Dell, eDiets, Disney, and thousands of other merchants. Figures 7-8 and 7-9 illustrate the Commission Junction and LinkShare e-business Web sites.

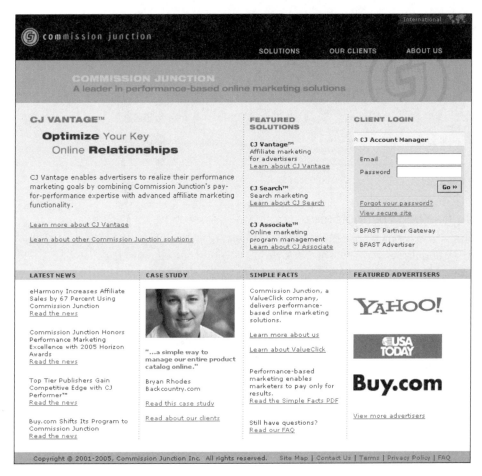

FIGURE 7-8 Commission Junction

FIGURE 7-9 LinkShare

Affiliate Tracking Technologies

For an affiliate program to operate effectively, each click-through must be tracked so that information about the source of the click-through can be passed on to the merchant and so that the affiliate can be credited for any action the customer takes. The tracking information that identifies the affiliate is included in links (URLs) the merchant provides to the affiliate for the affiliate's Web site. If the merchant is part of an affiliate management network, these URLs contain information that identifies both the affiliate and the destination merchant site. This information is either passed directly to the merchant's server or to the affiliate management network server, which instantly records the source of the click-through and directs the visitor to the correct merchant Web page.[32] For example, the Amazon.com Associate program uses information coded in the URL to flag the originating affiliate of the click-through. Figure 7-10 depicts a link from an Amazon.com associate, Books for Managers, to the Amazon.com Web page for a book titled *Business Planning for the Entrepreneur*.

The first part of the URL, "http://www.amazon.com/exec/obidos/ASIN/0324220979/" links to the Web page at the Amazon.com site for the book *Business Planning for the Entrepreneur*. The last part of the URL, "booksformanag-20", is the affiliate code for Books for Managers. When a visitor clicks through to Amazon.com using this link and makes a purchase, Amazon.com pays the Books for Managers associate a commission for referring the sale.

LinkShare, the affiliate management network, also uses information encoded in links to identify the originating affiliate as well as the merchant and destination Web page. Figure 7-11 depicts a link from the Books for Managers site to the LinkShare merchant client OfficeMax.

The first part of the URL, "http://click.linksynergy.com/fs-bin/click" is the LinkShare server address; the second part, "?id=4z3CHHjtAoo" identifies Books for Managers as the affiliate;

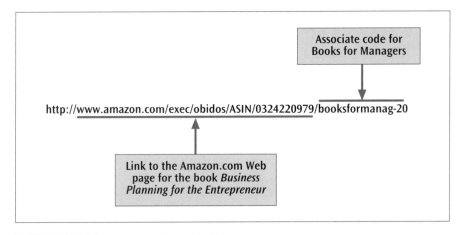

FIGURE 7-10 Amazon.com Associate link example

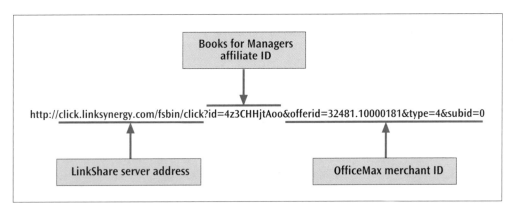

FIGURE 7-11 LinkShare affiliate management network link example

and the ending "&offerid=32481.10000181&type=4&subid=0" identifies OfficeMax as the merchant and directs the visitor to an OfficeMax page.

A merchant that participates in an affiliate management network can rely on the network's technologies to collect and track click-through information. A merchant that manages its own affiliate program must use some type of affiliate tracking technology. Some Web hosting companies provide this for their e-business clients, and some e-business software packages include tracking technologies.

Two other methods of acquiring tracking technologies are by installing third-party tracking software or by using an affiliate tracking service provided by an application service provider (ASP). The affiliate tracking technology from a third-party or an ASP does more than just track the click-throughs from the affiliate site to the merchant site. It also generates a cookie—a small text file that you learned about in Chapter 6—and stores the cookie on the visitor's computer. While the cookie does not contain any personal information about the visitor, the cookie's ID text can be used to identify a specific click-through as a single unique visitor, even if the user clicks more than once on the affiliate links.[33] If the visitor blocks cookies or deletes them, the affiliate tracking software uses other tracking methods to track the click-through, such as embedding JavaScript instructions in a URL.

AFFILIATE MARKETING RISKS AND CHALLENGES

Affiliate marketing is not without certain risks and challenges, primarily from unethical affiliates and the negative perceptions online consumers have about tracking technologies. Despite spelling out restrictions in an affiliate agreement, merchants still find it difficult to control the actions of any unethical affiliate determined to use methods such as cookie stuffing, parasiteware, and spam in order to get credit for unearned sales or to drive up Web site traffic. In addition, many online consumers are concerned about how tracking technologies might invade their privacy.

Cookie Stuffing and Parasiteware

Cookie stuffing is when an unethical affiliate places multiple cookies containing its commission codes on an unsuspecting visitor's computer during a single visit to the affiliate's site. These cookies, which can sometimes number in the hundreds, represent multiple merchant programs in which the affiliate participates. When the visitor directly accesses one of these merchants' sites and makes purchases, the cookies force the merchant's tracking software to assume the visitor clicked through from the affiliate's site (rather than directly), and the affiliate gets credit for the visitor's action.[34]

Spyware is a general term used to describe software that has been installed on a personal computer without the owner's permission. Users sometimes unknowingly download and install spyware when they download legitimate software such as games, screensavers, freeware utilities, and so forth. Once installed, spyware can then be used to gather personal information, track a user's online behavior, serve up ads, and perform other more malicious activities such as logging user keystrokes. **Parasiteware** is a kind of spyware that redirects affiliate links on a user's computer or replaces the content of the user's existing affiliate-tracking cookies. Some unethical affiliates use parasiteware to redirect traffic from the Web sites of other affiliates to their own Web site, or to hijack tracking cookies by replacing the commission codes of competing affiliates with their own commission codes—in other words, literally stealing the commissions of other affiliates.[35] The affiliate agreements for some merchants, such as Oreck (vacuum cleaners), now expressly prohibit the use of parasiteware by their affiliates.[36]

Spammers

Because the merchant has little control over what their affiliates do—the only recourse being to remove them from the affiliate program—there is a constant risk of an affiliate conducting itself in a manner that reflects poorly on the merchant or that even poses a potential legal risk to the merchant. One example is the use of unsolicited e-mail advertising or spam. Spam is a scourge both to consumers, who deplore it, and to legitimate marketers, who don't want their messages to get lost or ignored. Merchants also realize that spam generates a negative impression of their organizations in the minds of most consumers.

Unfortunately, some unethical, criminal, or simply naive affiliates continue to use spam. There are even reports that international criminals have combined e-mail addresses illegally harvested from Web sites with legitimate affiliate management network accounts to send affiliate marketing spam and generate income.[37, 38]

Merchants who operate an affiliate program should be aware that sending spam is not only unethical, it is now illegal. The CAN-SPAM Act of 2003 holds the *merchant* responsible for spam sent by its affiliates if the merchant should have known about the spam or failed to take reasonable precautions to monitor its affiliates, actions.[39, 40, 41] Therefore, most affiliate agreements now expressly prohibit the use of spam to generate Web site traffic.

QUOTES ON SUCCESS

"Our cases [Federal Trade Commission (FTC) actions against spammers] are sent with a message for companies to understand they are potentially liable if their affiliates violate the law. Certainly we are paying the most attention to entities causing the most injury, but quote-unquote legitimate marketers are not immune to FTC action if they are violating CAN-SPAM."

Steven Wernikoff, staff attorney for the FTC

For merchants, the best defense against abuses like cookie stuffing, parasiteware, and illegal affiliate spam is to monitor their affiliates' actions as carefully as possible and stay on the lookout for customer complaints and evidence of affiliate misbehavior.

Blocking Software

One of the biggest challenges facing merchants and affiliates is the use by consumers of software that blocks tracking cookies. Many consumers do not understand how tracking cookies work and equate them with spyware, viruses, worms, Trojan Horses, and other destructive software, sometimes called malware. According to a study by JupiterResearch, 38 percent of Web site viewers think cookies are an invasion of their privacy, and 44 percent believe they are protected by blocking or deleting cookies.[42]

Most cookies, like affiliate tracking cookies, are benign, and are used to store information such as Web site passwords, viewing preferences, and click-through identification. However, the software that detects, eliminates, and blocks malware frequently does the same to benign cookies. This can prevent a merchant or affiliate management network from properly tracking click-throughs.[43] According to the same JupiterResearch study, 58 percent of Web site visitors have deleted cookies from their computers and 39 percent delete cookies on a regular basis. These Web site visitors are most certainly deleting affiliate tracking cookies. Blocking or deleting affiliate tracking cookies severely compromises the accuracy of affiliate commission and referral fee reporting based on tracking cookies.[44] Merchants can begin to tackle this problem by educating online users about the usefulness of benign tracking cookies—and by using more than one type of tracking technology in their affiliate marketing programs. Indeed, educating site visitors about cookies and encouraging site visitors to read posted privacy policies might benefit both merchants and affiliates.

. . . SEASONAL PROFITS

The match between Jeffrey Rusinow, the retiring retailing executive, and Getz and Majdoch, the young entrepreneurs who started BuyCostumes.com, appears to have been a good one. Rusinow liked the idea of becoming an entrepreneur and Getz and Majdoch were happy to attract an investor and business partner with Rusinow's experience and talent to their young company. Once on board, Rusinow immediately began a program to find additional financing and to scale down operating costs. Under Rusinow's guidance, the business was able to raise a little more than $1 million in equity investment from 24 angel investors, and then split into two entities: BUYSEASONS, INC., the e-business portion of the business to be run by Getz, and GMI Inc., the brick-and-mortar stores to be run by Majdoch. As part of his cost-cutting program, Rusinow scrapped an original plan to spend $2.6 million for a traditional advertising campaign and replaced it with a public relations effort that was managed in-house, a portal advertising deal with MSN, e-mail promotions and catalogs sent to existing customers, and affiliate marketing.

This successful marketing tool mix allowed BuyCostumes.com to survive the dot.com bust, and by mid-summer 2001, BUYSEASONS, INC. and its BuyCostumes.com Web site were approaching profitability. In fact, BuyCostumes.com was heading toward becoming the largest costume e-business on the Web.[45] By 2003, BuyCostumes.com had sales of more than $10 million, and its affiliates were credited with driving 20–30 percent of these sales.[46, 47]

Today, BUYSEASONS, INC. operates two e-business sites—Deal.com and BuyCostumes.com (Figure 7-12). Both have affiliate programs managed by Commission Junction.

FIGURE 7-12 BuyCostumes.com

Chapter Summary

- Affiliate marketing is a revenue-sharing approach to promoting an e-business's products and services.

- An affiliate program is a marketing tool for the merchant who operates it and a source of revenue for the affiliate who participates in it.

- The advantages to a merchant of operating an affiliate program include the ability to measure marketing efforts and the benefit of paying only for results.

- The advantages to participating in an affiliate program include increasing site revenue for an affiliate e-retailer, or generating revenue for sites that offer information or entertainment or for sites that feature an e-mail newsletter or a blog.

- In a pay-per-click or cost-per-click affiliate program, the merchant pays the affiliate a set fee each time a visitor clicks through to the merchant's site whether or not the visitor takes any action at the site.

- In a pay-per-lead or cost-per-lead affiliate program, the merchant pays the affiliate a set fee for each visitor who clicks through and takes an action at the merchant's site, such as completing an online survey, registering at the site, or opting-in to receive e-mail.

- In a pay-per-sale or cost-per-sale affiliate program, the merchant pays the affiliate a percentage of the sale when a visitor clicks through to the merchant's site and makes a purchase. Some pay-per-sale programs have a residual component that enables the affiliate to continue to earn commissions on subsequent sales to the same visitor.

- Merchants who operate an affiliate program should treat affiliates as long-term business partners, be responsive to affiliate and potential affiliate inquiries, share important statistics with potential affiliates, and provide extra benefits to super affiliates.

- To participate in an affiliate program, an e-business should first carefully select the appropriate merchant(s) and program(s), fully understand the terms of the affiliate agreement(s) it is required to sign, update its Web site with merchant links and other content, and focus on drawing traffic to its site.

- For an affiliate e-business, selecting an affiliate program(s) that fits well with its products and/or services—or, in the case of information- or entertainment-based sites, with the content of its site, newsletter, or blog—is critical to successful participation.

- An affiliate agreement defines all aspects of the affiliate program and the relationship between the merchant and affiliate.

- Specially coded URLs and tracking cookies are used to track click-throughs from an affiliate site to a merchant site.

- An affiliate management network is a third-party e-business that recruits affiliates, manages the registration process, tracks and properly credits all of the referral fees and commissions, and arranges for payment.

- The risks and challenges of operating and participating in an affiliate program primarily originate from the activities of unethical affiliates (such as cookie stuffing, parasiteware, and spamming) and negative user perceptions about how cookies represent threats to their privacy (which lead to the use of software that blocks or deletes tracking cookies).

275

Checklist

Participating in an Affiliate Program:

❏ Does the affiliate program fit well with your products and services or the content of your Web site, newsletter, or blog?

❏ Is the merchant that operates the affiliate program a reputable merchant with a good business history?

❏ Does the merchant's Web site make shopping easy?

❏ Does the merchant offer good quality, competitively priced products and/or services that will be of interest to your site visitors?

❏ Does the merchant offer product guarantees, customer support, clearly stated return policies, a privacy statement, and other customer service benefits that will ensure that visitors to your site who click through and make a purchase at the merchant's site will be satisfied customers?

❏ Are the merchant's referral fees and commissions competitive with other similar programs?

❏ What types of affiliate member support services are available?

❏ Is it easy to download and install the merchant's links and affiliate content?

❏ Has the affiliate agreement been reviewed by an attorney, and are all terms fully understood?

❏ Have any affiliate program FAQs (frequently asked questions) been reviewed, and are the answers fully understood?

Key Terms

advertiser	cost-per-lead affiliate program
affiliate	cost-per-sale affiliate program
affiliate agreement	merchant
affiliate management network	parasiteware
affiliate marketing	pay-for-performance program
affiliate tracking system	pay-per-click affiliate program
associate	pay-per-lead affiliate program
associate program	pay-per-sale affiliate program
cookie stuffing	publisher
cost-per-click affiliate program	spyware

Review Questions

True/False Questions

1. An affiliate program is both a marketing tool and a revenue generator. True or False?

2. A pay-per-click program requires a visitor to make a purchase at the merchant's site in order for the affiliate to earn a referral fee. True or False?

3. A super affiliate is a one of a handful of high-performing affiliates that generate the majority of affiliate traffic to a merchant site. True or False?

4. An affiliate agreement may restrict the types of marketing tools an affiliate can use to drive traffic to its site. True or False?

5. Parasiteware is used by unethical affiliates to download hundreds of cookies from their Web sites to a visitor's computer. True or False?

Multiple Choice Questions

1. The affiliate program in which a click-through must result in a sale in order for an affiliate to earn a fee or commission is a:
   a. cost-per-lead program.
   b. cost-per-click program.
   c. pay-per-click program.
   d. None of the above.

2. The e-business that operates an affiliate program is called the:
   a. publisher.
   b. super affiliate.
   c. associate.
   d. merchant.

3. Which of the following items is usually defined in an affiliate agreement?
   a. dispute resolution
   b. fee and commission schedules
   c. restrictions on using a merchant's name, logo, and site content
   d. all of the above

4. An affiliate management network:
   a. provides network security services for its members.
   b. manages all aspects of an affiliate program for its members.
   c. sends spam on behalf of its members.
   d. blocks tracking cookies for its members.

5. Which federal action holds merchants responsible for e-mail advertising messages sent on the merchant's behalf?
   a. Advertising Providers Act of 2001
   b. MBPA Act of 2003
   c. Affiliate Program Monitoring Act of 2005
   d. CAN-SPAM Act of 2003

Exercises

1. Using links located on this text's student online companion or online search tools, locate the affiliate program pages, affiliate agreements, and affiliate program FAQ pages, if available, for Rackspace Managed Hosting, Amazon.com, Oreck, and 1-800-FLOWERS.COM. Answer the following questions for each program:

    a. What is the fee or commission structure?

    b. How frequently are payments made to affiliates?

    c. Are there any restrictions on the types of marketing tools an affiliate can use?

    d. What are terms under which the merchant can terminate the affiliate agreement?

2. Using links located on this text's student online companion or online search tools, locate three blogs that focus on affiliate marketing. Survey the blog postings to identify some current hot topics in the affiliate industry. Then meet with a group of your classmates to discuss: 1) the various types of affiliate industry topics that are under discussion at the blogs, 2) the effectiveness of the blogs as a marketing tool for the blog publishers and as a source of information for entrepreneurs thinking about participating in affiliate programs, and 3) whether the blogs themselves participate in affiliate marketing programs.

3. Using links located on this text's student online companion or online search tools, locate three affiliate management networks. Review the merchant information provided by each network and a list of its merchant clients. Use online search tools to search for articles, news stories, and blog postings that discuss each network. Then use your research to create a table or chart that illustrates the advantages and disadvantages of contracting with each of the three affiliate management networks.

4. Use online search tools to search for news reports, articles, and blog postings that describe unethical actions by affiliates. Select three instances of an unethical action by an affiliate and find at least two credible sources that describe that action. Report to a group of your classmates on the unethical actions in question and the ramifications of the actions.

5. Use online search tools to locate three e-businesses (not discussed in this chapter) that participate in one or more affiliate programs. Visit the merchant sites for each affiliate program of which each e-business is a member. Using the criteria outlined in this chapter, evaluate whether the program represents a good match between the merchant and the affiliate. Then discuss your evaluation with a group of your classmates.

Case Projects

1. You publish a blog for mystery novel aficionados and would like to use the blog to earn a little extra money. Using online search tools, locate five affiliate programs that you could consider joining. Then, using the information in this chapter, evaluate each program for how well it suits your blog's content and how effective it might be in generating revenue for your blog. Choose two of the programs, and present the results of your evaluation of your two choices to your classmates. Give the reasons for your choices.

2. You and your sister have just opened your own interior design business and created a Web site to promote and sell your design services. To increase revenues for your business, you want to participate in one or more appropriate affiliate programs. You decide you will have access to better quality and more appropriate programs if you become a member of an affiliate management network. Using links located on this text's student online companion, check out an affiliate membership in the Commission Junction or LinkShare affiliate management network. Decide which network best meets your needs. Discuss with your classmates the advantages and disadvantages of joining an affiliate management network and offer the reasons for your choice of networks.

Team Project

You and three classmates are working on the marketing tactics portion of a marketing plan for a new B2C e-business. As a group, create a name and description of the e-business and determine the major products or services your e-business will sell. Also, identify available URLs that you think will be effective in building your e-business's brand.

One of the tactics your e-business will use to acquire new customers and build sales is to operate an affiliate program. Work together to answer the following questions (you may make any assumptions about your e-business not explicitly stated here):

a. What types of affiliates should the program target?

b. What type of affiliate program will you create: pay-per-click, pay-per-lead, or pay-per-sale? Why?

c. What is a competitive referral fee or commission structure for your program?

d. What are the key points to cover in your affiliate agreement and affiliate program FAQ page?

e. Should you manage your affiliate program yourself, or should you contract with an affiliate management network to manage it for you? Why?

f. If you choose to manage your own program, what software or service options are available to help you track click-throughs and manage affiliate payments?

Using Microsoft PowerPoint or another presentation tool, create a 10-15 slide presentation describing your business, its name, its URL, and an outline of your proposed affiliate program using answers to the above questions as your guide. Give your presentation to a group of classmates selected by your instructor who will critique your business, name, URL, and affiliate program outline.

For Further Study

Here are some resources that might help you in further investigating the topics covered in this chapter.

Student Online Companion

Check out the *Creating a Winning E-Business, Second Edition* student online companion Web site for links to the sites discussed in this chapter and to other useful Web sites.

Articles and Books

AberdeenGroup. "Revisiting Affiliate Marketing: A New Sales Tier Emerges in the Digital Network: An Executive White Paper." www.linkshare.com/press/Aberdeen.pdf. September 2003.

Ericksen, Gregory K. *What's Luck Got to Do With It: Twelve Entrepreneurs Review the Secrets Behind Their Success.* New York: Wiley. 1997.

Laudon, Kenneth C. and Guercio, Carol. *E-commerce: Business, Technology, Society.* New York: Addison-Wesley. 2002.

Picarille, Lisa. "Stumped About Stopping Spyware." Revenue. www.revenuetoday.com/featurearticle_rev1.htm. 2005.

End Notes

[1] Hajewski, Doris. "Costume Investor." *Milwaukee Journal Sentinel.* www.buycostumes.com/ContentDisplay.aspx?siteid=1&pageid=102. July 26, 2000.

[2] Hajewski, Doris. "Successful Dot-Com About to Turn a Profit." *Milwaukee Journal Sentinel.* www.buycostumes.com/ContentDisplay.aspx?siteid=1&pageid=73. July 22, 2001.

[3] Maguire, James. "Case Study: BuySeasons/Microsoft Commerce 2000." Microsoft Corporation. www.buycostumes.com/ContentDisplay.aspx?siteid=1&pageid=62. June 28, 2002.

[4] Hajewski, Doris. "Costume Investor." *Milwaukee Journal Sentinel.* www.buycostumes.com/ContentDisplay.aspx?siteid=1&pageid=102. July 26, 2000.

[5] Housley, Sharon. "Understanding Affiliate Programs." Concept Marketing Group, Inc. www.marketingsource.com/articles/view/1802. 2005.

[6] Harris, Tom. "How Affiliate Programs Work." How Stuff Works. money.howstuffworks.com/affiliate-program.htm/. 2005.

[7] Gray, Daniel. *The Complete Guide to Associate and Affiliate Programs on the Net*; 3. New York: McGraw-Hill. 2000.

[8] Nielsen, Jacob. "Affiliates Programs: History of Affiliates Programs." Useit.com. www.useit.com/alertbox/990711_affiliates.html. July 1999.

[9] The-history-of.net. "The History of Affiliate Programs – Designed for Success." www.thehistoryof.net/history-of-affiliate-programs.html. 2005.

[10] Amazon.com Press Release. "Web Sites Big and Small Participating – From Film Critics to Chefs to Puppies." Amazon.com. phx.corporate-ir.net/phoenix.zhtml?c=97664&p=IROL-NewsText&t=Regular&id=643390&. July 18, 1996.

[11] Hostbyte.com Press Release. "Rackspace Managed Hosting Launches Worldwide Business Partner Program; Rackspace Signs Leading Integrators and Developers into New Program." Hostbyte.com. www.hostbyte.com/hosting-news/28/. September 29, 2003.

[12] Rackspace Managed Hosting. "Partner Programs." www.rackspace.com/aboutus/partner_programs.php. 2005.

[13] Ibid.

[14] Weston, Graham. E-Mail Summary: Referral Programs. August 5, 2005.

[15] Ibid.

[16] Barlas, Pete. "Sellers Stock Up on Affiliate Web Sites to Boost Marketing: Avon Calling, Calling, Calling; EBay, Apple, Amazon are Among Those Using Many Outlets to Get out the Word." *Investor's Business Daily*, A04. January 14, 2005.

[17] Ibid.

[18] Dilbert.com. "Biography of Scott Adams." dilbert.com/comics/dilbert/news_and_history/html/biography2.html. 2005.

[19] Canavor, Natalie. "1986: Flower Biz Blooms with 1-800 Number." *Long Island Business News*. www.libn.com/libnat50.cfm?id=4286. July 8-14, 2005.

[20] Ibid.

[21] Pellet, Jennifer and Schira, George. "This Bud's for You – 1-800-FLOWERS CEO and President Jim McCann." *Chief Executive* as reported by FindArticles. www.findarticles.com/p/articles/mi_m4070/is_n121/ai_19294274. March 1997.

[22] SAS. "Intimate Relationships in Bloom: 1-800-FLOWERS Sees Revenue Hikes After Implementing SAS® CRM Solution." www.sas.com/success/1800flowers.html. 2005.

[23] *Hoovers*. "1-800-FLOWERS.COM, Inc." www.hoovers.com/1-800-flowers.com/--ID__43451--/free-co-factsheet.xhtml. July 2005.

[24] 1-800-FLOWERS.COM "Welcome to the Commission+ Affiliate Program." affiliate.1800flowers.com/. 2005.

[25] Kohler, Cathy. "1-800-Flowers Develops BloomNet into a Wire Service." *SAF Wednesday E-Brief*. newsmanager.commpartners.com/safwed/issues/2005-04-06.html. April 6, 2005.

[26] Fiore, Frank. "Successful Affiliate Marketing: What Do Affiliates Want?" *Que* as reported by informIT. www.informit.com/articles/printerfriendly.asp?p=23009&rl=1. August 20, 2001.

[27] Wall, Thomas. "How to Select an Affiliate Program That Works for You." Concept Marketing Group, Inc. www.marketingsource.com/articles/view/1838. 2005.

[28] Ibid.

[29] Rackspace Managed Hosting. "Frequently Asked Questions About Our Affiliate Program." www.rackspace.com/aboutus/affiliate_faq.php. 2005.

[30] Amazon.com Associates Program. "Operating Agreement." associates.amazon.com/gp/associates/agreement/102-3933252-3817732. July 1, 2005.

[31] Harris, Tom. "How Affiliate Programs Work." How Stuff Works. money.howstuffworks.com/affiliate-program.htm. 2005.

[32] Ibid.

[33] DirectTrack. "DirectTrack FAQ Page." www.directtrack.com/faq.html. 2005.

[34] Holland, Anne. "Welcome to Affiliate Marketing Hell Week." Marketing Sherpa. www.marketingsherpa.com/sample.cfm?contentID=2750. June 24, 2004.

[35] Campanelli, Melissa. "Quit Bugging Me!" Entrepreneur.com. www.entrepreneur.com/Magazines/Copy_of_MA_SegArticle/0,4453,306756,00.html. March 2003.

[36] Oreck.com. "Affiliate Program: Parasiteware Policy." www.oreck.com/customer-service/affiliate_parasiteware_policy.cfm. 2005.

37 Kladko, Brian. "Spam by Association." NorthJersey.com. www.northjersey.com/page.php? qstr=eXJpcnk3ZjczN2Y3dnFlZUVFeXkyOSZmZ2JlbDdmN3ZxZWVFRXI5NjY1ODc1NiZ5- cmlyeTdmNzE3Zjd2cWVlRUV5eTI=. February 27, 2005.

38 Singel, Ryan. "Shady Web of Affiliate Marketing." *Wired News*. wired-vig.wired.com/news/print/ 0,1294,66556,00.html. February 10, 2005.

39 Wilson, Ralph F. "How to Comply with the CAN-SPAM Act of 2003." *Web Marketing Today*. www. wilsonweb.com/wmt9/canspam_comply.htm. January 15, 2004.

40 Singel, Ryan. "Shady Web of Affiliate Marketing." *Wired News*. wired-vig.wired.com/news/print/ 0,1294,66556,00.html. February 10, 2005.

41 U.S. Federal Trade Commission. "The CAN-SPAM Act: Requirements for Commercial Emailers." www.ftc.gov/bcp/conline/pubs/buspubs/canspam.htm. April 2004.

42 JupiterResearch. "Measuring Unique Visitors: Addressing the Dramatic Decline in the Accu- racy of Cookie-Based Measurement." www.jupitermedia.com/corporate/releases/05.03.14- newjupresearch.html. March 14, 2005.

43 Kawamoto, Wayne. "Affiliate Marketers: 'Blocking Software is Killing Us.'" E-commerce Guide. ecommerce-guide.com/solutions/affiliate/article.php/3440931. November 29, 2004.

44 JupiterResearch. "Measuring Unique Visitors: Addressing the Dramatic Decline in the Accu- racy of Cookie-Based Measurement." www.jupitermedia.com/corporate/releases/05.03.14- newjupresearch.html. March 14, 2005.

45 Hajewski, Doris. "Successful Dot.com About to Turn Profit." *Milwaukee Journal Sentinel*. www.buycostumes.com/ContentDisplay.aspx?siteid=1&pageid=73. July 22, 2001.

46 Winters, Chris. "Become Superman at a Click of the Mouse." GMToday. www.gmtoday.com/ content/LSW/2004/May/14.asp. September 18, 2004.

47 E-consultancy.com. "How to make Millions in E-retail Sales During a Short Window of Opportunity: 5 Tactics." www.e-consultancy.com/newsfeatures/360108/how-to-make-millions- in-eretail-sales-during-a-short-window-of-opportunity-5-tactics.html. January 11, 2005.

CHAPTER 8

DESIGNING YOUR WEB SITE

CLIENTS COME FIRST . . .

Fulbright & Jaworski L.L.P. is a large international law firm with more than 900 attorneys in 11 worldwide offices, including its primary office in Houston, Texas. In the mid-1990s, Fulbright & Jaworski launched its first Web site, which consisted of little more than a page of contact information for its various offices.[1] Over the next several years, the Fulbright & Jaworski Web site continued to evolve, adjusting to changes in both Web-related technologies and the expectations of Web site visitors, who were increasingly more experienced at using the Web. For example, the Web site began to incorporate graphics on its home page, and grew to include additional pages containing information about the areas of legal practice in which Fulbright & Jaworski specialized, attorney biographies, and current legal news and events.[2] By 2003, more and more Internet-savvy individuals, corporations, non-profit organizations, and government agencies were turning to the Web to connect with law firms that could provide the legal services they needed.[3] Was the Fulbright & Jaworski Web site up to the challenge of competing in this new environment in which a firm's Web site had become a key means of attracting clients?

WEB SITE PLANNING PROCESS

Whether your e-business Web site and its pages are created from scratch by someone in your business, or by a Web design and development service provider that you've hired as a contractor, you must make important decisions about how your site is organized, what pages to include, and how the pages should look and function. Making good decisions about your Web site's organization and page design begins with creating a plan.

In Chapters 3 and 6 you learned about business planning processes that involved defining your e-business's overall purpose, setting specific business goals, planning ways to market your e-business, and budgeting for each of these processes. Designing your e-business Web site involves a similar planning process. Before you begin to design your Web site, you must first identify the business objectives for the site, and then define its target audiences and how these audiences will use the site.

Web Site Business Objectives

Your e-business is likely to have a number of business objectives for its Web site. The primary objective for your site is to create profits, which it can do by helping build your e-business's image and brand, generate sales and revenues, reduce expenses, grow your customer base, and increase sales to repeat customers.[4] Although the particulars vary from e-business to e-business, your Web site's other business objectives may involve:

- educating consumers about your products and services
- providing technical support for your products and services
- providing customer support after the sale
- collecting information about current and potential customers
- offering a virtual community where consumers can interact with each other
- directing consumers to other useful Web sites
- recruiting talented employees

While these objectives may not focus directly on creating profits, they can contribute indirectly to your profitability. Therefore, identifying your Web site's many business objectives is an important first step in the Web design process. To do this, you might ask yourself questions such as:

- Is there anything about my e-business's products or services that customers may find unfamiliar or unusual?
- Do my customers need technical support to work with my e-business products or services?
- Do my customers expect support after they purchase one of my products or services?
- What features do my competitors offer at their Web sites?
- How will it help my bottom line if I solicit or collect information from my customers?

After you determine the business objectives for your site, the next critical steps are to identify your site's target audiences, determine why these audiences might want to visit your site, and identify any technological constraints they might experience when viewing your site.

Web Site Audiences

Before you begin designing your Web site (and its pages), you must gain a clear understanding of who will visit your site, what these visitors will want or need to accomplish at your site, and how they will do it. The primary audience for an e-business's Web site is its customers and potential customers. If you created a marketing plan as described in Chapter 6, then you can use the section in which you defined your target market to identify who these customers are and what they might expect to find at your Web site. The secondary audiences for your e-business's Web site may include any or all of the following: your vendors, strategic partners, investors, other stakeholders, and the general public.

After defining the audiences for your site, you should next consider why anyone from these audiences would want or need to visit your site. Remember that visitors to your site are not interested in helping you build your brand, reduce your expenses, or improve your e-business's bottom line—those are *your* business objectives. What will bring visitors to your site is the opportunity to satisfy *their* wants and needs—for example, their desire to gather information or make a purchase.[5] It is important, therefore, to focus your Web site planning not just on how you want the site to look and function, but also on how the site's organization and page design will enable visitors to fulfill their wants and needs. In short, your Web site must allow customers to find the information they need quickly and to make purchases easily.

QUOTES ON SUCCESS

"Nobody cares about you or your site. Really. What visitors care about is getting their problems solved."

Vincent Flanders, author, consultant, lecturer, and Web design guru

When planning your Web site's organization and its page design, you should take into account the tools your site's visitors will use to access your site and view its pages. For example, you shouldn't assume that all members of your target audiences are using the latest version of their Web browsers, high-speed broadband Internet connections, and the highest monitor resolution available. It would be wiser instead to plan and design for visitors who might be working under the most common technological constraints. To do this, you will need answers to the following questions.

- *Are your site's visitors likely to comprise experienced Internet and Web users, novice users, or a mix of both?* Although millions of people use the Internet and Web today, your site's visitors may still include some novice users. One way to assess this is to consider the products or services you offer. For example, if your e-business sells technical products or services, your site visitors will likely be experienced Internet and Web users who can handle a more sophisticated level of Web site complexity. If your e-business sells products or services to older adults, your site's visitors may include computer novices who would benefit more from a simpler Web site organization and page design.

- *Which browsers (and which versions of these browsers) will your site's visitors likely use: Internet Explorer, Firefox, Netscape, Opera, or some other browser?* As of this writing, the three most recent versions of the Internet Explorer Web browser continue to be the most widely used browsers by far. But other browsers, such as Firefox, are gaining in popularity.[6] You should be aware that Web pages designed specifically to be viewed in the most recent Internet Explorer browser version may not look the same or function as well when viewed in other browsers. To make certain your Web pages look and function properly in different browsers, it is wise to test your Web pages with several of the most popular browsers and most commonly used browser versions.

- *At what speed will your site's visitors connect to the Internet: dial-up, high speed dial-up, or broadband?* A 2004 research report by Nielsen//NetRatings indicates that almost 70 percent of e-retail purchases in November, 2004 were transacted over a broadband connection.[7] In 2005, there were more than 150 million broadband Internet subscribers worldwide, and this number is expected to grow to more than 400 million by 2009.[8] Although broadband usage continues to grow at a rapid pace, you should consider whether adding features that require a broadband connection for optimum viewing will cause problems for those visitors to your site who are still using slower dial-up connections.

- *At what screen resolution will your site's visitors view your Web pages: 640 x 480, 800 x 600, 1024 x 768, or higher?* While old computers with limited video memory and smaller screens are still around, current estimates indicate about 90 percent of today's monitors are set for 800 x 600 or higher resolution.[9, 10] But creating fixed-width Web pages that view best at 800 x 600 resolution may result in extra margin space at higher resolutions or may require horizontal scrolling at a lower resolution. Later in this chapter you learn about an alternative to the fixed-width design that will help you create a Web page suitable for multiple screen resolutions.

Once you have established your Web site's business objectives and identified its target audiences, their expectations, and the most common technologies they will use to view your Web pages, you are ready to organize the pages at your site.

WEB SITE ORGANIZATION

Organizing your Web site's pages in a way that visitors find logical and intuitive will help your visitors navigate the site and accomplish their goals. Organizing a Web site does not simply involve creating links between various Web pages. Instead, you should begin the process of Web site organization by anticipating the information viewers might seek out when they first visit your site, as well as the content they might be interested in exploring in-depth. Once you've begun identifying this sort of information, you can start creating the Web pages that correspond to it and determine how these Web pages best fit together. Figuring out how the information on your Web pages best fits together will, in large part, determine the linking relationships between the Web pages—and thereby the organization of your Web site. A good place to begin the process of Web site organization is to determine the pages you will include at your site.

Web Site Pages

Web sites are dynamic. What this means is that you can expect the number and type of pages at your e-business Web site to evolve over time as your e-business accommodates changes in your customers' wants and needs, shifts in your business objectives for your site, trends in the marketplace, and new technological developments. You should keep this in mind as you establish the initial pages of your site. Initial pages include a primary page (usually called a home page) and the main secondary pages your visitors need.

Home Page

Your site's **home page** is often the first page your site's visitors see. Because of this, the content of your home page must provide quick answers to these basic visitor questions.

- Who are you?
- What do you do?
- Where can I find what I want or need?
- Why should I be interested in your products and services?

Consequently, your home page should include information such as your e-business's name, business slogan, logo, or trademark; a description of your products or services; a brief explanation of how your products or services can benefit the visitor, or a link to a page that discusses these benefits; and easy-to-follow links to other pages at your site. The Rackspace Managed Hosting home page, as shown in Figure 8-1, illustrates content that answers these basic who? what? where? and why? questions.

FIGURE 8-1 Rackspace Managed Hosting home page

As you can see in Figure 8-1, the company's name and logo are highly visible in the upper-left corner of the home page. A nicely formatted, descriptive headline and trademark line is positioned near the top of the page to inform a visitor what Rackspace Managed Hosting does. A row of links between the company name and logo and the descriptive headline tell a visitor where to find additional information about what the company sells (Hosting Solutions and Managed Services), the benefits provided by the company's services (Fanatical Support™), the technologies the company uses (Superstructure), and company background and management (About Us).

Secondary Pages

The secondary pages at your site provide the visitor with more detailed information and features that support your home page. The type and number of secondary pages will depend on the types of products and services you sell and to whom, the business objectives for your site, and your visitors' wants and needs. Your secondary pages may include (but are not limited to) the following pages.

- A customer login page that allows customers to log in, if necessary, to view other pages on the Web site.
- Products or services pages that provide information about the individual products or services you sell and how visitors can purchase them.
- A "shopping cart" page that lists items the visitor wishes to purchase plus any related information the visitor needs in order to "check out" or pay for these purchases and then track their delivery.
- Shipping and return policy pages that describe shipping options and how your e-business handles product returns.
- Customer account information pages that allow your customers to review and edit their billing address, shipping address, and other information.
- Customer support pages that provide your customers with information about the support you provide for the products or services you sell.
- A contact information page that provides your mailing address, telephone and fax numbers, and e-mail addresses.
- "About Us" pages that provide additional company information, such as your e-business's history, professional biographies for the key members of your management team, and community outreach efforts.
- Forms pages that use text boxes, option buttons, drop-down lists, and other features to gather information from your viewers.
- A privacy statement page that explains what visitor information you gather and how you use it.
- An acceptable use policy page that outlines the rules and restrictions visitors must follow when using your Web site and its content.
- Frequently asked questions (FAQ) pages that itemize and provide answers to questions visitors or customers frequently ask about your company, its products and services, and Web site features.
- An employment opportunity page that lists job openings at your e-business, if any, and how potential employees can apply for them.
- "What's new?" pages that discuss new products or services, introduce new Web site features, or discuss important changes taking place at your e-business.
- Customer stories or case studies pages that illustrate how your current customers have benefited from using your products and services.
- Affiliate program pages that explain how your affiliate program works, present your affiliate agreement, and outline how e-businesses can join your program.
- Help pages that tell visitors how to navigate your site or use the special features of your site.

After determining which secondary pages you should include at your site, you need to think about how those pages should be organized to best benefit your site's visitors.

Web Site Structure

The organizational structures commonly used to build Web sites are a linear (or sequential) structure, a non-linear (or webbed) structure, a pure hierarchical structure, and a mixed hierarchical structure that combines a hierarchical structure with cross-referencing links between pages as necessary.[11, 12] The simplest organizational structure is the **linear structure**, in which visitors view a series of pages in sequential order. Web pages with content that must be viewed in order—such as pages with online instructions that must be followed in sequence, as shown in Figure 8-2—are organized in a linear structure. Visitors start at Step 1, click a link to the next page in the sequence, Step 2, and so on until they reach the last page. In contrast, a non-linear or **webbed structure**, shown in Figure 8-3, allows pages to be linked to each other without regard to how the content of these pages fits together logically. While this structure exploits the power of links to tie pages together, it may be the most confusing structure for site visitors to attempt to navigate.[13]

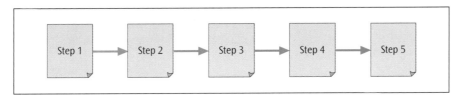

FIGURE 8-2 Linear structure

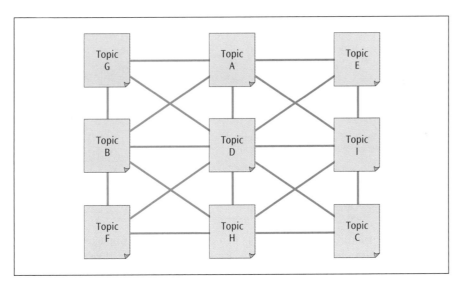

FIGURE 8-3 Webbed structure

A **hierarchical structure** presents carefully organized information in different levels, beginning with a top level (general information) followed by multiple levels of increasingly more detailed information. You may already be familiar with other instances of information organized in a hierarchical structure. One common example is a business's management team organization chart, such as the one depicted in the business plan in Chapter 3. A hierarchical Web site structure allows for more complexity than a linear structure but is easier to understand and navigate than a webbed structure. A Web site that is built in a pure hierarchical structure starts with the home page at the very top level of the hierarchy. Additional pages are organized by topic or category below the home page in multiple levels; the deeper the level, the more detailed the page content.[14] Figure 8-4 illustrates a Web site organized in a pure hierarchical structure with four levels, beginning with the home page.

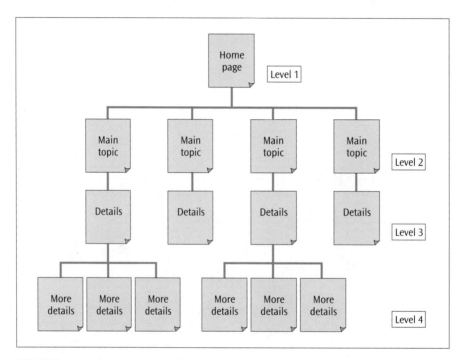

FIGURE 8-4 Pure hierarchical structure

A pure hierarchical structure in which visitors can move from page to page only by proceeding down or up the hierarchy can be restrictive for Web sites with multiple related pages. These sites use instead a **mixed hierarchical structure** that allows for cross-linked pages within a hierarchy, as shown in Figure 8-5. Some studies indicate that visitors remember more of the information at a site when they navigate a mixed hierarchical structure.[15]

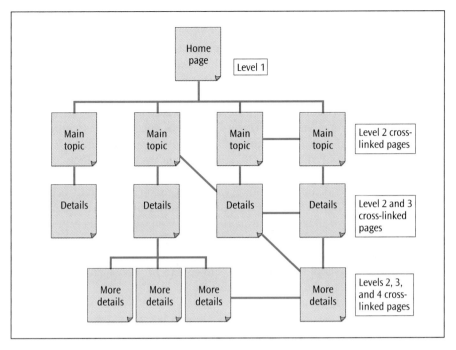

FIGURE 8-5 Mixed hierarchical structure

Another issue to consider is how shallow or deep you should make your hierarchy of pages. The number of levels will depend on the number of pages and the level of detail your Web site needs to present its content. Some simple Web sites might feature a single level of separate and unrelated pages to which visitors link directly from the home page, as shown in Figure 8-6. Although it's easy for visitors to navigate, a flat structure can be uninteresting.

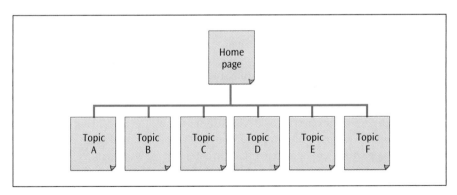

FIGURE 8-6 Flat hierarchical structure

Other, more complex Web sites use a mixed hierarchical structure with a deeper hierarchy that features multiple layers of pages. A structure with too deep a hierarchy can also be problematic in that it may require visitors to click through a frustratingly large number of pages to make a purchase or find the information they need. One way to achieve some balance—that is, to avoid creating a Web site structure that is either too flat or too deep—is to include as much important information as possible in the first few levels of the hierarchy, which, you'll notice, amounts to creating a structure similar to Figure 8-5.

One rule of thumb, sometimes called the **Three-Click Rule**, suggests that visitors should be able find useful information or make a purchase in no more than three clicks from the home page. The reasoning behind the Three-Click Rule is that visitors are likely to get frustrated and move on to another site if they can't accomplish their goals at your site within three or fewer clicks. The basic idea here—leading visitors to actionable content as quickly as possible—is indeed useful. But more important than following an arbitrary rule about adhering to a specific number of clicks is ensuring that the organization of the pages at your Web site will make logical sense to visitors and be easy for them to navigate.[16, 17]

After determining which initial pages you need at your site and considering how those pages might be organized, you should test your Web site's organizational plan. For this, you can borrow a common practice from the movie industry.

Web Site Storyboard

A good way to test your Web site's organizational plan is to use a technique called storyboarding. A **storyboard** is a blueprint for a design of a narrative, and is used in movie and television production to show the copy, dialogue, and actions corresponding to various important moments (or scenes) in the narrative. A storyboard generally consists of a board or panel that displays a series of small drawings or sketches that roughly depicts a sequence of actions.[18] The storyboarding process can be used to good effect when testing your site's organization because it allows you to visualize the linking relationships between pages.

To create a Web site storyboard, you begin by summarizing the planned content of each Web page on an individual sticky note, index card, or sheet of paper, depending on the level of detail you want to show for the page. Next, you group the notes, cards, or sheets by major categories, such as products or services, customer accounts, or "About Us" topics. Then you fasten these dummy Web pages onto a wall, whiteboard, or bulletin board, arranging them in a hierarchical order of importance beginning with the home page at the top of the hierarchy. One important aspect of this hierarchy is that you must assemble the pages according to the perspective of a Web site *visitor*, not your e-business's priorities.

The second level of the hierarchy typically represents the major categories of information at your site that a visitor would be interested in viewing. The remaining levels then contain the pages that provide details for each major category. As you arrange the pages, keep your visitors in mind and consider what they will want, need to know, and hope to accomplish at each level in the hierarchy. After your dummy Web pages are arranged, draw or create temporary lines from page to page to represent the linking relationships among them. Once you've done this, you will see a visual representation of how a visitor might navigate from page to page at your site.

Storyboarding helps you identify potential navigation problems a visitor might experience, such as important pages that require too many clicks to access quickly. This technique can also help you identify other types of organizational problems, such as orphan pages (pages that do not currently fit into the hierarchy) or missing pages (pages that you discover you need to add). Figure 8-7 illustrates a Web site storyboard in progress.

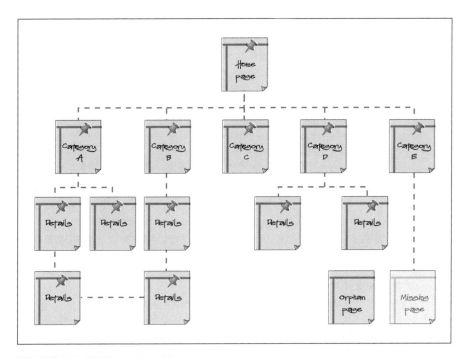

FIGURE 8-7 Web storyboard in progress

You should reposition any orphan pages, add any missing pages, and continue to rearrange your notes, cards, or sheets until you have a satisfactory organization that your visitors can navigate easily. After you complete the storyboarding process, you should then create a more formal chart of your Web site's organizational plan using flowcharting software or the drawing tools found in business productivity software, such as Microsoft Office.

With a formal Web site chart in hand, you are ready to plan the overall design of your Web site's pages. You should start by considering the general design issues that can make your Web pages more (or less) useful and attractive to your visitors.

USEFUL AND ATTRACTIVE WEB PAGES

Good Web page design creates useful and attractive Web pages that support your e-business's Web site message rather than detracting from it. For example, if visitors to your site are spending time thinking about how "cool" your pages are, they may be missing your site's primary message—buy our products or services! Over the past few years, the practices associated with Web page design have evolved to take advantage of advances in technology and keep up with the increased sophistication of the average Web page viewer.

Given this constant evolution, the upcoming sections focus on general design guidelines that will be useful for you to consider as you plan your Web pages. Two important principles of good Web page design are accessibility and usability.

Accessibility and Usability

Many people with disabilities are able to use computers with the help of special tools called assistive technologies. Assistive computer technologies include voice recognition software and screen reader software. Voice recognition software recognizes spoken commands and text entries for users who, for example, cannot use a keyboard. Screen reader software uses speech synthesizers to read aloud content on the computer screen for users who are visually impaired. The principle of Web **accessibility** involves designing Web pages so that Web resources are available to people with disabilities.

The World Wide Web Consortium (W3C) is an international group that oversees and develops protocols and guidelines for the Web. Directed by the inventor of the World Wide Web, Tim Berners-Lee, the W3C has been active in developing Web accessibility guidelines since the late 1990s. In general, the W3C's Web accessibility design guidelines seek to ensure that:[19]

- visual or auditory content is supported by corresponding alternative content; for example, for images, alternative text is assigned that enables screen readers to describe aloud what is shown in the image; similarly, visual text is added to supplement audio content
- information is not conveyed only through the use of color; for example, not relying on color alone to indicate a link
- background and foreground colors provide adequate contrast
- text and objects that move, scroll, or blink can be turned off without losing information or navigation
- navigational links are presented clearly and consistently
- page content is consistent across all pages
- text is simply worded and easy to understand

The W3C guidelines also specify a number of technical standards for using HTML and other technologies to create accessible Web pages that support this list of accessibility guidelines. You'll learn more about HTML and other Web page technologies in Chapter 9.

QUOTES ON SUCCESS

"The power of the Web is in its universality. Access by everyone regardless of disability is an essential aspect."

Tim Berners-Lee, director of W3C and inventor of the World Wide Web

Another design issue that is closely related to Web accessibility is Web site and page usability. While accessibility guidelines focus on ensuring that people with disabilities have equal access to Web resources, Web site and page **usability** guidelines are designed to help all Web site visitors accomplish their goals quickly and easily. Designing for accessibility gives you a good start toward designing for usability.

Since the mid-1990s, scientists, engineers, and psychologists have conducted a number of research studies to learn how humans interact with different Web site and Web page elements in order to identify ways to optimize site and page usability.[20] These studies looked at a variety of elements that can affect Web site/page usability, including:

- organizational structures
- frames and splash pages
- the way Web page text is written
- color, fonts, font sizes, and font styles
- different types and arrangement of links
- page scrolling actions
- images and multimedia
- the time it takes to download pages in a browser

In recent years, the results of some of the older studies have been mitigated or undercut by transformations such as the increasing level of sophistication of Web site visitors, changes in programming standards, greater access to high-speed Internet connections, and advances in the technologies used to create and view Web pages. For example, recommendations to create fixed-width pages for a specific monitor resolution have largely been offset by recommendations to create Web pages that resize with the browser window size. In other instances, however, the findings of these studies still inform Web page design today. For example, the recommendation to use a consistent page layout and design across all the pages of a Web site has become common, if not standard, practice.

An important result of these research studies was that many Web designers began adhering to their findings, and this, over time, led to the creation of a number of de facto guidelines for Web page layout. The application of these guidelines has, in turn, led to Web site visitors having a particular set of design expectations. For example, viewers are now accustomed to seeing a company's name and/or logo in the upper-left corner of a home page and finding links to other pages at the site near the top of a page and/or on the left side of the page. Earlier in this chapter, you learned that studies have shown that a mixed hierarchical organizational structure contributes to a more usable Web *site*. In the next several sections, you will learn about guidelines for designing Web content to enhance the usability of your individual Web *pages*.

Web Access for All

In the late 1990s, many scientists and researchers involved with Web development and design became increasingly concerned that as Web pages evolved and began to rely more on graphics, animation, and other special effects, it was actually becoming more difficult for people with disabilities to access all the content on a Web page. A movement led by Tim Berners-Lee and other scientists at the World Wide Web Consortium (W3C) that called for full Web accessibility gained momentum. In 1997, the W3C brought together government agencies, IT industry representatives, organizations that represented the disabled, disability researchers, and other stakeholders to create its Web Accessibility Initiative (WAI). The mission of the WAI is to develop international standards for Web accessibility and to make available materials and resources that will help Web designers, software engineers, and others involved with Web technologies to understand and implement those standards.[21] In May 1999, the WAI published its Web Content Accessibility Guidelines (WCAG 1.0) for Web content and Web technology developers. As of this writing, the WCAG 1.0 guidelines are still in effect. However, an updated set of guidelines, WCAG 2.0, is under review by WAI members.

In 1998, the U.S. Congress joined the Web accessibility movement by amending the U.S. Rehabilitation Act to add Section 508, the final standards of which became effective on June 25, 2001.[22] Section 508 mandates that all federal agencies eliminate barriers to public electronic information for people with disabilities. Section 508 covers a host of information technology standards, from standards for software procured by government agencies to those for making government Web pages accessible. The standards for government Web page accessibility under Section 508 are based on the WAI published standards.[23] Because of the procurement power of the U.S. government, many colleges, universities, and companies that do business with government agencies are taking steps to make certain their products and Web sites are Section 508 compliant.

To learn more about Web accessibility guidelines and standards, use the links on the student online companion to this text to view the complete WAI Web Content Accessibility Guidelines at the W3C Web site and find information about Section 508 of the Rehabilitation Act at the Section 508 Web site.

TIP

Designing your Web pages with accessibility and usability guidelines in mind makes good business sense. For one, having pages that are accessible by people with disabilities increases the number of visitors to your Web site, and thus the number of potential customers for your e-business. The idea is if your pages are easy to use, you are more likely to convert a wider range of site visitors into customers. Also, following accessibility and usability guidelines—such as using alternative text for images and animated objects, and writing clear and simply worded text—may help optimize your pages for search engine indexing.[24]

Consistency Across Site Pages

When planning the general design of your pages, make sure that all the pages are consistent in design and function. Although your home page is likely to be the starting point for many visitors, there is no guarantee which of your pages a visitor will access first, especially if he or she arrives at your site by clicking through from a search results list. Additionally, a visitor may click through to a specific page other than the home page when he or she uses an affiliate link. To avoid disorienting a visitor who enters your site at a page other than your home page, you should design of all your pages so that they present a common look and feel. For example, you might consider placing your e-business's name and logo in the same location on every page, positioning the same style navigational elements in the same location on every page, and using the same color scheme and font styles on every page. This sort of consistency in your design will help reassure visitors that they are still browsing pages at your site and have not inadvertently moved to pages at a different site.

The Rackspace Managed Hosting site provides an excellent example of design consistency across pages at a Web site. Figure 8-1 illustrated the Rackspace home page. Figures 8-8 through 8-10 illustrate three additional pages at the Rackspace Managed Hosting site: the Managed Hosting page, the Network page, and the Leadership page. Note the consistent use of the company name and logo, navigational elements, font styles, and other design elements on all four pages. Anyone who enters the Rackspace site at any of these pages and then browses to other pages at the site should easily understand where they are.

FIGURE 8-8 Managed Hosting page

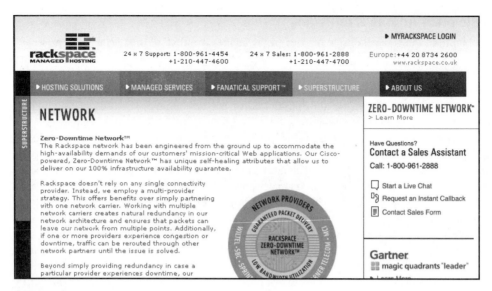

FIGURE 8-9 Network page

FIGURE 8-10 Leadership page

According to U.S. Census Bureau projections, 15 percent of the U.S. population—approximately 49.9 million people—will be age 62 and over by 2010.[25] Web sites with content directed to an audience of older adults should consider incorporating special design features that accommodate visitors whose physical abilities may have changed due to aging. The student online companion to this text contains links to Web pages that provide information and tips for designing Web pages for such audiences.

Navigational Elements

An **internal link** is a connection between two pages at the same Web site. You can create internal links using either text or images, but remember to always include complementary text links if you choose to use image links so that the links are accessible to all your site's visitors. You can use a number of different styles for your site's internal links, including text links, navigation bars, menus, navigation tabs, breadcrumb trails, and site maps. Most sites use a combination of these styles. For example, you are likely to see embedded text links in paragraphs, a graphic navigation bar across the top of a Web page, and corresponding simple text navigation bar at the bottom of the same Web page. Whichever link style you use, remember to apply that style consistently across the pages at your site. This helps your visitors clearly identify links as they move from page to page. Because visitors do not necessarily enter your site at the home page, remember also to include a link to the home page on all other pages at your site. The upcoming sections explore some navigational elements and styles commonly found on Web pages.

Embedded Text Links

An **embedded text link** is a link positioned inside a text paragraph. When creating an embedded text link (or, for that matter, any text link), make certain that the text you use for the link clearly describes what page viewers will see when they click the link. For example, text links Home or Privacy Statement clearly identify the pages to which the links refer. If you embed a text link in a paragraph, the link's text should fit into the context of the surrounding text. This enables both humans and screen readers to visually scan for and identify the links. For example, you should use phrasing like "Prices on the latest ZAX phones" instead of "Click here to view our prices on ZAX phones." It is easy to identify the "ZAX" phones link when scanning the paragraph for relevant links, but the "Click here" link means nothing unless you read the remainder of the sentence.[26]

Earlier in this chapter, you learned that Web page viewers have come to expect a Web page to contain certain design elements. For example, most Web users are familiar with the change in appearance that text links undergo when they have been accessed—namely, the use of underlining and a dark color (often blue) to indicate a text link that has not yet been clicked, and the use of underlining and a purple or maroon color to indicate a text link that has already been clicked. Both of these commonly recognized formats for text links also serve to distinguish the links from any surrounding text. As already noted, these formats also make the process of visually scanning a Web page for links easier. If you decide to stray from this common practice, remember to pay particularly close attention to accessibility and usability issues. Make certain the alternative format you use allows visitors to quickly and easily distinguish your text links from their surrounding text. Finally, don't forget to avoid using color alone to indicate a link.

Clickable Table of Contents and Top-of-Page Links

You can help viewers navigate between topics on a long Web page by adding a **clickable table of contents** section with text links to individual sub-topics throughout the page. A clickable table of contents allows visitors to click a subtopic link in the table of contents to quickly "jump" to that section of the page. You should also add a **top-of-page link**—a link to a position at the top of the Web page—at the end of each subtopic to allow viewers to quickly return to the top of a page. The text "Top" or "Back to top" or "Top of Page" is often used for top-of-page links.

Combining a clickable table of contents and with top-of-page links enables your viewers to read just the subtopic in which they are interested without having to scroll through a long page.[27] Figure 8-11 illustrates the clickable table of contents at the top of the very long page on Tanzania at the Central Intelligence Agency's The World Factbook site. Figure 8-12 show a top-of-page link that appears further down on the same page on Tanzania.

FIGURE 8-11   The World Factbook – Tanzania (top of page)

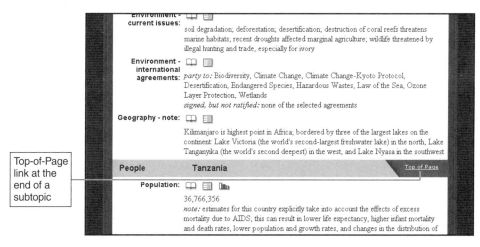

Top-of-Page link at the end of a subtopic

FIGURE 8-12 The World Factbook – Tanzania (middle of page)

TIP

A link that directs a visitor outside your Web site is called an **external link**. You can create external links using text or images. External links should open a new browser window to display the link's target page so that it is clear to visitors that they have exited your Web site.

Navigation Bars, Menus, and Tabs

A **navigation bar** is a series of graphic or text-based internal links that send visitors to the various major pages of a given Web site. Using navigation bars can help you create a common look for links across all the pages at your site. Remember to place the navigation bars at the same location on each of your pages.[28] Figure 8-13 and 8-14 show the graphic and text navigation bars on the home page of a Web site that specializes in advertising government jobs, USAJOBS.

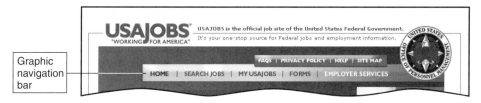

Graphic navigation bar

FIGURE 8-13 Top of the USAJOBS page

Designing Your Web Site

Text navigation bar

FIGURE 8-14 Bottom of the USAJOBS page

A **navigation menu** is a list of internal links, similar to a list of commands from a software program menu. To make more efficient use of Web page space, some complex Web sites use drop-down navigation menus near the top of the page, expandable navigation menus on the left side of the page, or **navigation tabs**, similar to file folder tabs. Figure 8-15 illustrates a drop-down navigation menu near the top of the Rackspace Managed Hosting home page. Figure 8-16 shows an expandable navigation menu and navigation tabs on the home page of the U.S. government's Medicare program.

Drop-down navigation menu

FIGURE 8-15 Rackspace Managed Hosting

Navigation tabs

Expandable navigation menu

FIGURE 8-16   Medicare

Breadcrumb Trails

Another way to help visitors better navigate your site is to use a hierarchical navigational outline, called a **breadcrumb trail**, which shows all the levels of links between the page currently being viewed and the Web site's home page (or another major page). A breadcrumb trail can provide a site's visitors with feedback on where they are at a site and how they got there.[29] In addition, visitors at a site can use the breadcrumb trail to quickly move up or down in the hierarchy and more easily understand the relationship of the page they are viewing to the page from which they started. Although breadcrumb trails are great visual cues, some site visitors may be unfamiliar with them. Therefore, a breadcrumb trail should be used in conjunction with other more familiar navigational links such as navigation bars. Figure 8-17 illustrates a breadcrumb trail at the U.S. government's official Web portal, FirstGov.gov. The breadcrumb trail begins with a Home page link (where the visitor started), continues with the Agencies page link (which was accessed via the visitor's next click), and ends with a link to the Federal Executive Branch page (which the visitor is currently viewing). The visitor can move back through their viewing history by clicking the Agencies or Home page link.

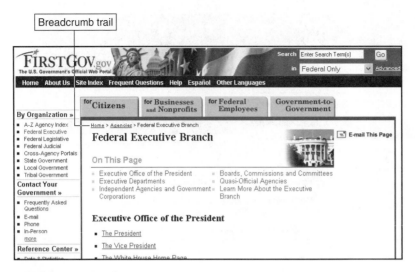

FIGURE 8-17 FirstGov.gov

Site Maps

A **site map**, sometimes called a **site index**, is a Web page that shows a summary of all the linked pages at a Web site and depicts how those pages fit into the site's organizational structure. A site map can be very useful to viewers of a complex Web site that has many pages. Figure 8-18 illustrates the site map at the Federal Communications Commission (FCC) Web site.

FIGURE 8-18 Federal Communications Commission (FCC) site map

Rollover Links

Despite design guidelines that discourage the practice of designing pages with confusing or hard to find links, some Web sites still use **rollover links**, animated graphics with links that appear or disappear as a visitor moves the mouse pointer over the graphic. When designing your Web pages, remember that the whole point of your pages' links is to quickly get your visitors to the information, products, or services they want and need. Links that are pretty and animated, but hard to locate and use, may frustrate your visitors and drive them to your competitors' Web sites. If you must use rollover links in your design, remember to provide alternative text links, navigation bars, menus, or tabs.

> **TIP**
>
> For a fun (but not necessarily funny) look at poor Web design techniques, including Web sites with confusing or hidden navigational elements such as rollover links, use a link on the student online companion to this text to check out the examples at Vincent Flanders' Web site.

Page Layout

Some studies have shown that viewers generally look first at the top of page, then they look left and right—in that order—before reading down the page.[30] With this in mind, you should consider placing important or critical information near the top center, top left, and top right areas of the page. For example, because quick telephone contact with sales and customer support staff is an important element of the services provided by Rackspace Managed Hosting, critical sales and customer support phone numbers are positioned near the top-center of each page, along with the company name and logo (positioned at the top left) and a link to the client login page (top right), followed immediately by navigation menus centered near the top of the page (as was shown in Figure 8-1).

Fixed-width Pages vs. Liquid Design

Because most people will view your pages at a resolution of 800 x 600 or higher, some Web designers suggest creating pages that view best at the 800 x 600 resolution. However, creating a fixed-width page—a page whose width does not vary—for the 800 x 600 resolution does have some drawbacks. When the page is viewed at the 1024 x 768 or higher resolution, a blank area will appear at the page margins; when the page is viewed at a lower resolution, viewers will have to scroll the page horizontally to see all the content on the page. Usability guidelines generally discourage requiring viewers to scroll pages horizontally.[31]

Because you have no control over the screen resolutions of your visitors' computers, a more important design issue to consider is how your pages will look in differently sized Web browser windows. Instead of designing your pages for a specific screen resolution and so that they fit within a fixed viewing area (for example, a maximized browser window), you should consider using liquid design techniques to create your pages. **Liquid design** techniques allow your pages to automatically resize to fit the browser window in which they are being viewed. Two tools for liquid design are tables and cascading style sheets.

A **table** is a layout of columns and rows within which text or data or images can be positioned. You may be familiar with using tables to organize text in a word processing document or data in a spreadsheet. By some estimates, more than 90 percent of all Web pages designed today use tables to organize page content.[32] You can use tables to create fixed-width pages by specifying the width of each column to be limited to an exact number of pixels (a pixel is a single point of light on a monitor's screen). But you can also use tables in a liquid design technique by specifying a table's width to be 100 percent of the browser window's viewing area instead of occupying a specific number of pixels.[33]

An alternative to using percentage-width tables to lay out a page in liquid design is to use cascading style sheets. **Cascading style sheets (CSS)** are files that contain the rules or codes that define style issues—fonts, color, and item positioning—for all the pages at a site. CSS codes can be inserted internally in a Web page or can be placed in a separate document that is referenced by a browser as it displays the page. [34, 35]

Page Length

Early Web usability studies indicated that Web page visitors sometimes became frustrated or disoriented when they had to scroll vertically to view the content on very long Web pages. In general, it is a good idea to break up a very long page into several shorter pages. However, as with other page layout considerations, the length of a page depends on what viewers are doing at the page.[36] For example, if a page has content that is likely to be printed and read offline, such as the individual country pages at The World Factbook site shown in Figures 8-11 and 8-12, then creating a long page with a clickable table of contents and intermittent top-of-page links is preferable. If a page is to be read primarily online and can be logically broken into multiple shorter pages, doing so helps your site visitors avoid unnecessary vertical scrolling.

Splash Pages and Frames

Two older Web design techniques that are less commonly used today, but still worth discussing, are splash pages and frames. As you learned earlier, the first page or home page at your Web site is very important and should make clear to a viewer who you are and what you do. A **splash page**, sometimes called an **entry page**, is a Web page that is used to create a showy entrance to a Web site, which it does by featuring big, flashy, sometimes animated graphic images (and occasionally sound effects). After the animated graphics and sounds finish playing, the Web site home page automatically loads in the Web browser. To allow visitors to move on, most splash pages contain a link that the visitor can click to bypass the animation and get right to the site's home page. For site visitors eager to find information or make a purchase, a splash page can be annoying; many design guidelines suggest avoiding splash pages unless they are essential to conveying your site's message.

A Web browser's display area can be divided into separate sections called **frames** in which different Web pages appear. For example, the home page of a Web site could consist of three frames: a top frame containing a page with the company name and logo, a left frame containing a page with navigational links, and a main frame in which pages containing content appear. Unfortunately, frames may make a page look cramped and cluttered and may cause navigation problems for viewers.[37] Additionally, as you learned in the search engine optimization section of Chapter 6, frames may cause problems for search engine robot or spider programs trying to index the site's pages. As in the case of splash pages, design guidelines generally suggest avoiding frames.

QUOTES ON SUCCESS

"The golden rule of doing business on the Web is 'Don't do anything [when designing your Web pages] that gets in the way of the sale.'"

Vincent Flanders, author and Web design guru

Search Function and Forms

Most Web users are very familiar with using a search tool, such as Google, to find information on the Web. It may be helpful, therefore, to add a **search function** to your site so that visitors can perform keyword searches for information on products and services at your site. If you add a search function, you should include the function on all pages at your site where visitors might want to use it; and you should place the search feature where it will be easy to find and use. Alternatively, you can create a separate search page. If you create a separate search page, be sure to place a link to it on all the other pages where users might need it. Also, be sure to include "how to search" instructions and examples for both the basic search feature and any advanced search options that are available.[38]

Another Web page element that you will likely include at your site is forms. You use Web page **forms** to collect information from viewers. Forms can consist of text labels and the related input boxes, option buttons, drop-down lists, or check boxes. Forms allow a viewer to enter specific information on the Web page and then send that information to the e-business's e-mail address or Web site database. Forms are used for many functions, such as providing site feedback, registering for approval to use Web site functions, and ordering products online. You can also allow customers to download your forms so they can print and complete the forms offline. Figure 8-19 illustrates both a search function and downloadable forms function on the Internal Revenue Service (IRS) Forms and Instruction page.

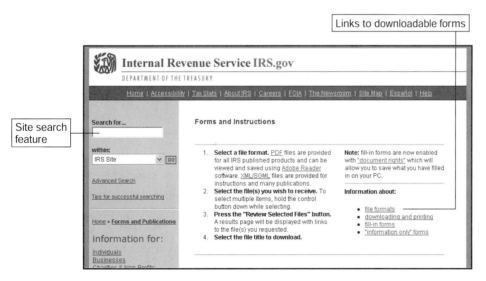

FIGURE 8-19   Internal Revenue Service

Writing for the Web

Usability studies indicate that there are important differences between writing for the Web and writing for the printed page. For example, the results of one early study by Dr. Jakob Nielsen at Sun Microsystems indicated the following:

- 70 percent of viewers scanned Web page content instead of reading it word for word.
- Reading text on a screen was 25 percent slower than reading printed text.
- Web page text should have 50 percent of the word count it would have if printed on a paper.[39]

Thus, early guidelines for writing for the Web emphasized writing short, succinctly written paragraphs and using bulleted or numbered lists instead of densely worded text to enhance the visitor's ability to scan Web page text. More recent guidelines for writing for the Web continue to recommend writing scannable text for most pages; but they also recognize that some Web page text is designed to be read word-for-word and not scanned.[40]

The "Research-Based Web Design & Usability Guidelines" report developed by the U.S. Department of Health and Human Services offers some useful guidelines for writing text to be read online.[41, 42, 43, 44]

- Use simple and direct language, and write in active voice.
- Start each paragraph with a descriptive sentence; write succinctly by limiting the number of words in a sentence and the number of sentences in the paragraph.
- Avoid industry jargon, and make sure that your site's target audiences will understand any acronyms or abbreviations you use.
- To add emphasis, use boldface formatting; avoid using italic formatting, which is hard to read online; and never use underlining for emphasis, as this signals a text link.
- Place related items in a bulleted or numbered list for easy scanning; introduce each list with a descriptive heading.
- Use a dark text color on a light colored background, such as black text on a white background.
- Use a familiar font, such as Arial, Times New Roman, Georgia, Helvetica, or Verdana.
- Use a font size no smaller than the equivalent of a 12-point font for the printed page.

Finally, it is very important that all the your Web page text be checked for spelling errors and professionally proofread and edited for grammatical and stylistic errors.

"It is up to your organization to speak the language of your customer, not the other way around. If you publish organization-speak on your website you will end up talking to yourself."

Gerry McGovern, author, lecturer, and consultant on Web content management

Color

Choosing the right colors for a Web site can be critical, because color is one of the first things a viewer sees as a Web page loads in his or her browser. Also, color can quickly set the tone (good or bad) for a viewer's experience at your Web site. Consider choosing a color scheme of no more than three or four colors and applying this scheme consistently to the page elements across all your Web pages.[45] Make certain that the scheme you choose allows for a good contrast between text color and the background color.

Color evokes both a cultural and emotional response; therefore, your Web page color scheme should reflect the values and expectations of your site's target audience(s). For example, the Web page color scheme for a financial services company should use colors that evoke feelings of trust, competency, and professionalism, such as white, blue, brown, and gray. In contrast, the color scheme for an entertainment site should evoke feelings of fun and excitement, so colors like red, orange, and black might be more suitable.

309

TIP

To have a little fun while experimenting with color on a Web page, use a link on this text's student online companion to check out the Mundi Design Studio's Web Color Theory page, where you can drag colors from a color grid to a simulated Web page.

According to some reports, more than 75 percent of all Internet users live outside the United States. This means that your e-business Web site will, whether you intend for it to or not, most certainly have an international audience.[46] You may, therefore, want consider how people from different cultures around the world will respond to the colors you choose for your Web site. For example, the colors red and blue may evoke very different cultural and emotional responses in Eastern cultures than they do in Western cultures. Working with a design professional can help you select the most appropriate color scheme for your site.

Critical First Impressions

In the late spring of 2000, the international retailer of luxury goods LVMH (Moët Hennessey - Luis Vuitton) jumped into the online marketplace by launching eLUXURY.com, an upscale fashion site designed to appeal to the "current buyer of luxury goods and services, as well as the aspirational buyer."[47] After the launch, the eLUXURY.com management team paid very close attention to customer feedback. What they found was surprising. The feedback indicated that the Web site's audience was younger than anticipated, with more customers falling into the 25–35 age group, instead of the older age range of the typical consumer of luxury goods who shopped in the company's brick-and-mortar stores. By October, a scant 3 1/2 months after the launch, the company did a major redesign of the Web site to target these younger, online consumers.[48]

The redesign included replacing the dominant black-and-white color scheme with bolder colors and using bolder typefaces to suggest a fresher, younger attitude. It also included more product illustrations and photographs of a wider product assortment to place a greater emphasis on a fashion-forward approach. In addition, the redesign effort involved adding fashion show video clips to the Web site's online magazine, and modifying the Web site's navigational structure to add links to the online magazine from any page at the Web site. The immediate result of the redesign was a doubling of the pages viewed by visitors—and an increase in sales.[49]

Now catering to a more youthful but affluent market, eLUXURY.com styles itself as the "ultimate destination for what's hip, hot, and hard to find" and sells a variety of prestigious luxury goods brands such as Luis Vuitton, Christian Dior, and Baccarat (Figure 8-20).[50]

FIGURE 8-20 eLUXURY.com

Images and Multimedia

Images, audio, and video can enhance your site's design and effectively communicate your Web site's message. Your logo image helps brand your site, and pictures of your products help your customers make good buying decisions. Audio and video clips and other animation add interest and excitement to a site. For example, the use of fashion show video clips at the eLUXURY.com site supported the site's fashion-forward message. News sites usually include audio and video clips to supplement their news stories. While the use of images and multimedia clips can make your Web site stand out, you should never include an image or multimedia clip simply because it looks or sounds great. Just like your other Web page design elements, the images and multimedia you choose must support your site's business objectives and satisfy your target audiences' expectations. As you go about making these decisions, you should be aware of the various design guidelines for image and multimedia clips.

QUOTES ON SUCCESS

"Never confuse eye-candy with content. Graphics should only ever be used to support the main purpose of your site: to get people to buy what you have to sell."

Corey Rudl, author of Insider Secrets to Marketing Your Business on the Internet *and founder of The Internet Marketing Center*

Image File Size

Image file size is an important factor to consider when designing Web pages because the larger the image file size, the longer a Web page takes to download. To improve download times, Web page image files should be compressed or reduced in size. There are two primary types of compressed images used on Web pages—Graphic Interchange Format or **GIF images** (usually pronounced "jif") and Joint Photographic Experts Group or **JPEG images** (pronounced "jay-peg")—each with its own properties and uses. Web page bitmap images, such as logos and icons that require no more than 256 colors, are compressed as GIF files because the compression does not result in a loss of image quality. Photographs, which have more detail, compress better into JPEG images. Although there is some loss of detail when a photograph is compressed, the loss is generally not noticeable.[51, 52] In addition, GIFs have some interesting features that JPEGs do not. GIFs can be animated in that they can be shown as a rapid sequence of frames, like a very short movie or flipbook animation. GIFs can also have a transparent color, which allows the Web page background to show through.

Because the use of large images can increase the amount of time it takes for a Web page to load in a browser, some Web pages feature a small version of an image, called a **thumbnail**, in place of the larger image. Thumbnails help reduce download time, and once the Web page has downloaded, a viewer can click the thumbnail image to see the larger version of the image. E-businesses often use thumbnail images to illustrate their product catalogs.

Background Images

While it is possible to add a background image behind other Web page content, most design guidelines recommend that background images be used sparingly, if at all.[53] Background images may obscure Web page text and slow page download time. Instead of a background image, a neutral background color (with an appropriately contrasting text color) is usually preferable for an e-business Web site.

Image Links

Images links are often used in addition to text links. For example, an image of a shopping cart might be used as a link to an e-retailer's shopping cart page, or the image of an envelope on a company's contact information page might signal a link to an e-mail message window. If you use image links, you should make certain that you assign alternative text to each image link so that screen readers can identify the link. You should also remember to include nearby alternative text links in case viewers have a problem recognizing the image as a link.

> **TIP**
>
> E-business Web sites that are designed to attract viewers and then convert those viewers into customers are referred to as external Web sites. In contrast, internal Web sites are published on a company's intranet, and thus are designed for a different audience. To learn more about external vs. internal Web site design issues, use a link on the student online companion to this text to view the Site Design chapter of the *Web Style Guide, 2nd Edition*.

WEB DESIGN OUTSOURCING

When you are planning your Web site's design, another important issue to consider is your budget. The decision whether to develop your Web site in-house or to outsource some or all of the work will have a major impact on how much it will cost you to design and build the site. Related to this is the decision you make about whether or not to outsource your

Web hosting needs—this will determine how much it will cost to operate your site. Estimating design and development costs can be difficult. Depending on the circumstances, costs may range from a few hundred dollars to several thousand dollars.

In-house Web development may require the addition of technical personnel, software, hardware, and office space. You may need other equipment such as a digital camera to photograph products in a digital format or a scanner to scan existing photos into a digital format. You may also need software such as Adobe Photoshop to clean up and enhance the digital photos you decide to use, or you may need a Web authoring tools such as FrontPage or Dreamweaver to create your Web site. (You learn more about Web authoring tools in Chapter 9.) Finally, your e-business's staff must expend time and effort to participate in the design, testing, and maintenance of the Web site.

Outsourcing Web design work can save a startup e-business time and money by eliminating the costs associated with recruiting and hiring in-house Web design professionals and purchasing additional equipment and software. Outsourcing Web design also gives you greater access to experienced design specialists who are familiar with the current best practices and trends and the latest technological advances. Outsourcing all or part of your Web site development will, however, require you to enter into contracts with Web designers, programmers, and testers. Before selecting an outsourcing contractor, you should thoroughly review several outsourcing candidates to get answers to the following questions:

- What services do they provide?
- What are their staff capabilities, and what portion of the design work, if any, will they themselves subcontract?
- Can they provide references and examples of their work?
- What is their track record for completing projects on schedule?

Any contract to outsource Web design must also address the important issues of: (1) who is responsible for updating and maintaining the site, (2) what happens if updates are not made on a timely basis, and (3) who owns the Web site content. You must be sure that you are not giving away copyright ownership of the resulting Web site to the designer. If in-house employees do the Web design and work within the scope of their duties, the copyright falls to the e-business. If you outsource the Web design to an independent contractor, however, the copyright ownership may remain with the creator of the design unless you have it formally transferred in writing to your e-business. Therefore, it is important to clarify copyright ownership in writing as part of any Web design agreement with an outside designer or developer. And, as with all other contractual relationships, it is a good idea to have your attorney review the terms of the written agreement.

One considerable benefit of outsourcing your Web design is that it may allow you to take good advantage of usability analysis. For example, many Web design firms employ usability analysts, specialists in human-to-computer interactions who work directly with clients to fine-tune the clients' Web site plans into usage scenarios and process flow diagrams. These scenarios and diagrams are then passed on to Web designers and technicians, who use them to fine tune the Web site's design.

In this chapter, you learned some guidelines for designing useful and attractive Web pages and organizing those pages at your site. In Chapter 9, you will learn about the technologies you can use to create Web pages and how you can measure your Web site's effectiveness at attracting visitors and converting visitors into customers.

. . . CLIENTS COME FIRST

By 2003, the executive committee at Fulbright & Jaworski realized that more and more of the law firm's corporate clients were turning to the Internet for information about almost everything—including legal services. To meet client expectations for a more comprehensive and useful experience at the Fulbright & Jaworski site, the committee decided that it was time to give the Web site a major overhaul and redesign. This project was so important to Fulbright & Jaworski that a senior partner led the in-house team in the redesign effort. The primary focus for the in-house team was to develop an understanding of what kind of information Fulbright's domestic and international clients wanted and needed to find at the Fulbright & Jaworski Web site.[54]

As a result of this client-oriented focus, Fulbright & Jaworski revamped the organization, color scheme, links, and other elements at its site. Additionally, several new features were added, including a feature that translates key site content into different languages (German, Chinese, Japanese, Spanish, Portuguese, and French); downloadable articles and papers that viewers might find relevant and interesting; online registration for seminars; and a vCard feature that enables clients to download an individual attorney's contact information into their electronic address books.[55] By August 2003, the redesigned Fulbright & Jaworski site was launched, and in September 2004, *Law Office Computing* magazine awarded the Fulbright & Jaworski site its 2004 Best Large Law Firm Web Site Award.[56]

QUOTES ON SUCCESS

"We approached our web site the way we approach our clients' legal matters. First, we listened. Then, we focused on the needs of both our domestic and international clients, and designed a site that addressed those needs. Client insight was essential to creating a meaningful web tool."

Linda L. Addison, technology partner and member of the executive committee of Fulbright & Jaworski, L.L.P.

Chapter Summary

- Before you create your Web site, you should identify your site's business objectives, target audiences, and how these audiences will use your site.

- Your Web site will have many business objectives that both directly and indirectly contribute to your e-business's profitability.

- The primary audience for your Web site is the customers and potential customers you identified in the target market assessment portion of your e-business marketing plan. Secondary audiences include vendors, strategic partners, investors, and the general public.

- Your Web site and page design decisions should be geared toward meeting the wants and needs of visitors to your site—in particular, fulfilling their desire to find information or make a purchase.

- You should try to anticipate any technological constraints your viewers might experience when accessing the pages at your site, viewing page content, or using page features.

- Your Web site will consist of a primary page, also known as a home page, and as many secondary pages as necessary to meet your visitors' wants and needs. Home page content should readily provide a visitor with the answers to four basic questions about your e-business: who you are, what you do, where the visitor can find what he or she wants or needs, and why the visitor should be interested in your products and services.

- The organizational structure of your site determines the linking relationships between its Web pages. Studies have shown that a mixed hierarchical structure with necessary cross-linked pages improves a site's usability.

- A storyboard is a helpful planning tool for visualizing your site's organizational structure and the linking relationships between the pages.

- Good Web page design creates useful and attractive Web pages that support your e-business's Web site message rather than detracting from it.

- Designing your site and pages with Web accessibility guidelines in mind makes your Web site and pages available to people with disabilities.

- Designing your site and pages with Web usability guidelines in mind can help ensure that visitors can quickly and easily find information or make a purchase at your site.

- It is very important that the design of your Web pages, including elements such as page layout, navigational elements, and color scheme, be consistent across all the pages at your site.

- Embedded text links should use descriptive text and be formatted in such a way that the links stand out from the surrounding text.

- In general, it is a good idea to break very long pages into multiple shorter pages. Some pages, however, such as those designed to be printed and read offline, can remain long. To help viewers navigate very long pages, provide a clickable table of contents and intermittently placed top-of-page links.

- Navigation bars, menus, and tabs can save page space and help you create a common look for the links across all your Web site's pages. A breadcrumb trail shows visitors where they are at your site in relation to where they started.

- A site map is a summary of all the categories of pages and page links at your site.

- Important information, such as your e-business's name, logo, primary navigation elements, and other critical information, should appear at or near the top of each page.

- Fixed-width Web pages display best in a maximized browser window at a specific screen resolution. Liquid design allows a Web page to fit within different sized browser windows.

- Avoid any Web design elements (such as splash pages, frames, and unnecessary images or multimedia) that make it more difficult for visitors to find information or make a purchase.

- Writing for the Web is different from writing for the printed page in that online text should be succinctly worded in simple, direct language; related items should be placed in bulleted or numbered lists for easy scanning; and the text should be formatted in a familiar font and easily read font size.

- Color generates different emotional and cultural responses in various audiences; therefore, select your Web site color scheme carefully and make certain it supports your Web site's message.

- When used appropriately, images and multimedia elements can enhance a Web site's design and effectively communicate information. To optimize the performance of these elements, you should become familiar with the different file sizes and try to use the smallest files possible (to avoid overly long download times). Avoid background images and be sure to include alternative text links for any image links.

- Outsourcing Web design work can save your startup e-business time and money by eliminating the costs of recruiting and hiring in-house Web design professionals and purchasing additional equipment and software. Outsourcing Web design work also gives you access to experienced design professionals who are familiar with current best practices and latest technologies as well as access to usability testing.

Checklist

Does Your E-Business Web Site Pass the Test?

❏ Does the overall design of your e-business's Web site adequately support its business objectives and your target audiences' expectations?

❏ Does your Web site design use text, color, images, and multimedia in a way that provides viewers with a clear message about your e-business?

❏ Is your Web site and Web page content accessible to visitors with disabilities?

❏ Does your Web site organization and page design follow the current generally accepted usability guidelines?

❏ Does content on your home page answer the basic visitor questions of who? what? where? and why?

- ❏ Is information that is critical and common to all pages (company name, logo, navigational elements) positioned at or near the top of each page?
- ❏ Is the page layout and color scheme consistent across all the pages at your site?
- ❏ Are embedded text links easy to identify?
- ❏ Are navigational bars, menus, tabs, and breadcrumb trails easy to find and consistently used across all the pages at your site?
- ❏ Is your page text easy to read and scan, if appropriate?
- ❏ Do very long pages have a clickable table of contents and top-of-page links?

Key Terms

accessibility	mixed hierarchical structure
breadcrumb trail	navigation bar
cascading style sheets (CSS)	navigation menu
clickable table of contents	navigation tabs
embedded text link	rollover links
entry page	search function
external link	site index
forms	site map
frames	splash page
GIF images	storyboard
hierarchical structure	table
home page	Three-Click Rule
internal link	thumbnail
JPEG images	top-of-page link
linear structure	usability
liquid design	webbed structure

Review Questions

True/False Questions

1. A Web site should be designed around both the site's business objectives and its audiences' needs and wants. True or False?

2. Web sites designed to work successfully with high-speed connections and the latest Web browser technologies pose no problems for site visitors using older technologies and slower connection speeds. True or False?

3. The storyboarding process can help you plan for an optimum Web site structure and linking relationships between the site's pages. True or False?

4. Visitors to your Web site are primarily interested in generating profits for your e-business. True or False?

5. The terms "accessibility" and "usability" mean the same thing in good Web site and page design. True or False?

Multiple Choice Questions

1. The first step in creating an e-business Web site is to:

    a. select a color scheme.

    b. determine the site's business objectives and its target audiences' expectations.

    c. create a storyboard.

    d. decide which pages to include.

2. Which of the following Web site organizational structures creates a more interesting and, at the same time, more usable site?

    a. linear structure

    b. webbed structure

    c. pure hierarchical structure

    d. mixed hierarchical structure

3. Which of the following navigational elements provides feedback to visitors about where they are at the site in relation to where they started?

    a. navigation tab

    b. embedded text link

    c. breadcrumb trail

    d. rollover link

4. Which of the following is not a good design practice when writing Web page text?

    a. using a dark text color on a light background

    b. writing succinct sentences and paragraphs

    c. using industry jargon

    d. using a familiar font such as Times New Roman or Arial

5. Outsourcing your startup e-business's Web design and development can:

    a. save time and money.

    b. provide access to experienced design professionals who are knowledgeable about current design guidelines and technologies.

    c. provide access to Web usability experts.

    d. All of the above.

Exercises

1. Following the usability guidelines outlined in this chapter, create a checklist to assess the usability of a Web site and page. Next, use online search tools to locate five e-businesses that sell the same or similar products. Use your checklist to evaluate each of the e-businesses' Web sites to see which guidelines each site follows. Browse the pages at each site. Then try to find the same or similar product at all the sites. Based on your checklist and browsing experience, rate each site's usability on a scale of 1 (least usable) to 5 (most usable).

   a. Which site was the least usable? Why?

   b. Which site was the most usable? Why?

   Discuss your finding with a group of classmates.

2. Using online search tools and the keywords "Web accessibility statement," locate five Web sites that post Web accessibility statements at their site. Review the statements. Report to your class on the type of sites that tend to post Web accessibility statements and which accessibility standards these sites tend to reference.

3. Review the home pages of five e-business Web sites of your choice. Look for elements on the home page that answer the visitor questions who?, what?, where?, and why?.

   a. Which home page does the best job of quickly answering these questions? Why?

   b. Which home page does the worst job of quickly answering these questions? Why?

   Discuss your findings with a group of classmates.

4. Select five e-business Web sites of your choice and review the color scheme for each site. Evaluate how well you think each color scheme fits its Web site's message. Write a one-page report describing the sites and evaluating the effectiveness of each site's color scheme.

5. Select three e-business Web sites of your choice and review their different navigational elements, such as navigation bars, menus, tabs, breadcrumb trails, embedded text links, and image links. Try out the different navigational elements at each site.

   a. Which navigational elements did you find most useful? Why?

   b. Which navigational elements did you find least useful? Why?

   Discuss your research with a group of classmates.

Case Projects

1. You are the assistant to the human resources manager for a Web design firm. You have been asked to prepare a job posting for a usability analyst. As this is a new position at your e-business, you will need to learn more about usability analyst positions. To do this, you decide to check out job postings listed by other Web technology firms. Using online search tools or other relevant sources, research the kind of work done by a usability analyst (the job title may also be referred to as a usability engineer, information architect, or information designer). Locate information on current job openings for a usability analyst, including educational requirements and salary ranges. Then write a description for the new position that could be posted on your company's Web site. Remember to follow usability guidelines for writing for the Web.

2. You and your partner are starting a new B2C e-business that sells custom-designed educational toys. You want your Web site to have a light-colored background with bright primary color accents, and your partner wants to use a black background with red and blue accent colors. The two of you are meeting for lunch tomorrow to make the final decision on the color scheme. Using online search tools or other relevant sources, review design issues relating to the use of color on a Web site. Then create a list of talking points that support your color scheme choice that you can take to the lunch meeting.

3. The executive assistant to the vice president of sales needs to put a section of the printed customer support manual on the company's Web site. He asks for your help. Create a checklist that the executive assistant can use to ensure that the text from the printed document is easy to read online and easy to understand.

Team Project

You and three classmates are planning your Web site for your new B2C e-business startup. Working together, create a description of the e-business, set the business objectives for your site, and define the target audiences for your site and their expectations. (You may assume any information about your e-business not explicitly stated here.) Then complete the following tasks:

- Using the storyboarding process, design the Web site's structure and page linking relationships.
- Create a formal chart for the site's organization.
- Develop a color scheme for the Web site.
- Design the layout of your home page, including the use of fonts, images, and navigational elements.
- Create a mock-up of your home page in a word processor, graphics program, Web design program, or other available software.

Use Microsoft PowerPoint or another other presentation tool to create a 5–10 slide presentation describing your e-business, your site's organizational structure, and the design of the home page. Give your presentation to a group of classmates, who will critique the structure and home page mock-up using the following criteria:

- Will the site's structure enhance or detract from the site's usability?
- Do the home page color scheme, page layout, navigational elements, text, images, and multimedia (if used) support the Web site message and provide for visitor usability?

For Further Study

Here are some resources that might help you in further investigating the topics covered in this chapter.

Student Online Companion

Check out the *Creating a Winning E-Business, Second Edition* student online companion Web site for links to the sites discussed in this chapter and to other useful Web sites.

Articles and Books

Bernard, Michael. "Criteria for Optimal Web Design (Designing for Usability)." Wichita State University—SURL. psychology.wichita.edu/optimalweb/structure.htm. 2003.

Brinck, Tom et al. *Usability for the Web: Designing Web Sites that Work*. San Francisco: Morgan Kaufmann. 2001.

Cohen, Judith. *Unusually Useful Web Book*. Indianapolis: New Riders Press. 2000.

Dunne, Danielle. "A Discussion with Jakob Nielsen and Vincent Flanders." *CIO Magazine*. www.cio.com/archive/120101/online.html. December 1, 2001.

Flanders, Vincent and Peters, Dean. *Son of Web Pages That Suck: Learn Good Design by Looking at Bad Design*. San Francisco: Sybex, Inc. 2002.

Flanders, Vincent and Willis, Michael. *Web Pages That Suck: Learn Good Design by Looking at Bad Design*. San Francisco: Sybex, Inc. 1998.

Kalbach, James. "The Myth of 800x600." Dr. Dobb's Software Tools for the Professional Programmer. www.ddj.com/documents/s=2684/nam1012432092/. March 16, 2001.

Krug, Steve. *Don't Make Me Think: A Common Sense Approach to Web Usability*. Indianapolis: New Riders Press. 2000.

Lynch, Peter J. and Horton, Sarah. *Web Style Guide, 2nd Edition*. New Haven: Yale University Press. 2002.

Maguire, James. "Site Design Tips to Improve Your Sales." Ecommerce Guide. www.ecommerce-guide.com/solutions/customer_relations/article.php/3390731. August 4, 2004.

Marcus, Aaron. "Are You Cultured? Global Web Design and the Dimensions of Culture." *New Architect*. March 2003.

McGovern, Gerry et al. *The Web Content Style Guide*. United Kingdom: Financial Times Prentice Hall. 2001. Excerpts available at www.gerrymcgovern.com/guide_design_1.htm.

Moss, Trenton. "How To Sell Accessibility." Sitepoint. www.sitepoint.com/article/sell-web-accessibility. April 6, 2004.

Nielsen, Jacob, et al. "Writing for the Web." Sun Microsystems. www.sun.com/980713/webwriting/. 2002.

Nielsen, Jacob. *Designing Web Usability: The Practice of Simplicity*. Indianapolis: New Riders Press. 1999.

Porter, Joshua. "Testing the Three-Click Rule." User Interface Engineering. www.uie.com/articles/three_click_rule/. April 16, 2003.

Poynter Institute. "Eyetrack III: Online News and Consumer Behavior in the Age of Multimedia." www.poynterextra.org/eyetrack2004/index.htm. 2004.

Rudl, Corey. "Capturing Your Site Visitors' Attention." Entrepreneur.com. www.entrepreneur.com/article/print/0,2361,308838,00.html. May 19, 2003.

Usborne, Nick. *Net Words: Creating High-Impact Online Copy*. New York: McGraw-Hill. 2001.

Wright, Steven. "11 Tips for Making Your Web Site Work Harder." Entrepreneur.com. www.entrepreneur.com/article/print/0,2361,311209,00.html. September 29, 2003.

End Notes

[1] "Fulbright & Jaworski L.L.P" as archived by the Internet Archive Wayback Machine. December 22, 1996.

[2] "Fulbright & Jaworski L.L.P" as archived by the Internet Archive Wayback Machine. December 12, 1998 and January 24, 2002.

[3] Paperstreet. "TouchPoint Metrics 2003 Report." www.paperstreet.com/facts.htm. 2003.

[4] Knemeyer, Dirk. "Jared Spool: The InfoDesign Interview." InfoDesign. www.informationdesign. org/special/spool_interview.php. April 2004.

[5] Flanders, Vincent. "Vincent Flanders Presents: The Biggest Web Design Mistakes of 2004 (Part 1 of 2)." www.webpagesthatsuck.com/biggest-web-design-mistakes-in-2004. 2005.

[6] JohnHaller.com. "Browser Statistics by Rendering Engine." johnhaller.com/jh/useful_stuff/ browser_statistics/. July 2005.

[7] Nielsen/NetRatings Press Release. "More Than Two Thirds of Online Retail Purchases are Transacted via Broadband, According to Nielsen//Netratings." www.nielsen-netratings.com/pr/ pr_050119.pdf. January 19, 2005.

[8] IMS Research Press Release. "Broadband Subscribers Surge Past 150 Million." www.imsresearch.com/members/pr.asp?X=180. March 3, 2005.

[9] *Browser News*. "Resolution Trends." www.upsdell.com/BrowserNews/stat_trends.htm#R3. July 23, 2005.

[10] theCounter.com. "Resolution Stats." www.thecounter.com/stats/2005/June/res.php. June 30, 2005.

[11] Lynch, Peter J. and Horton, Sarah. "Site Design." *Web Style Guide, 2nd Edition*. www.webstyleguide.com/. July 12, 2005.

[12] Bernard, Michael. "Criteria for Optimal Web Design (Designing for Usability)." Software Usability Research Laboratory, Department of Psychology, Wichita State University. psychology. wichita.edu/optimalweb/structure.htm. March 31, 2003.

[13] Lynch, Peter J. and Horton, Sarah. "Site Design." *Web Style Guide, 2nd Edition*. www.webstyleguide.com/. July 12, 2005.

[14] Ibid.

[15] Bernard, Michael. "Criteria for Optimal Web Design (Designing for Usability)." Software Usability Research Laboratory, Department of Psychology, Wichita State University. psychology. wichita.edu/optimalweb/structure.htm. March 31, 2003.

[16] Porter, Joshua. "Testing the Three-Click Rule." User Interface Engineering. www.uie.com/ articles/three_click_rule/. April 16, 2003.

[17] Eleniak, Marta. "Essential Navigation Checklists for Web Design." Sitepoint. www.sitepoint.com/ article/checklists-web-design. May 29, 2003.

[18] Kodak. "Glossary of Motion Picture Terms." www.kodak.com/US/en/motion/students/handbook/ glossary13.jhtml?id=0.1.4.9.6&lc=en. 2005.

[19] World Wide Web Consortium (W3C). "Web Content Accessibility Guidelines 1.0." www.w3.org/ TR/WAI-WEBCONTENT/#conventions. May 1999.

[20] Bernard, Michael. "Criteria for Optimal Web Design (Designing for Usability)." Wichita State University – SURL. psychology.wichita.edu/optimalweb/structure.htm. 2003.

[21] World Wide Web Consortium (W3C). "WAI Mission and Organization." www.w3.org/WAI/about. html. July 25, 2005.

[22] Thatcher, Jim. "Accessibility Is a Serious Issue." www.jimthatcher.com/index.htm. 2005.

[23] "Summary of Section 508 Standards." Section 508. www.section508.gov/index.cfm?FuseAction=Content&ID=11#web. August 15, 2002.

[24] Moss, Trenton. "Secret Benefits of Accessibility Part 2: Better Search Ranking." Sitepoint. www.sitepoint.com/article/accessible-search-friendly-site. October 19, 2004.

[25] U.S. Cencus Bureau, Population Division. "Intern State Population Projections, 2005. Table B1." www.cencus.gov/population/projections/SummaryTabB1.pdf. April 21, 2005.

[26] Eleniak, Marta. "Essential Navigation Checklists for Web Design." Sitepoint. www.sitepoint.com/article/checklists-web-design. May 29, 2003.

[27] U.S. Department of Health and Human Services. "Research-Based Web Design & Usability Guidelines: Navigation." Usability.gov. usability.gov/pdfs/chapter7.pdf. June 2003.

[28] Ibid.

[29] Ibid.

[30] U.S. Department of Health and Human Services. "Research-Based Web Design & Usability Guidelines: Page Layout." Usability.gov. usability.gov/pdfs/chapter6.pdf. June 2003.

[31] U.S. Department of Health and Human Services. "Research-Based Web Design & Usability Guidelines: Scrolling and Paging." Usability.gov. usability.gov/pdfs/chapter8.pdf. June 2003.

[32] Mardiros Internet Marketing. "CSS Layouts vs. Table Layouts – Alternate Browsers and Accessibility Issues." www.mardiros.net/css-layout.html. 2005.

[33] Sanchez, Mario. "Usability and Design – The Web Page Width Dilemma." AlteredImpressions. www.alteredimpressions.com/Usability101/Web_Page_Width_Dilemma.htm. 2005.

[34] SitePoint Glossary. "CSS." www.sitepoint.com/glossary.php?q=C. 2005.

[35] Marencin, Bob. "CSS Tutorial." HTML Center. www.htmlcenter.com/tutorials/tutorials.cfm/58/CSS/. 2004.

[36] U.S. Department of Health and Human Services. "Research-Based Web Design & Usability Guidelines: Scrolling and Paging." Usability.gov. usability.gov/pdfs/chapter8.pdf. June 2003.

[37] Usability.gov. "Design Considerations." http://usability.gov/guidelines/. 2005.

[38] U.S. Department of Health and Human Services. "Research-Based Web Design & Usability Guidelines: Search." Usability.gov. usability.gov/pdfs/chapter17.pdf. June 2003.

[39] Nielson, Jakob and Fox, Jonathan. "Writing for the Web." Sun Microsystems. www.sun.com/980713/webwriting/. 1994.

[40] U.S. Department of Health and Human Services. "Research-Based Web Design & Usability Guidelines: Headings, Titles, and Labels." Usability.gov. usability.gov/pdfs/chapter9.pdf. June 2003.

[41] Ibid.

[42] U.S. Department of Health and Human Services. "Research-Based Web Design & Usability Guidelines: Text Appearance." Usability.gov. usability.gov/pdfs/chapter11.pdf. June 2003.

[43] U.S. Department of Health and Human Services. "Research-Based Web Design & Usability Guidelines: Lists." Usability.gov. usability.gov/pdfs/chapter12.pdf. June 2003.

[44] U.S. Department of Health and Human Services. "Research-Based Web Design & Usability Guidelines: Writing Web Content." Usability.gov. usability.gov/pdfs/chapter15.pdf. June 2003.

[45] Clukey, Tim. "5 Steps to a Stand-Out Website." WebDesign & Review at DT&G. www.graphic-design.com/Web/5_Steps.html. 2003.

[46] Internet World Stats. "Internet Usage Statistics – The Big Picture." www.internetworldstats.com/stats.htm. July 23, 2005.

[47] Lee, Lydia. "Spend Shamelessly.com." Salon.com. dir.salon.com/business/feature/2000/07/10/eluxury/index.html. July 10, 2000.

[48] Seckler, Valerie. "First Impressions Prompt a Swift ELuxury Makeover." *Women's Wear Daily*, p. 1. October 2, 2000.

[49] Ibid.

[50] eLUXURY.com. "About ELUXURY: OVERVIEW." www.eluxury.com/help/help2.jhtml?PageName=about_eluxury&ContentID=4. 2005.

[51] Wikipedia. "GIF." en.wikipedia.org/wiki/GIF. August 6, 2005.

[52] Wikipedia. "JPEG." en.wikipedia.org/wiki/JPEG. August 8, 2005.

[53] U.S. Department of Health and Human Services. "Research-Based Web Design & Usability Guidelines: Graphics, Images, and Multimedia." Usability.gov. usability.gov/pdfs/chapter14.pdf. June 2003.

[54] Fulbright & Jaworski. "Fulbright Wins Best Law Firm Web Award for 2004." www.fulbright.com/index.cfm?article_id=4129&fuseaction=news.detail. September 29, 2004.

[55] Ibid.

[56] Addison, Linda. "Eight Steps to a Great Law Firm Website." *Texas Bar Journal*. www.fulbright.com/images/publications/techpage_nov1.pdf. November 2004.

CHAPTER 9

UNDERSTANDING WEB TECHNOLOGIES

LEARNING OBJECTIVES

In this chapter, you will learn to:

- Identify Web site and Web page development tools
- Explain the importance of Web site testing
- Define Web site benchmarking
- Describe ways to measure Web site ROI using Web analytics
- Identify Web analytics software and service vendors

MONITORING VISITORS' BEHAVIORS . . .

Originally founded in the mid-1980s, the Resort Sports Network (RSN) is a national cable television network with 29 television station affiliates broadcasting adventure sports and travel programming into more than 70 resort-destination markets in North America.[1] RSN provides information such as weather conditions, where to eat, where to shop, what to do with the kids, and so forth for more than 100 mountain, beach, and golf resort areas.[2] Rated the number one network in its market by Nielsen//Netratings, RSN launched its Web site, RSN.com, in 1996.[3] By 2000, RSN.com's management wanted to revamp the site by adding interactive features that would allow visitors to personalize and customize content at the site. How could the management at RSN.com know that interactive personalization and customization features were what its site visitors wanted and needed?

WEB SITE AND PAGE DEVELOPMENT TOOLS

As you learned in Chapter 8, whether you have in-house employees develop your e-business Web site or you outsource the site's design and development, you still are responsible for making important decisions about how your site looks and functions. For example, in order to hire an in-house programmer or outside Web developer, it may be useful to have some knowledge of the tools that person might use to create your Web pages. Some of the more common and popular Web page development tools are:

- markup languages
- text editors
- HTML/XHTML editors
- Web authoring software
- multimedia tools and scripting languages (which are used to add images, animation, sound, and video content to your site)

Markup Languages

A **markup language** is a set of rules and instructions, also called **tags**, that are embedded in an electronic document in order to define various aspects of the document's content and appearance. These markup tags are codes enclosed in angle brackets (for example <body> and </body>). They can describe a document's data and specify how document and Web page elements, such as text and images, are formatted and positioned on a page by computers.[4] They can also be used to describe the actual data (for example, purchase order or invoice data) contained in a document that businesses exchange across an extranet.

About 40 years ago, scientists began working on a generic coding system that would allow documents created for one computer system to be correctly displayed and printed when transported to a different computer system. The **Standard Generalized Markup Language (SGML)** was the result of that work. More precisely, SGML is a standard that defines a set of rules for encoding documents so that the documents are transportable among computer systems. The SGML standard is approved by the **International Organization for Standardization (ISO)**, an international body of worldwide organizations that set standards for numerous industry and government products and services.[5, 6] The SGML standard (ISO 8879:1986) can be used to create other markup languages. For example, three markup languages based on the SGML standard that are commonly used on the Web are Extensible Markup Language (XML), Hypertext Markup Language (HTML), and Extensible Hypertext Markup Language (XHTML).

Extensible Markup Language (XML)

The **Extensible Markup Language (XML)** is a streamlined subset of SGML developed in 1996 by the XML Working Group of the World Wide Web Consortium (W3C). XML is used to describe data transmitted over the Web.[7, 8] The XML tags used for a given document are defined in a specification—known generally as a document type definition specification (DTD) and specifically as an **XML schema**—that is either part of or attached to the document.[9] Developers using XML create customized tags, which can be used to describe or identify any type of data—the sender and receiver of the data must simply agree on the tags and on what the tags describe. Due to this versatility, XML is often used by businesses to exchange a range of data, including company-specific data related to various business processes such as shipping, receiving, and billing. Figure 9-1 shows an example of how customized XML tags can be used to define purchase order data exchanged between two businesses.

```
<message>
    <to>Company A</to>
    <from>Company B</from>
    <topic>Purchase Order Number</topic>
    <body>Purchase order data exchange</body>
</message>
```

FIGURE 9-1 Sample XML tags

Another subset of SGML, the Hypertext Markup Language (HTML), was developed to control how data looks when viewed in a Web browser.

Hypertext Markup Language (HTML)

The **Hypertext Markup Language (HTML)** is an easy-to-use markup language that was originally developed by Tim Berners-Lee, the inventor of the Web.[10] Berners-Lee created HTML to allow documents with headings, text, tables, bulleted or numbered lists, images, links, and other such style elements to be published on the Web and displayed properly in a Web browser.[11] When creating HTML in the early 1990s, Berners-Lee relied, in part, on SGML tag formulation, but HTML did not become an SGML-based standard until the mid-1990s.[12]

Unlike XML, which is used to describe data, HTML is used to display and format data for viewing in a Web browser. Suppose that purchase order data is to appear on a page titled "Purchase Order." HTML tags could then be used to specify that this heading text should be displayed in, for example, the center of the Web page between the left and right margins, and that it should be formatted with the Arial font and the bold font style. Another difference between XML and HTML is that whereas developers using XML can create their own custom tags to describe data, developers using HTML can use only predefined tags to lay out and format their Web pages. These predefined HTML tags can, however, be used to communicate a range of information to a Web browser, including:

- where to start and stop a Web page
- where to find special heading information, such as a page title or meta tag description and keywords
- where to place paragraph breaks

- how to format heading text
- where bulleted or numbered lists begin and end
- what text is bolded or italicized for emphasis
- which images to insert and where the images are stored
- which text or image should function as a hyperlink

Like XML tags, HTML tags are placed inside angle brackets. For example, the tags that specify the beginning and end of a Web page are <html> and </html>, respectively. (Note that, as in XML, the closing tag of a pair of HTML tags includes a forward slash.) Between the <html> and </html> tags are placed other HTML tags that display Web page elements, such as heading text, paragraph text, images, and hyperlinks. Some basic HTML tags and their respective elements are as follows.

- The <head> and </head> tags indicate the beginning and end of a Web page's heading section, which generally contains the page title and meta tags you learned about in Chapter 6.
- The <title> and </title> tags indicate the text that appears in the title bar of the Web browser.
- The <body> and </body> tags indicate the beginning and end of the body of a Web page, which generally consists of text, images, and hyperlinks.
- The <p> and </p> tags indicate the beginning and end of a paragraph.
- The <h1> and </h1> tags indicate an HTML heading style—in this case, heading 1—that should be applied to the text these tags surround.
- The <b> and </b> tags indicate that the bold font style should be applied to the text these tags surround.

Figure 9-2 illustrates these basic HTML tags (which were created in the software program Notepad) as they might be used to format and display a very simple Web page. Figure 9-3 shows this page displayed in a Web browser (in this case, the Firefox browser).

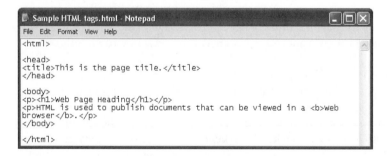

FIGURE 9-2 Basic HTML tags used to create a simple Web page

FIGURE 9-3 Simple Web page appearing in a Web browser

The World Wide Web Consortium (W3C), which you learned about in Chapter 8, sets HTML standards and adds, replaces, or deletes tags in the markup language on an ongoing basis. Since its conception, HTML has evolved through several versions. As of this writing, the current W3C standard is HTML version 4.01.[13] One example illustrating how HTML has evolved over the years is the issue of how Web page text is formatted. Whereas older versions of HTML used tag pairs and their properties to indicate how text should be formatted (for example, to specify the type of font and font size to be used or how the text is horizontally aligned on the page), the HTML 4.01 standard is deprecating (that is, phasing out) formatting tags in favor of using cascading style sheets (CSS), which you learned about in Chapter 8, to control the formatting of text across multiple Web pages.[14, 15]

In the past, Web browsers have been forgiving of sloppy HTML coding practices and would try to display a page despite some coding errors or omissions, such as the omission of the paragraph closing tag </p> at the end of a paragraph. Web browsers would also interpret HTML tags whether they were written in uppercase, lowercase, or a combination of both. For example, an opening and closing pair written as <HTML> and </html> would be accepted. However, coding standards are evolving, and the HTML 4.01 coding standards are stricter than the standards for previous HTML versions. For example, unlike earlier HTML versions, HTML 4.01 requires developers to write tags and their properties only in lowercase. Tags that in earlier versions were single tags, such as the <img> tag used to indicate the insertion of an image, must now have a closing tag, </img>. This evolution of HTML toward using CSS and toward adopting stricter coding standards is preparing Web developers to convert to the "next generation" of HTML—the Extensible Hypertext Markup Language, or XHTML.

TIP

For more information on using HTML 4.01 tags and their properties and on how CSS can be used to structure and format more complex Web pages, use links on the student online companion to this text to check out Web sites that offer tutorials on HTML 4.01 and CSS.

Extensible Hypertext Markup Language (XHTML)

The **Extensible Hypertext Markup Language (XHTML)** version 1.0 reformulated HTML 4.01 into XML. XHTML 1.0 is in fact very similar to HTML 4.01.[16] While Web developers still create pages using HTML 4.01, some developers stress the importance of migrating to XHTML from HTML in order to stay current with changing technologies and W3C standards. In addition to complying with W3C standards, the benefits of using XHTML include stricter coding rules to help eliminate coding errors and omissions, better structured documents that display in a Web browser more quickly, the flexibility of creating custom tags, and more control over how Web pages are viewed by wireless devices and screen readers.[17, 18] XHTML 1.1, which added markup features for Asian languages, became a W3C standard in May 2001; as of this writing, developers are discussing a yet newer version, XHTML 2.0.[19, 20]

QUOTES ON SUCCESS

"XML is a new standard—suite of standards—coming out of W3C. XML doesn't replace HTML directly—it replaces the SGML family of which HTML was one. XHTML is equivalent to HTML but in the XML family. XML is simpler than SGML. It makes creating new languages (like XHTML) easier."

Tim Berners-Lee, director of the W3C and inventor of the World Wide Web

TIP

Two other W3C markup language standards for the Web are the **Synchronized Multimedia Integration Language (SMIL)**, which is used to synchronize multimedia elements on a Web page, and the **Mathematical Markup Language** (**MathML**), which is used to structure mathematical expressions on a Web page.

Text Editors and Web Authoring Software

You can create a Web page from scratch by typing the HTML or XHTML tags and related text in a simple **text editor**, software that allows you to create text documents but lacks the special document creating and formatting features found in a word processor. The Notepad software installed with the Windows operating system, and shown in Figure 9-2, is an example of a text editor.

You also can use a text editor designed specifically for creating HTML/XHTML documents. HTML/XHTML editors provide features that make coding a Web page easier and faster, such as line numbering, tag menus, and a familiar working environment with menus and toolbars. Some inexpensive HTML/XHTML editors are TextPad (Helios Software), BBEdit for the Macintosh (Bare Bones Software), HomeSite (Macromedia), and HTMLPad 2005 (Blumentals Software). Figures 9-4 and 9-5 illustrate how the HTMLPad 2005 text editor is used to create a Web page using HTML 4.01 and XHTML 1.0, respectively.

FIGURE 9-4 Example of HTML 4.01 tags in a text editor

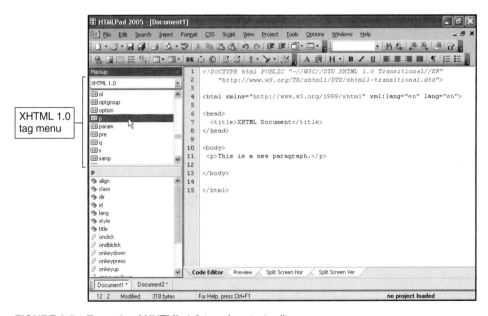

FIGURE 9-5 Example of XHTML 1.0 tags in a text editor

Web authoring software provides a "What You See Is What You Get" or **WYSIWYG** environment in which you create Web pages in much the same way you create documents in word processing software. Like HTML/XHTML editors, Web authoring software provides a familiar working environment with menu commands and toolbar buttons that enable you to build a Web page by keying and formatting text, inserting images, and creating hyperlinks. But unlike HTML/XHTML editors, Web authoring software can also *automatically* insert the appropriate markup tags and their properties (values that further define the tags) as you enter and format Web page elements. Two popular Web authoring software packages are FrontPage (Microsoft) and Dreamweaver (Macromedia).

FrontPage

Microsoft **FrontPage** Web authoring software allows you to create a Web site folder, called a FrontPage Web, and then create the site's individual Web pages, which are saved in the folder. In addition, FrontPage shares features like the Clip Organizer, common menu commands, toolbar buttons, and task panes with Microsoft Office programs, such as Microsoft Word. Also like Microsoft Office, FrontPage provides "wizards" (step-by-step instructions) and templates (model documents) for creating a new FrontPage Web and new Web pages.

FrontPage offers a number of views and sub views that you can use to create your Web pages. You use the Design sub view in Page view to see the text, images, and links you insert on a page; the Code sub view to track the markup code being inserted automatically as you work; the Preview sub view to see how your page looks in a Web browser; and the Split sub view to see both the Design sub view and Code sub view at once. The appendix to this book discusses other FrontPage views.

FrontPage also offers color-coordinated visual themes that, like Microsoft PowerPoint's Design templates, contain graphics, bullets, and font formatting that you can apply to one or more pages. In addition, FrontPage supports both table layout and CSS to help you maintain a consistent look across your Web site. Because of its familiar working environment and competitive pricing (less than $200), FrontPage may be a good choice for creating and maintaining a non-business or small business Web site. Figure 9-6 shows a Web page created in FrontPage's WYSIWYG environment and displayed in Design sub view. Figure 9-7 depicts the Split sub view, in which you can see both the Web page (in a WYSIWYG environment) and its corresponding HTML tags, which were automatically inserted by the software. Figure 9-8 shows just these HTML tags in Code sub view. Figure 9-9 illustrates the Preview sub view and thus shows the Web page as it looks in a Web browser.

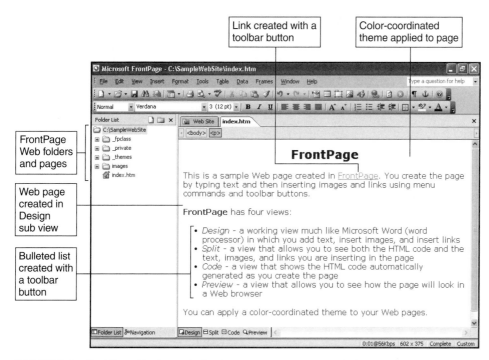

FIGURE 9-6 FrontPage Design sub view

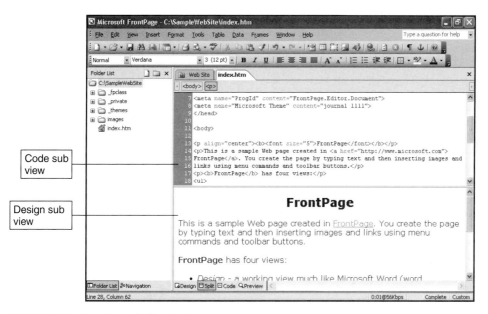

FIGURE 9-7 FrontPage Split sub view

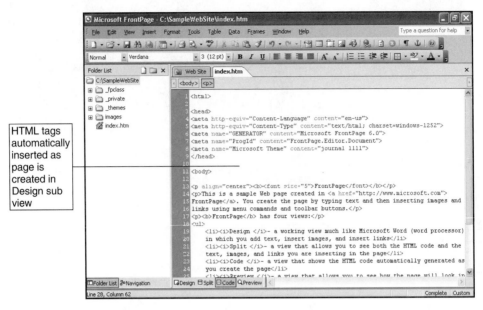

FIGURE 9-8 FrontPage Code sub view

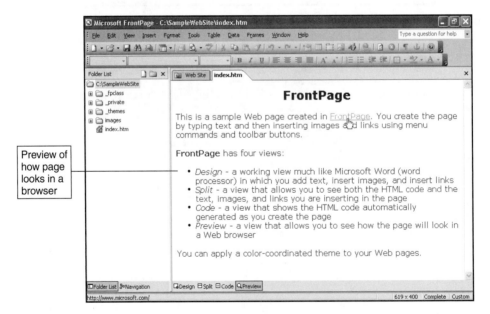

FIGURE 9-9 FrontPage Preview sub view

Dreamweaver

Macromedia **Dreamweaver** is considered by many Web developers to be the "industrial strength" version of Web authoring software. More powerful and rich in features than FrontPage, Dreamweaver is also much more complex and may, in fact, be difficult for Web authoring novices to learn to use. Like FrontPage, Dreamweaver offers different views: a WYSIWYG Design view (Figure 9-10), a markup Code view (Figure 9-11), and a split view that shows both the Code and Design views (Figure 9-12).

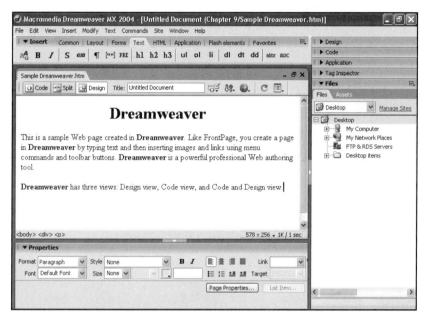

FIGURE 9-10 Dreamweaver Design view

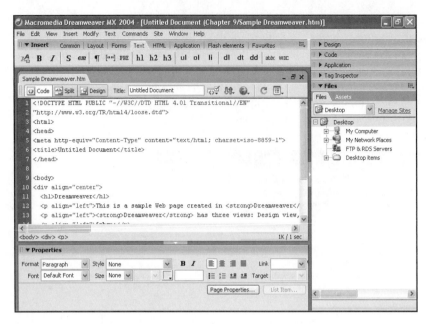

FIGURE 9-11 Dreamweaver Code view

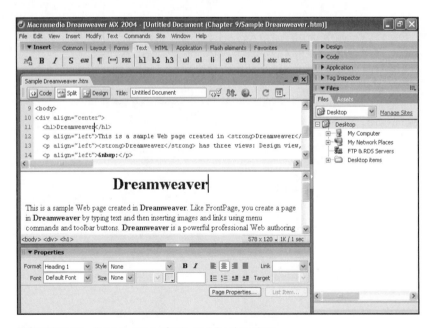

FIGURE 9-12 Dreamweaver Code and Design view

In Dreamweaver, you use a menu command to launch a Web browser and preview the page currently being developed. Like FrontPage, Dreamweaver also supports table layout and CSS. Dreamweaver is more expensive (about $400) than FrontPage, and, because of its complexity, is generally regarded as a Web authoring tool for professional Web developers.[21]

Multimedia Tools

As you learned in Chapter 8, the appropriate use of images, multimedia, and animation can add interest to your Web site. Several popular tools for creating and editing images, multimedia, and animations include Photoshop (by Adobe Systems), Flash, Fireworks, and FreeHand (all three developed by Macromedia).

Photoshop and Fireworks

Adobe **Photoshop** is image editing software that you can use to create the animated GIFs you learned about in Chapter 8, draw vector graphics, and edit bitmap graphics. **Vector graphics** are created by drawing lines, curves, and polygons to create images. Vector graphics file formats include the Windows Metafile (.wmf), the PostScript (.ps), and the Flash (.swf) formats.[22] **Bitmap graphics**, also called **raster graphics**, are created by defining the color of individual pixels in a bitmap. Raster file formats include the Windows .bmp format as well as the .gif and .jpeg formats you learned about in Chapter 8.[23]

Using Photoshop, you can scale and rotate a raster graphic, adjust its colors, and correct flaws, such as scratches. You can draw vector shapes (lines, rectangles, polygons, and so forth) to create a drawing and then add layers of special effects to the drawing. Figure 9-13 shows a picture (in the .jpg format) of a boy being edited in Photoshop.

FIGURE 9-13   Photoshop

Like Photoshop, **Fireworks** is used to draw vector graphics, edit bitmap graphics, and create animated GIF files. Figure 9-14 illustrates the same picture of a boy being edited in Fireworks.

FIGURE 9-14 Fireworks

Flash

As you have learned, vector graphics are created with lines and curves. **Flash** is a multimedia program originally designed to provide animation for vector graphics. Flash animation is often used in Web page ads or online product tutorials. Some Web authoring software, such as FrontPage and Dreamweaver, have a feature you use to incorporate Flash animation into Web pages. Additionally, developers can use Flash to create interactive Web sites with animation, sound effects, and other fun elements. A visitor to a site that has Flash content must have a Flash player plug-in installed in his or her Web browser in order to view the Flash content. Figure 9-15 illustrates vector graphic animation being created in Flash.

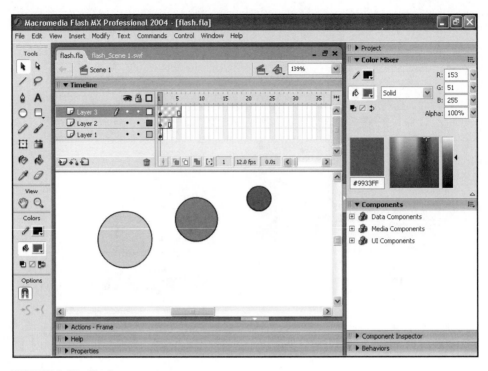

FIGURE 9-15   Flash

FreeHand

FreeHand is a graphic design program for planning the layout of any document that contains graphics, such as a brochure, product catalog, or a Web page. FreeHand can also be used to create a Web site storyboard. Figure 9-16 shows a new document with a perspective grid layout being created in FreeHand.

You can explore all four of these multimedia tools by reviewing information at the respective vendor's Web sites and by downloading trial versions of the software.

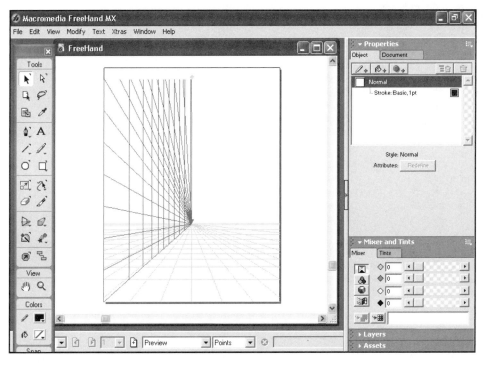

FIGURE 9-16　FreeHand

341

> **TIP**
>
> Both Adobe Systems and Macromedia offer software packages that combine their Web authoring soft-ware with their multimedia tools. Both companies have grown into Web technology powerhouses. On April 18, 2005, Adobe announced it was acquiring Macromedia in a stock transaction valued at approximately $3.4 billion.[24]

A Multimedia Vision

Today's Internet users constantly encounter images and multimedia—from family photographs taken with a digital camera, edited on a home computer, and posted to a Web-based photo album, to complex animation, video, and sound on e-business Web pages. But it wasn't always so. In fact, early Web pages were composed of little more than static text and images. But in the mid-1980s, an outspoken, colorful entrepreneur with a multimedia-based vision for the future was instrumental in helping develop the technologies that would ultimately change Web pages from static to dynamic documents, and thus lay the groundwork for the multimedia-enriched Web pages we know today.

In the mid-1980s, to add animation and sound to video games, you had to write a computer program in a language such as the C programming language.[25] In 1984, Marc Canter and two other video game developers, Jay Fenton and Mark Pierce, combined resources to form a company named MacroMind. The mission of Canter and his partners at MacroMind was to develop video game authoring software that would make it easier for nonprogrammers—artists, designers, and musicians—to add animation and sound to video games.[26] Calling themselves a "software rock and roll band," the MacroMind team began developing software for Macintosh computers, such as MusicWorks (which enabled game developers to add music and sounds to CDs) and VideoWorks (used to create and play multimedia animation). MusicWorks failed to catch on, but VideoWorks was licensed to Apple Computers, who used it to create training discs that were shipped with every Macintosh computer and featured "guided tours" of the Mac and its software. By the late 1980s, two of the original founders of MacroMind (Fenton and Pierce) had moved on to other projects. Canter remained with MacroMind, where he helped guide the evolution of its primary software product, VideoWorks, into Director, a very popular multimedia software program for the Macintosh that even nontechnical people could use to author multimedia.

In the early 1990s, a number of changes were taking place at MacroMind. First, in 1991, MacroMind merged with another multimedia developer named ParaComp to form MacroMind-ParaComp. Then, in 1992, MacroMind-ParaComp merged with a third multimedia company, Authorware, Inc., to create a new multimedia powerhouse named Macromedia.[27] In the meantime, the popularity of the Director software (now known as Macromedia Director) continued to grow. By the mid-1990s, the software was also supported in the Windows operating system environment, and was being used to create more than 70 percent of multimedia CD-ROMs.[28]

Over the next several years, Macromedia became a huge player in the multimedia software market. In addition to its success with Director, the company developed many other popular software products, including Flash (animation), Dreamweaver (Web authoring), Fireworks (graphics), FreeHand (page layout), and ColdFusion (Web page/database interactivity). In 2005, Macromedia reported $4.4 billion in sales, announced a net income of $42 million, and was purchased by one of its main competitors, Adobe Systems, for $3.4 billion.

continued

But what happened to Marc Canter? Sometimes called the "father of multimedia," Canter left Macromedia in 1992, shortly after the merger that created it, and, in his own colorful words, "took the 90s off." But Cantor remains a vocal supporter of multimedia technologies. Today, he is the CEO of Broadband Mechanics, a company that is developing technologies to allow users to keep track of their digital collections (music, photos, videos, and so forth) and then publish their own content based on items in their collections.[29]

QUOTES ON SUCCESS

"Nowadays everybody and his mother is a multimedia expert. All they have to do is go buy an iPod, digital camera, PC, Mac, even a videogame console—and they've got the power of computers, graphics, digital video and audio—at their disposal. That wasn't possible in 1985. Somebody, or group of people had to conceptualize what was possible and make it all happen. I was one of those people."

Marc Canter, co-founder of MacroMind, which became Macromedia

Tools for Interactive Web Page Elements

As you learned in Chapter 8, Web pages can contain forms and other interactive content such as clickable buttons and animation. Developers can add interactive Web page elements using a variety of technologies, including Java applets and servlets, Active X controls, JavaScript and JScript, PHP, CGI, and Active Server Pages (ASP).

Java Applets and Servlets and ActiveX Controls

Developed by Sun Microsystems, **Java** is an object-oriented programming language that is used to create applets and servlets.[30] (Object-oriented programming languages for the Web control interactive Web page elements such as clickable buttons and scroll bars.) An **applet** is a small program embedded in a Web page that executes when the page is loaded in a Web browser.[31] Java applets can be used to get information from databases, perform interactive Web page animation and calculations, or carry out other such actions without requiring the browser to send a request back to the server on which the Web page is stored.[32] A **servlet** is a small program that is similar to an applet but is executed at the server, before the Web page is loaded in a browser.[33] Servlets are often used to process data from online forms and authenticate user names and passwords, as well as create interactive content.[34]

ActiveX technologies are object-oriented technologies that were developed by Microsoft Corporation to compete with the Java programming language. **ActiveX controls** are small programs, similar to Java applets, that are used to provide Web page interactivity primarily for pages viewed in the Microsoft Internet Explorer Web browser.[35, 36]

JavaScript, JScript, PHP, CGI, and ASP

JavaScript, JScript, and PHP are all examples of scripting languages. While the details are beyond the scope of this book, briefly, a **scripting language** is a programming language used to embed run-time instructions in a Web page. Run-time instructions are commands that do not have to be translated or compiled into a machine-readable format before they execute—which means that the commands can execute very quickly and efficiently. Despite its name, **JavaScript** was developed independently of Sun Microsystems' Java programming language; the Netscape Corporation developed JavaScript for use with its Netscape browser.[37] Like Java, JavaScript has its own Microsoft equivalent, **JScript**. Both JavaScript and JScript are used to add Web page interactivity and dynamic content, such as the opening of pages in pop-up windows.[38]

PHP: Hypertext Preprocessor, or simply **PHP**, is a scripting language that can be embedded in HTML and can be used to build Web pages "on the fly" with elements—such as text, images, audio, and video—stored in databases. PHP is used primarily with Web servers running the UNIX operating system.[39]

Another technology, the **Common Gateway Interface (CGI)** standard, is a set of rules that determine two-way communication between a Web server and a Web browser. For example, when a user submits information entered in a Web-based form, a CGI script can provide the instructions to add the information to a database. CGI scripts can be written in a number of programming languages, including C++ and Perl.

Active Server Page (ASP) technologies were developed by Microsoft to embed server-side controls such as JScript or ActiveX programs in a Web page. ASP is an alternative to PHP and CGI for pages stored on a Web server running the Microsoft Internet Information (IIS) Web server program in a Windows operating system environment.[40]

It is very likely that your e-business Web pages will be created using some of the Web technologies discussed here. After your Web pages are completed, you should test them thoroughly.

TIP

Dynamic HTML (DHTML) uses HTML tags, style sheets, and a scripting language, such as JavaScript, to make animations and interactive features (such as a scrolling menu or banner) available in the browser window after a page is downloaded from the server.[41]

E-CASE

"Smarts, Aggressiveness, and Good Timing"

Growing up in Berkeley, California, Kim Polese was always interested in science, computers, and owning her own business. These days, Polese is a successful serial Web technology entrepreneur. During a 1998 interview for the Women of NASA's "Take Our Daughters to Work Day" chat forum, she described her ambition of becoming an entrepreneur this way: "It has always been a dream of mine to start my own business. I have known ever since I was a little kid that in some way I [would] make this dream happen."[42]

continued

Polese got a chance to make her dream come true while she was working for Sun Microsystems, a company she had joined in the late 1980s. By the summer of 1993, she was the product marketing manager for a new programming language named Oak, which was designed to replace the C++ programming language and be used to program "smart" appliances, such as coffee makers and microwave ovens. By mid-1994, however, the focus of Oak development had changed. The Oak development team realized that the Internet was on its way to becoming an interactive "super highway" and was the perfect cross-platform environment in which to exploit the features of its new programming language. Instead of focusing on how to use Oak to program a coffee maker to make a morning cup, the team set out to discover how the language might be used to add interactivity to static Web pages.[43]

As part of this re-focusing effort, the Oak programming language was renamed Java and the rest, as the saying goes, is history. As product manager, Polese's job included using her entrepreneurial skills to create the business model for marketing Java, develop a business plan, brand and position Java in the marketplace, and develop business partnerships with complementary technology providers, such as Netscape Communications. Polese's efforts paid off and the Java programming language became the hot technology for creating Web page interactivity.

Buoyed by the success of Java, Polese and three of her Java group co-workers (Sami Shaio, Jon Payne, and Arthur van Hoff) left Sun Microsystems in 1996 to establish their own Web technology company named Marimba. Each of the four invested $15,000, and then the group raised $4 million in startup funds from the VC firm Kleiner Perkins Caulfield & Byers.[44, 45] Marimba focused on developing and marketing technologies that pushed software patches and updates to computers over corporate networks. While serving as CEO of Marimba, Polese's long-time dream of becoming an entrepreneur was finally realized. In 2004, the co-founders harvested Marimba by selling it to BMC software for $239 million.[46]

Currently, Polese's desire to run her own business is being realized through other means. She is CEO of the technology startup SpikeSource, a company that provides support servers to businesses whose IT departments are using open-source software.[47]

QUOTES ON SUCCESS

"Java was the first language that made it possible to transport programs along with a Web page...Suddenly you could play checkers or get live sports scores or stock quotes instead of just seeing dried, dead text."

Kim Polese, original product marketing manager at Sun Microsystems for Java products and serial Web technology entrepreneur

WEB SITE TESTING

Before your Web site is published to its final destination Web server, and thus made available to the public, you should have your Web pages tested for accessibility and usability. All the features on each Web page should be checked to make sure they work as expected. One way to do this testing is to publish your Web site to a temporary server, called a **staging server**, and then perform a variety of tests to ensure that:

- links work correctly
- data submitted by visitors on Web-based forms update the appropriate database
- dynamic or interactive elements function properly

Additionally, you should have your site undergo a "stress test" to ensure that it can handle a heavy load of customer activity. If you develop your site in-house, you may lack the resources for a thorough testing process. If this is the case, it is a good idea to use the products or services of professional Web site testing companies, such as UsableNet, Segue Software, and Empirix.

After your new Web site is thoroughly tested and all the issues discovered are resolved adequately, you are ready to publish your Web pages to their final destination server. But publishing your Web pages doesn't mean that all your work on the Web site is finished. Because of the dynamic nature of the Web and the constant advancement in Web technologies, over time you may have to adjust how various pages look and function. You should also keep your Web site current by soliciting and evaluating feedback about your site from visitors and then using this feedback to make appropriate changes to page content or features. Finally, remember to retest your site at intervals as your business evolves.

After your Web site has been tested and is in operation, you must determine the success or failure of your site in achieving its business objectives by carefully evaluating the site's performance.

EVALUATING WEB SITE PERFORMANCE

In Chapter 8, you learned that determining your Web site's business objectives is a critical first step in developing your site. After your site "goes live" you should regularly measure the site's performance against these previously established business objectives. This evaluation not only determines the success or failure of your site in meeting these objectives, but it helps you learn how to improve your e-business's and your Web site's operations. More to the point, success or failure in most businesses, including e-businesses, is determined by the ability to learn from experience and make improvements. To evaluate your site's performance, begin by setting appropriate benchmarks and evaluating those benchmarks using the appropriate Web analytics measurements.

Web Site Benchmarking

To measure your Web site's performance, you must first set **benchmarks**, or performance-based goals, for the site. You can develop benchmarks by observing the actual performance of similar e-businesses or by reviewing and comparing your e-business's performance to industry averages. If your e-business has an operating history, you may also use historical data to set your benchmarks. For example, if you are operating a brick-and-click e-business, and thus have access to historical brick-and-mortar performance data, you may set benchmarks based on annual increases in sales over historical sales. If your e-business is new, you may need to look at similar brick-and-mortar businesses as well as similar e-businesses to establish your site's performance benchmarks.

Once you set your Web site's benchmarks, you must then compare the actual site results to these benchmarks to determine if the Web site performed as well as, better than, or worse than the benchmarks. Next, you must arrive at meaningful conclusions on why the benchmarks were or were not met or exceeded. Because it is possible to establish benchmarks that are either unattainable or too easily achieved, you may periodically need to reevaluate your original business objectives and benchmarks to determine if they are reasonable. Only through reevaluation can you do a better job of setting future business objectives and establishing appropriate benchmarks upon which to measure your site's performance.[48]

For evaluating the performance of a Web site, the measurements that are typically benchmarked are the number of site visitors during a specific period of time or the number of visitors who become customers during a specific period of time. Such measurements are grouped together in a system of performance measurements known as Web analytics.

Web Analytics

However you choose to develop, market, and operate your e-business Web site, it will be an expensive undertaking. One common method you can use to evaluate your Web site's performance is to identify its **return on investment** or **ROI**—the benefit your e-business gets in return for the investment in its Web site. The ROI that most e-businesses are looking for is an increase in sales. They want to know that the capital they are spending is generating additional sales and, they hope, increased profits. While increases or decreases in sales and profits may be easy to quantify, other factors that influence your Web site's ROI may not be as easy. For example, one of the primary factors that can affect your site's ROI is the increase (or decrease) in customer satisfaction with interactions at your site, but this sort of customer satisfaction or dissatisfaction may not be immediately measurable in terms of sales or profits. Determining your site's ROI can also be challenging because some online marketing activities may not be immediately measurable. For example, as you learned in Chapter 6, you can participate in industry-specific newsgroups or publish a blog to indirectly market your e-business, but it may be difficult to trace the direct impact on your site's ROI from these types of activities.

Web Analytics Performance Measures

When trying to determine your site's ROI, you should be aware that no single measure can define it. That is, the ROI is likely to depend on a number of factors. While the nature of these factors can vary, many e-businesses have found that analyzing the behaviors and actions of visitors at their Web sites can yield a generally useful, reliable ROI. Some visitor behaviors and actions that are useful to track include whether a visitor is visiting your site for the first time or is a repeat visitor; the last page the visitor accessed before exiting your site; the keywords visitors use to search within your Web site; and so forth. These and other visitor behaviors and actions like them are referred to as **Web analytics**, sometimes called **Web metrics**, as they can be measured, evaluated, and subsequently used to help determine your site's performance and ROI.[49]

TIP

Although the term "Web metrics" has been used in the past to describe the analysis of individual Web site visitor activity and site performance, some Web marketers and analysts now prefer to use "Web metrics" to describe measures of overall activity on the Internet, such as the number of people online, the number of people using broadband connections, online demographics, and so forth.[50]

Some commonly used Web analytics performance measurements are:[51]

- *Visit or session*: This is a measure of the continuous requests for Web pages at a site by a single user's Web browser that hasn't been inactive for a specific period of time, for example, 30 minutes. Requests from the same browser after 30 minutes of inactivity are counted as a new visit. This measurement indicates how long visitors are staying at the site.
- *Unique visitors (users)*: This is a measure of an individual visitor to a site, generally based on a combination of IP address and browser identification or cookies. The number of unique visitors can help you evaluate the success of your search engine optimization efforts and site advertising or promotion efforts.
- *Repeat visitors (users)*: This measurement identifies unique visitors who visit a Web site more than once during a specific time period. A large number of repeat visitors can indicate visitors are getting their wants and needs met at your site; a small number of repeat visitors can indicate problems at your site.
- *Page views or impressions*: This measurement indicates the number of times a specific page is viewed. Page views information tells you which pages your visitors are (and are not) viewing at your site.
- *Page views per visitor*: This measurement is developed by dividing the number of page views by the number of unique visitors, and it therefore measures how deep visitors are going into your Web site. A larger number of page views per visitor usually means that visitors are spending more time at your site.
- *IP addresses*: This measurement identifies the origin of a unique visitor. Analyzing IP addresses enables you to approximately identify countries from which your Web site visitors are coming, as well as from which networks.

- *Referring URLs*: This measurement indicates how visitors reached your Web site—whether by keying in the URL directly, clicking a link, or using a search tool. Reviewing the referring URLs can help you evaluate the effectiveness of online advertising links.
- *Browser type*: This measurement indicates what type of Web browser a visitor is using. This information can be used to make certain your site's pages operate correctly in your target audience's browsers.
- *Click-stream analysis*: This measurement identifies the path visitors take when navigating through the pages at your Web site. A click-stream analysis shows when and at what page a visitor entered your site, provides a list of all pages the visitor viewed in the order in which the pages were accessed, tracks the amount of time a visitor spends at each page, and indicates when a visitor leaves your site and from which page. A click-stream analysis tells you what visitors at doing at your site.
- *Conversion rate*: This measure indicates the rate at which visitors are becoming buyers. You can calculate the conversion rate by dividing the number of online orders by the number of unique visitors.
- *Shopping cart abandonment*: This measurement calculates the rate at which online shoppers fail to complete a purchase of the items in their shopping cart. A high rate of shopping cart abandonment may indicate that your site's checkout process is too complex or difficult.

Server log analysis and page tagging are two important tools used to develop these and other Web analytics.

Web Analytics Tools: Server Log Analysis and Page Tagging

Web servers record all of the events that occur on the server in **server log files**. For example, every time a browser requests a Web page or an image, or submits a form, the Web server log file records this event. It should come as no surprise, therefore, that you can obtain a treasure trove of information by analyzing these log files, which contain data such as date, time, the IP address of the computer making the request, browser type, referring URL, a myriad of other visitor information, and error messages. Figure 9-17 illustrates raw data recorded in the server log for the Books for Managers e-business you learned about in Chapter 7.

349

FIGURE 9-17 Web server log entries

Obviously, raw server log data such as that shown in Figure 9-17 may be difficult to interpret. Here is an explanation for the first entry in this server log:

- 65.54.188.51: This is the IP address of the computer visiting the Books for Managers Web site (in this case, the IP address tracks back to the msbot.msn.com domain).
- [02/Aug/2007:18:23:12 -700]: This is the date and time of the visit, including a time zone offset from Greenwich Mean Time (-700).
- "GET /cgi-bin/ae.pl?asinsearch=0142000280 HTTP/1.0": This is a request for a specific page at the Books for Managers site and the access method to be used (HTTP).
- 200 13730: This line contains the event status code (200), which indicates that the event was successful, and provides the size of the data returned in bytes (13730).
- "msnbot/1.0 (+http://search.msn.com/msnbot.htm)": This indicates that the entity accessing the Books for Managers site is an MSN Search spider.

The analysis of server log data can be performed in-house by installing server log analysis software or by contracting with an application service provider (ASP) that provides server log analysis services.

An alternative to server log analysis is page tagging. **Page tagging** involves adding hidden JavaScript images or tags to each of your Web pages. Each time a page is viewed in a visitor's browser, the data about the visitor and the page is sent to a remote server, where it is stored in a database. The database is then accessed to create a variety of reports that provide an analysis of visitors' behaviors.[52] Figure 9-18 illustrates the process for gathering visitor information using page tagging.

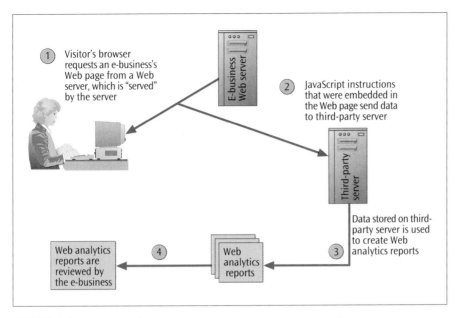

FIGURE 9-18 Page tagging process

The page tagging solution for developing Web analytics is generally provided by a third-party ASP service and may also involve the use of tracking cookies like those you learned about in Chapter 7.

Accuracy of Web Analytics

Measurements of visits, unique visitors, and other visitor actions at your Web site are based on the interaction between your visitors' Web browsers and the Web servers on which your pages are stored; the availability of cookies on the visitors' computers; the accuracy of human programmers; and many other variables. In other words, a number of factors can compromise the accuracy of Web analytics measurements. To gain a better understanding of what can go wrong, let's take a look at the problems that can arise when an e-business tries to measure the number of unique visitors at its Web site.

As described earlier in this chapter, a unique visitor is identified by combining IP addresses and browser identification information. One way this measurement can be flawed is that many organizations, for security reasons, substitute a common IP address for each of their users' internal IP addresses and also mandate that the same browser and browser version be used across all workstations in the organization. Consequently, all visitors to an e-business's Web site from an organization that has implemented these security measures would have the same IP address and browser identification—and thus would be counted as the same unique visitor. What about visitors that are assigned dynamic IP addresses (IP addresses that change with each session), such as visitors who are traveling and working from their laptops? In this case, a visitor who has been assigned dynamic IP addresses might appear to be a *different* unique visitor each time he or she visits your site.[53]

But the difficulty of obtaining an accurate count of unique visitors does not end there. If your e-business uses page tagging, there's a chance that human errors were introduced during the development of the JavaScript instructions, and this might also distort results. Alternatively (or in addition), your e-business may use tracking cookies to identify unique visitors. As you learned in Chapter 7, however, many people delete tracking cookies from their computers in an attempt to protect their privacy, and this, too, can compromise your e-business's ability to measure the number of unique visitors to your site.[54]

Analyzing visitors' behavior at your site is useful and important, but neither server log analysis nor page tagging can produce 100% accurate data about visitors' behaviors at your site. However, you can still glean useful estimates and trends from both of these methods of analysis.

QUOTES ON SUCCESS

"Web analytics is growing more sophisticated. We're developing methods for understanding customers, predicting trends, and assessing ROI. Every month analytics gurus amaze you with the latest revelations to sharpen your focus and tune your spend[ing]. What no one is telling you is that all these systems and numbers are based on inaccurate numbers. The god of web analytics has feet of clay—100 percent accuracy is impossible."

Brandt Dainow, CEO of ThinkMetrics, a Web analytics consulting firm

TIP

In the early days of e-business, a common measure of Web site performance was a hit. A hit is a recorded event in a Web site's server log for each element of a Web page downloaded to a viewer's browser. For example, if a viewer loaded a Web page with four graphics in his or her browser, the Web server would record five hits (one for the page and one for each of the four graphics). Although early industry measurements were largely devoted to the number of hits at a Web site, a hit actually bears no relationship to the number of pages viewed or visits to a site. Indeed, a count of unique visitors is more meaningful than the number of hits.

Whether you choose to use server log analysis or page tagging or a combination of both, you can access the Web analytics products or services you need from a number of software and service vendors.

Web Analytics Vendors

If you choose to perform server log analysis in-house, you will find it helpful to purchase some software that is designed to interpret log event entries and create meaningful, easy-to-understand reports from those entries. Examples of Web analytics software vendors include Sawmill and WebTrends. If you choose a page tagging solution, you will probably need to contract with an ASP such as DEEPMETRIX to gather, store, and report the data. Some vendors offer both server log analysis and page tagging services for developing Web analytics.

Monitoring visitors' behaviors and actions can be expensive. According to Jupiter Research, many large e-businesses spend thousands annually to analyze visitors' behaviors and actions at their Web sites, and e-businesses are projected to spend almost $1 billion on Web analytics software and services by 2009.[55, 56] It is a good idea, therefore, to consider several factors as you determine which monitoring process, software, or service to use.[57, 58]

- Carefully evaluate your Web site's business objectives and then determine which Web analytics measures would best be able to help you determine if your e-business is succeeding at or failing to meet those objectives.
- Determine what types of Web analytics reports you really need. Accumulating too many reports can result in information overload; too few reports may not be useful. Decide if your e-business would benefit from periodic (weekly, monthly) reporting or if you require real-time reporting.
- Many Web analytics vendors offer product tours, sample reports, and free trials of their software or services, plus free online seminars. Take advantage of these freebies to learn more about your Web analytics options.
- Look carefully at each of the vendors that offer the Web analytics software or ASP services in which you are interested to determine the company's history and commitment to customer training and support.
- Consider your budget constraints. Whereas an in-house software solution might give you what you need for a one-time cost, contracting with an ASP may involve hefty monthly fees.

No matter which Web analytics process or vendor you choose, your final challenge, of course, is to draw useful conclusions from the Web analytics reports and then to implement the changes your Web site might need.

. . . MONITORING VISITORS' BEHAVIORS

Before revamping its Web site, the management at RSN.com, the Web site for the RSN cable network, wanted to be certain the plans to add personalization and customization features to the site were on the right track. They implemented a customer relationship management (CRM) solution to create and maintain customer profile data, and they installed Web analytics software to monitor the behaviors of visitors to their Web site. It wasn't long before an analysis of this data made clear that the primary information customers and site visitors wanted to find at the site was not what RSN.com's management expected. Instead of checking out multiple resorts in a given geographic area, RSN.com's customers and visitors wanted to see more in-depth information on the specific resort in which they were interested. Armed with this new information, RSN.com scrapped its plans to add personalization and customization features to its site and concentrated on providing more in-depth information on each of its resort partners.[59]

continued

As part of this initiative, RSN.com phased out its CRM solution and concentrated on using its Web analytics software to gather the information it needed to improve its site. Continued analysis of visitors' behaviors led to more site improvements. For example, RSN.com's management had originally supposed that most of its visitors entered the RSN.com Web site through the site's home page; but Web analytics indicated that most visitors bypassed the home page and went directly to other pages at the site. RSN.com's management had also thought that around 70 percent of its site visitors were repeat visitors. Web analytics indicated that repeat visitors represented only about 20 percent of total site visitors. On the basis of this information, the RSN.com site was again revamped in 2004 to improve navigation and content.[60]

Today, visitors to RSN.com can view the Web site's Insider's Guide to check out happenings at major resort areas, book travel packages, and buy hiking, camping, and skiing gear. For a monthly fee, RSN's resort partners can have a "minisite" at the RSN.com Web site where they can post a resort profile, add weather reports, and post travelers' reviews.[61] RSN.com continues to use Web analytics to refine its site, and it shares these analyses with its resort partners to help them improve the effectiveness of their minisites. Annual sales at RSN.com, which are reported to approach $1.3 million, now make up almost 15 percent of Resort Sports Network's total annual revenues.[62]

QUOTES ON SUCCESS

"In the process of implementing the [Web analytics] software, we realized one of those great organizational myths. The customer we thought we were serving was actually different from what we were serving."

Richard Bilodeau, vice president of marketing, research, and technology for RSN.com

Chapter Summary

- A markup language is a set of rules and instructions called codes or tags, which are embedded in a document. Markup languages are used to describe the data in a document, specify the document's structure and layout, and format its contents. Examples of markup languages for the Web include XML, HTML, and XHTML.

- The Standard Generalized Markup Language (SGML) is an ISO standard for encoding documents so that they are transportable among computer systems. The common markup languages used on the Web today are based on the SGML standard.

- The Extensible Markup Language (XML) is a subset of SGML used to describe the data in a document. XML is focused on defining the contents of a document rather than its layout or format.

- The Hypertext Markup Language (HTML) was developed by Tim Berners-Lee as an easy-to-use language for publishing information on the Web. HTML is used to specify the layout and format of headings, text, tables, images, and other Web page elements. Unlike XML, HTML is focused on defining the layout and format of a document's contents.

- The Extensible Hypertext Markup Language (XHTML) is a marriage between HTML and XML and is the current W3C standard for creating Web pages.

- Web pages can be created from scratch by manually entering text and HTML or XHTML tags in a text editor such as Notepad, TextPad, HTMLPad, and HomeSite.

- Web pages can also be created in a "What You See Is What You Get" (WYSIWYG) environment. WYSIWYG-based Web authoring tools such as FrontPage and Dreamweaver allow Web page developers to work in an environment that contains familiar menu commands and toolbar buttons. As the Web developer creates a Web page (by adding text, images, and other elements), the Web authoring software automatically enters the corresponding HTML or XHTML tags.

- Popular multimedia tools that can be used to lay out Web pages, edit images, and insert animation in Web pages include Photoshop, Fireworks, FreeHand, and Flash.

- Interactive Web page elements are created using Java applets and servlets, which are small programs that run in a browser or on a server when a Web page is loaded. ActiveX controls are also used to create interactive elements for Web pages.

- JavaScript and JScript are two scripting languages used to add interactive elements to a Web page. PHP: Hypertext Preprocessor, Common Gateway Interface (CGI) scripts, and Active Server Page (ASP) are technologies that allow Web developers to create Web page and database interactions.

- After your Web pages are developed, they must undergo thorough testing before they are published and made available to the public. This testing includes making sure that all the links on the Web pages work, that data submitted through forms update the appropriate database, that all interactive elements function properly, and that the site passes a "stress test" to ensure that it can handle a heavy load of customer activity.

- After your site "goes live," you must then evaluate its actual performance against its business objectives. Begin by setting measurable benchmarks for the site's performance that support its business objectives. Then evaluate the site's actual performance against these benchmarks, and, if necessary, make adjustments to fine-tune both the Web site and future benchmarks.

- Web analytics involves measuring and analyzing the behaviors and actions of visitors at your site. Common Web analytics measurements include visits, unique visitors, repeat visitors, and page views. The data for Web analytics can be gathered from Web server logs or by page tagging.

- Because of technological constraints and the possibility of human error, the data gathered for Web analytics analyses is not 100 percent accurate; however the results of such analyses can still provide useful and important information about estimates and trends.

- Analysis of raw server log or page tagging data can be done in-house using Web analytics software or by contracting with an ASP that provides Web analytics gathering and analysis services.

Checklist

Evaluating a Web analytics software or service:

❏ What aspect of your Web site's performance does your e-business need to measure?

❏ Which Web analytics measurements will best enable you to measure actual site performance against its business objectives?

❏ What types of Web analytics reports do you need?

❏ How often do you need these reports: periodically or in real-time?

❏ Have you reviewed the online product tours, sample reports, and free online seminars offered by the Web analytics software or service vendors you are considering?

❏ Are the Web analytics vendors' products or services easy to use, and are the resulting reports easy to read and understand?

❏ What type of training and customer support do the Web analytics vendors offer?

❏ How much can you afford to spend for Web analytics software or services?

Key Terms

Active Server Page (ASP)
ActiveX controls
ActiveX technologies
applet
benchmarks
bitmap graphics
Common Gateway Interface (CGI)
Dreamweaver
Dynamic HTML (DHTML)

Extensible Hypertext Markup Language (XHTML)
Extensible Markup Language (XML)
Fireworks
Flash
FreeHand
FrontPage
Hypertext Markup Language (HTML)
International Organization for Standardization (ISO)

Java

JavaScript

JScript

markup language

Mathematical Markup Language (MathML)

page tagging

Photoshop

PHP: Hypertext Preprocessor

raster graphics

return on investment (ROI)

scripting language

server log files

servlet

staging server

Standard Generalized Markup Language (SGML)

Synchronized Multimedia Integration Language (SMIL)

tags

text editor

vector graphics

Web analytics

Web authoring software

Web metrics

WYSIWYG

XML schema

Review Questions

True/False Questions

1. SGML is the ISO standard for markup languages. True or False?

2. XML is a markup language used to define data, while HTML is a markup language used to lay out and format data. True or False?

3. The Notepad program is a WYSIWYG-based Web authoring tool. True or False?

4. The data gathered from Web server logs or page tagging can give you a 100 percent accurate analysis of your Web site visitors' behaviors. True or False?

5. A Web site benchmark is a performance-based goal. True or False?

Multiple Choice Questions

1. The benefit an e-business gets in return for the time and money spent to plan, develop, publish, and operate its Web site is called:

   a. benchmarking.

   b. Web analytics.

   c. return on investment.

   d. page tagging.

2. Which of the following markup languages is the current W3C standard for creating Web pages?

   a. SMIL

   b. XHTML

   c. HTML

   d. XML

3. Which of the following is a WYSIWYG-based Web page authoring tool?

   a. HTMLPad

   b. Dreamweaver

   c. ActiveX controls

   d. Java

4. Which of the following multimedia tools can be used to lay out Web pages or create a Web site storyboard?

   a. Photoshop

   b. Flash

   c. FreeHand

   d. Fireworks

5. Which of the following tools are used to create Web page interactivity with databases?

   a. Active Server Page (ASP)

   b. PHP: Hypertext Preprocessor

   c. Common Gateway Interface (CGI) script

   d. all of the above

Exercises

1. Using a link on the student online companion to this text, visit the Web sites for NetTracker, BlizzardTracker, Urchin, DEEPMETRIX, and WebTrends, and take the product tours at each site. Create a table that compares the products and services offered by each company; then use the table to guide a discussion with a group of classmates about the types of Web analytics products and services available to e-businesses.

2. Using a link on the student online companion to this text or online search tools, locate free online tutorials for HTML 4.01 and XHTML 1.0 or 1.1. Take the tutorials and make notes about what you learn. Then, using your notes, discuss how you can use these markup languages to create Web pages. How are the languages alike? How are they different? If you were developing an e-business Web site, which markup language would you use to create its pages? Why?

3. Using online search tools, research the contributions that Tim Berners-Lee has made to the development of the World Wide Web, HTML, and other Web standards. Who is Berners-Lee? What is his educational and professional background? What has he done that has had such a major impact on today's world? When did he do it? What is his ongoing role in Web standards development today? Create a profile of Berners-Lee based on your research results. Use the profile to discuss Berners-Lee and his contributions with a group of classmates.

4. The rapid advances in Web technologies provide virtually unlimited opportunities for e-business entrepreneurs who focus on creating and marketing new technologies. Using online search tools, identify at least three e-business entrepreneurs who have been successful in creating and marketing new technologies. Create an outline describing each entrepreneur and the technologies he or she created and marketed. How have (or will) these technologies change peoples' lives? Share your research with a group of classmates.

Case Projects

1. You and your business partner operate a small B2C e-business and want to better understand how your Web site is performing relative to the business objectives you established as part of the Web site design process. Your business partner thinks page tagging is the best way to develop the Web analytics you need to assess the site's performance, but you think server log analysis might be a better approach. Create a description of your e-business, set the business objectives for your site, and define the target audiences for your site and their expectations. (You may assume any information about your e-business not explicitly stated here.) Then, using links on the student online companion to this text plus online search tools and other relevant sources, research the advantages and disadvantages of using page tagging versus server log file analysis to develop the Web analytics information for your Web site. Create a table comparing the two approaches and use the table as a springboard to conduct a discussion with your business partner.

2. You and your business partner have finished planning your new e-business Web site and are ready to begin creating it. Now you need to either hire an in-house programmer or outsource the work. You want to be certain that whomever you hire to do the work is knowledgeable of and experienced with the latest Web development technologies. Working together with a classmate, use information from this chapter and other relevant sources to create a checklist you can use during interviews to assess an interviewee's knowledge and experience using current Web technologies.

Team Project

You and three classmates are planning your Web site for your new C2C e-business startup. Working together, create a description of the e-business, set the business objectives for your site, and define the target audiences for your site and their expectations. (You may assume any information about your e-business not explicitly stated here.) Then complete the following tasks:

- Using the storyboarding process, design the Web site's structure and page linking relationships.
- Create a formal chart for the site's organization.
- Develop a color scheme for the Web site.
- Design the layout for your home page, including the use of fonts, images, and navigational elements.
- Create a mock-up of your home page in a word processor, graphics or Web authoring software, or another available tool.
- Select the Web technologies you will use to create the pages.
- Select a Web analytics vendor to provide ongoing analyses of site visitors' behaviors that you can use to evaluate the site's performance.

Use Microsoft PowerPoint (or other presentation tool) to create a 5–10 slide presentation describing your e-business, your site's organizational structure, the design of its home page, and your technology and Web analytics vendor choices. Give your presentation to a group of classmates, who will critique your site and technology choices.

For Further Study

Here are some resources that might help you in further investigating the topics covered in this chapter.

Student Online Companion

Check out the *Creating a Winning E-Business, Second Edition* student online companion Web site for links to the sites discussed in this chapter and to other useful Web sites.

Articles and Books

Burby, Jason. "Barriers to Using Web Analytics Data for Optimization." ClickZ Network. www.clickz. com/experts/crm/analyze_data/article.php/3496486#bio. April 12, 2005.

Carey, Patrick. *New Perspectives on HTML and XHTML, Comprehensive*. Boston: Course Technology. 2004.

Dainow, Brandt. "Things That Throw Your Stats (Part 1)." iMedia Connection. www.imediaconnection.com/content/5184.asp. March 3, 2005.

Dainow, Brandt. "Things That Throw Your Stats (Part 2)." iMedia Connection. www.imediaconnection.com/content/5192.asp. March 7, 2005.

Doyle, Bronwyn and Miller, Helen. "Choose Your Web-Authoring Software." WorkZ. www.workz. com/content/view_content.html?section_id=500&content_id=5241. February 5, 2005.

Eisenberg, Bryan. "Calculate Your Online Conversion Rate." ClickZ Network. www.clickz.com/experts/article.php/3393741. August 13, 2004

Eisenberg, Bryan. "Prioritize Usability Testing and Web Analytics." ClickZ Network. www.clickz. com/experts/crm/traffic/article.php/3483671. February 18, 2005.

Farah, Samar. "Analyze This: Software Helps Marketers Tweak Sites, Drive Conversion Rates." *CMO Magazine*. www.cmomagazine.com/read/050105/pt_analyze_this.html. May 2005.

Hart, Kelly and Geller, Mitch. *New Perspectives on Macromedia Dreamweaver MX 2004, Comprehensive*. Boston: Course Technology. 2004.

Howard, Niles. "Maximizing Your Website's ROI." Inc.com. pf.inc.com/articles/2004/03/webroi. html. March 2004.

Hunt, Lachlan. "The Future: HTML or XHTML." SitePoint. www.sitepoint.com/print/future-html-xhtml. April 14, 2005.

K'necht, Alan. "The Dollars and Sense of Building to Standards." *Digital Web Magazine*. digital-web.com/articles/building_to_standards/. February 9, 2005.

Lopez, Luis. *New Perspectives on Flash MX 2004, Comprehensive*. Boston: Course Technology. 2004.

Patton, Susannah. "Web Metrics That Matter." *CIO Magazine*. www.cio.com/archive/111502/matter.html. November 15, 2002.

Peterson, Eric T. *Web Analytics Demystified: A Marketer's Guide to Understanding How Your Web Site Affects Your Business*. Portland: Celilo Group Media. 2004.

Peterson, Eric. T. *Web Site Measurement Hacks*. Cambridge: O'Reilly. 2005.

Raggett, Dave. "Getting Started with HTML." World Wide Web Consortium (W3C). www.w3.org/MarkUp/Guide/. May 24, 2005.

Sane Solutions, LLC. "NetTracker Log File Analysis vs. Page Tagging: A Guide for Comparing Web Analytics Methodologies." businessintelligence.ittoolbox.com/documents/document.asp?i=2787. 2003.

Sol, Selena. "History of XML." Web Developer's Journal, Virtual Library. www.wdvl.com/Authoring/Languages/XML/Tutorials/Intro/history. March 8, 1999.

WebTrends, Inc. "Survey Finds Most Businesses Are Not 'Very Confident' in Measuring Web Marketing and Have Not Adopted Accuracy Best Practices." CRMToday. www.crm2day.com/news/crm/114571.php. June 9, 2005.

Wikipedia. "Web Analytics." en.wikipedia.org/wiki/Web_analytics. August 18, 2005.

End Notes

[1] RSN.com. "About RSN." www.rsn.com/aboutrsn/welcome.html. 2005.

[2] Ibid.

[3] *Ski Press Magazine.* "Interview: RSN President Jeff Dumain Talks Ski TV and the Vegas Show." www.skipressworld.com/us/en/daily_news/2005/01/interview_rsn_president_jeff_dumais_talks_ski_tv_and_the_vegas_show.html?cat=Snowlife. January 21, 2005.

[4] Library of Congress. "EAD Application Guidelines for Version 1.0, Appendix F: Glossary." Society of American Archivists. www.loc.gov/. 1999.

[5] International Organization for Standardization (ISO). "ISO in Brief." www.iso.org/iso/en/aboutiso/isoinbrief/isoinbrief.html. March 2005.

[6] International Organization for Standardization (ISO). "Information Processing—Text and Office Systems—Standard Generalized Markup Language." ISO Catalog. International standard confirmed August 13, 2001. www.iso.org/iso/en/CatalogueDetailPage.CatalogueDetail?CSNUMBER=16387. 2005.

[7] World Wide Web Consortium (W3C) Recommendation. "Extensible Markup Language (XML) 1.0 (Third Edition)." www.w3.org/TR/REC-xml/. February 3, 2004.

[8] Ibid.

[9] World Wide Web Consortium (W3C). "XML Schema." www.w3.org/XML/Schema. April 2000.

[10] World Wide Web Consortium (W3C) Recommendation. "Introduction to HTML 4." www.w3.org/TR/REC-html40/intro/intro.html. 2005.

[11] Library of Congress. "EAD Application Guidelines for Version 1.0, Appendix F: Glossary." Society of American Archivists. www.loc.gov/. 1999.

[12] Bryan, Martin. *SGML and HTML Explained: Chapter 1.* New York: Addison Wesley Longman. 1997.

[13] World Wide Web Consortium (W3C) Recommendation. "HTML 4.01 Specification." www.w3.org/TR/REC-html40/. December 24, 1999.

[14] World Wide Web Consortium (W3C) Recommendation. "Alignment, Font Styles, and Horizontal Rules." www.w3.org/TR/REC-html40/present/graphics.html. 2005.

[15] W3Schools. "HTML 4.01 Reference." www.w3schools.com/tags/default.asp. 2005.

[16] Underwood, Lee. "Why Switch to XHTML?" WebReference.com. www.webreference.com/authoring/xhtml/. June 7, 2004.

[17] Ibid.

[18] New York Public Library. "NYPL Online Style Guide." www.nypl.org/styleguide/xhtml/benefits.html. March 2002.

19 World Wide Web Consortium (W3C) Recommendation. "XHTML 1.1—Module Based XHTML." www.w3.org/TR/xhtml11/Overview.html. May 31, 2001.

20 Wikipedia. "XHTML." en.wikipedia.org/wiki/XHTML. August 15, 2005.

21 Stevens, Susan G. "Macromedia Dreamweaver MX 2004." CNET Reviews. reviews.cnet.com/Macromedia_Dreamweaver_MX_2004/4505-3637_7-30521301-2.html?tag=tab. August 25, 2003.

22 Wikipedia. "Vector Graphics." en.wikipedia.org/wiki/Vector_graphics. August 14, 2005.

23 Wikipedia. "Raster Graphics." en.wikipedia.org/wiki/Raster_graphics. August 15, 2005.

24 Adobe Systems Incorporated. "Adobe to Acquire Macromedia." About Adobe: Investor Relations. www.adobe.com/aboutadobe/invrelations/adobeandmacromedia.html. April 18, 2005.

25 Canter, Marc. "The Birth of MacroMind." www.aec.at/en/archives/festival_archive/festival_catalogs/festival_artikel.asp?iProjectID=12317. 2003.

26 Ibid.

27 TechEncyclopedia. "Macromedia." www.techweb.com/encyclopedia/defineterm.jhtml?term=Macromedia. 2005.

28 Wikipedia. "Macromedia Director." en.wikipedia.org/wiki/Macromedia_Director. August 6, 2005.

29 Userland Blogs. "Marc's Voice: Digital Lifestyle Aggregation." blogs.it/0100198/stories/2004/03/26/digitalLifestyleAggregation.html. March 26, 2004.

30 Indiana University Knowledge Base. "What is Java?" kb.iu.edu/data/acwo.html. July 22, 2005.

31 Indiana University Knowledge Base. "In Java, What are Applets and Servlets?" kb.iu.edu/data/anqj.html. July 19, 2005.

32 Microsoft Corporation. "Glossary of Networking Terms for Visio IT Professionals." www.microsoft.com/technet/prodtechnol/visio/visio2002/plan/glossary.mspx#E1AA. November 1, 2002.

33 Indiana University Knowledge Base. "In Java, What are Applets and Servlets?" kb.iu.edu/data/anqj.html. July 19, 2005.

34 Sun Developer Network. "New to Java Programming Center - Unraveling Java Terminology." java.sun.com/developer/onlineTraining/new2java/programming/learn/unravelingjava.html. 2005.

35 Indiana University Knowledge Base. "What is ActiveX?" kb.iu.edu/data/afoi.html. July 20, 2005.

36 Indiana University Knowledge Base. "What are ActiveX Controls?" kb.iu.edu/data/afqb.html. July 20, 2005.

37 Wikipedia. "JavaScript." en.wikipedia.org/wiki/JavaScript#Java.2C_JavaScript.2C_and_Jscript. August 13, 2005.

38 Wikipedia. "JScript." en.wikipedia.org/wiki/Jscript. August 3, 2005.

39 eSeeHosting Glossary. "PHP." www.eseehosting.com/support/glossary.php. 2005.

40 Indiana University Knowledge Base. "What are ASP and ASP.NET?" kb.iu.edu/data/anqf.html. July 19, 2005.

41 Wikipedia. "Dynamic HTML." en.wikipedia.org/wiki/DHTML. August 13, 2005.

42 Women of NASA. "Kim Polese." quest.arc.nasa.gov/women/TODTWD98/polese.bio.html. April 23, 1998.

43 O'Connell, Michael. "Kim Polese Talks Java." JavaWorld. www.javaworld.com/javaworld/jw-04-1996/jw-04-polese_p.html. April 1996.

44 BusinessWeek. "The Next Generation: Kim Polese--CEO, Marimba." www.businessweek.com/1997/34/b354164.htm. August 14, 1997.

45 Silicon Valley Radio. "Transcript of Kim Polese Interview." www.transmitmedia.com/svr/vault/polese/polese_transcript.html. 2005.

46 LaMonica, Martin. "BMC Snaps Up Marimba." news.com.com/BMC+snaps+up+Marimba/2100-1012_3-5201902.html. April 29, 2004.

47 Red Herring. "Women in Tech: Kim Polese. Former Java Darling Kim Polese Wants Open Source to Work for Big Companies." www.redherring.com/Article.aspx?a=12411&hed=Women+in+Tech%3A+Kim+Polese. June 6, 2005.

48 Anfuso, Dawn. "Moving from Metrics to Results." iMEDIA Connection. imediaconnection.com/content/5756.asp. May 12, 2005.

49 Sterne, Jim. "Web Metrics Versus Web Analytics." MarketingProfs.com. www.marketingprofs.com/preview.asp?file=/4/sterne14.asp. March 16, 2004.

50 Ibid.

51 OPENTRACKER. "Clickstream or Clickpath Analysis." www.opentracker.net/en/articles/clickstream-analysis.jsp. 2005.

52 Wikipedia. "Web Analytics." en.wikipedia.org/wiki/Web_analytics. August 18, 2005.

53 Dainow, Brandt. "Things That Throw Your Stats: (Part 1)." iMedia Connection. www.imediaconnection.com/content/5184.asp. March 3, 2005.

54 Dainow, Brandt. "Things That Throw Your Stats (Part 2)." iMedia Connection. www.imediaconnection.com/content/5192.asp. March 7, 2005.

55 Bannan, Karen J. "Web Analytics Now a Marketing Must." BtoB. www.btobonline.com/article.cms?articleId=23827. April 4, 2005.

56 Farah, Samar. "Analyze This: Software Helps Marketers Tweak Sites, Drive Conversion Rates." CMO Magazine. www.cmomagazine.com/read/050105/pt_analyze_this.html. May 2005.

57 Eisenberg, Bryan. "How to Choose a Web Analytics Solution." ClickZ Networks. www.clickz.com/experts/crm/traffic/article.php/2174241. April 4, 2003.

58 Eisenberg, Bryan. "In Search of Web Analytics Excellence." ClickZ Networks. www.clickz.com/experts/article.php/3374451. July 2, 2004.

59 Ohlson, Kathleen. "Case Study: Web Analytics Helps RSN Create Happy Trails for Outdoor Types." ADTmag.com. www.adtmag.com/article.asp?id=11210. June 1, 2005.

60 SPSS Inc. "Customer Success List: RSN (Resort Sports Network)." www.spss.com/success/template_view.cfm?Story_ID=152. 2004.

61 Ibid.

62 Ohlson, Kathleen. "Case Study: Web Analytics Helps RSN Create Happy Trails for Outdoor Types." ADTmag.com. www.adtmag.com/article.asp?id=11210. June 1, 2005.

CHAPTER 10

SECURING YOUR E-BUSINESS

LEARNING OBJECTIVES

In this chapter, you will learn to:

- Describe the risk management process
- Describe business continuity planning
- Discuss the importance of business records management
- List the security risks and remedies associated with networks and Web sites
- Discuss the value of a security audit and network penetration testing

OUT OF DISASTER: AN E-BUSINESS OPPORTUNITY. . .

In the late 1990s, Phil Gilmour was managing partner of Armstrong Gilmour, a Walnut Creek, California accounting and consulting firm. A significant portion of the firm's business involved administering private pension funds for individual clients. Clients' pension and investment data was stored in a company database, and clients could go online to access the database and check the status of their 401(k) accounts.[1]

The portion of the business dedicated to managing the pension funds was growing fast—too fast, as it turned out, for the database that supported it. Unable to handle the increasing traffic, the database crashed. Fortunately, the company's computer system had a magnetic tape backup and so clients' data was safe—or so Gilmour thought. When the company attempted to restore the pension data from the backup tapes, it found the data on the tapes was corrupted.[2] The only option left was an expensive and agonizing one. Employees at the firm had to spend long hours over the next several weeks recreating the database manually. But the total cost of the disaster to Armstrong Gilmour may have been much greater than just the employee time and thousands of dollars consumed by the database restoration process. Gilmour, who eventually sold the firm, believes the company lost credibility with some of its clients after the database crash, and that this loss of credibility ultimately resulted in millions of dollars of lost value in goodwill—a loss that was reflected in a lower selling price.[3, 4]

continued

As painful as this disaster was for all concerned, there was one bright spot—Gilmour began thinking about the other small- and medium-sized businesses that relied on magnetic tape to back up their critical data. He suspected that many of these businesses, like Armstrong Gilmour, lacked the technical expertise and time to make certain their backup tape systems were functioning properly.[5] Could there be a new e-business opportunity in Gilmour's unfortunate experience?

RISK MANAGEMENT

As you learned in Chapters 1 through 9, becoming an e-business entrepreneur can be both rewarding and demanding. Some of the entrepreneurial challenges you will face in setting up your e-business and getting it off the ground include:

- identifying a viable e-business idea
- creating a business plan
- securing funding
- resolving legal and taxation issues
- planning adequate office space
- hiring key employees
- providing secure transactions for your customers and protecting their data
- marketing your products and services
- making the right technology decisions for your e-business and its Web site

Operating your e-business will pose still more challenges. For example, you must be prepared to protect your e-business's future by managing both the general risks inherent in operating any business and the specific risks associated with operating an e-business. Securing your e-business to protect not only its physical assets but also its reputation and longevity is a serious issue. Consider the variety of potential threats that your e-business faces:

- natural or human-made disasters such as fire, flood, hurricane, earthquake, or terrorist attack
- physical theft of your equipment and data storage media or electronic theft of your customers' data
- business interruptions caused by vandalism of your Web site or outside attacks on your network
- litigation and settlement costs associated with the inappropriate use of e-mail and the Internet by your employees
- product or service claims against items advertised and sold via your Web site
- lawsuits resulting from infringements of Web site-related copyrights, trademarks, and patents

Such threats can result not only in immediate loss of revenue, but they can also compromise future revenue and may result in compensatory payments to others for damages. In short, losses from business risks such as these could threaten the very survival of your e-business. You must, therefore, have some mechanism in place for managing the risk of potential losses. That mechanism is risk management.

Risk management is a process that identifies a risk of a business loss, assesses the risk's potential impact on the business, and then determines how to handle the risk.[6] You can use the risk management process, depicted in Figure 10-1, to protect your e-business' assets.

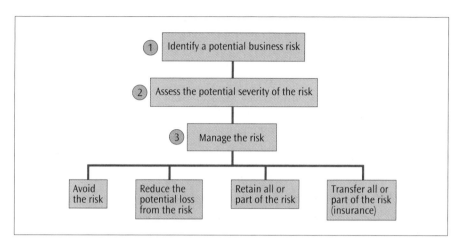

FIGURE 10-1   Risk management process

Suppose, for example, that you have just purchased vital and costly computer equipment, including networking equipment and individual workstations. Let's see how you can use the risk management process to protect your computer equipment assets against physical risk.[7]

Step 1: Identify the physical risk to the new equipment. Physical risks typically include fire, flood, and theft.

Step 2: Assess the potential impact on your e-business if you were to lose the new equipment as a result of fire, flood, or theft. The most obvious cost to your e-business is the cost of replacing the equipment. But other less obvious costs will also be incurred, such as the loss of software and data files, *plus* the employee-related costs lost to the process of restoring the software and data.

Step 3: Manage the risk. Avoiding the risk completely by not purchasing the equipment in the first place or by allowing your e-business to retain all of the risk are likely not viable options in this situation; therefore, the solution should probably involve a combination of risk management actions. For example, in addition to taking steps to reduce the potential loss from fire, flood, and theft, you might also retain a small part of the risk and transfer the remaining risk to someone else.

Reducing your potential loss might involve such actions as operating the network equipment from a secure, locked room; installing smart card or biometric identification devices that employees must use to enter their work areas; and installing a fire suppression system. Retaining part of the risk and transferring the remaining portion of the risk to someone else might involve purchasing fire, flood, and theft insurance with a deductible (your retained loss) that you can afford.

Risk management can help you not only protect your physical assets (cash, computer equipment, office furniture, and buildings) from physical damage or theft, but also protect your nonphysical assets (copyrights, trademarks, brand names, employees' knowledge, and goodwill) from the network-related risks to which your e-business may be exposed. Network-related risks can be complex in nature and their repercussions potent and damaging. For example, what if an intruder breaches your network and steals sensitive customer information, and then uses this information to commit credit card fraud? Alternatively, what if the intruder attempts to extort money from your e-business by threatening to expose the theft?[8] A comprehensive risk management program will include strategies for reducing the potential loss from network risks such as these. Those strategies may include sound security policies and procedures, network and Web site security and intruder detection programs, antivirus protection, firewalls, physical barriers to entry of the work area, a strong employee security education program, and a risk transfer program (insurance).

As you have learned, an important component of a comprehensive risk management program is the transfer of part or all of a risk to someone else, primarily through the purchase of insurance coverage. Table 10-1 illustrates types of traditional business insurance coverage, and Table 10-2 shows some of the different kinds of insurance coverage you should consider for your e-business.

TABLE 10-1 Traditional insurance coverage options

Traditional Insurance	Coverage
Employment practices liability	Protects employers from workers' claims of discrimination
Directors' and officers'	Protects corporate assets and the personal assets of directors and officers against wrongful acts such as mismanagement, fiscal irresponsibility, or security law violations
Product liability	Covers risks of third-party bodily injury or property damage from products sold
Business interruption	Mitigates revenue losses due to an interruption of business resulting from a malfunction of computer systems or other external or internal event
Crisis communication	Provides funds to hire professional public relations experts and others to handle damage control in a crisis
Crime loss	Protects against electronic theft of funds
Electronic data processing (EDP)	Covers hardware and software replacement and extra expenses related to hiring technical experts and others to recapture lost data

TABLE 10-2 E-business insurance coverage options

E-Business Insurance	Coverage
Computer virus transmission	Protects against losses that occur when employees open infected e-mail attachments or download virus-laden software
Extortion and reward	Responds to Internet extortion demands and/or pays rewards to help capture saboteurs
Unauthorized access / unauthorized use	Covers failure to protect against third-party access to data and transactions
Specialized network security	Responds to breach of network security and resulting losses
Media liability	Protects against intellectual property infringement losses
Patent infringement	Covers defensive and offensive costs when battling over patent infringement issues
Computer server and services errors and omissions	Protects e-businesses against liability for errors and omissions when their professional advice causes a client's financial loss

Insurance companies, such as Zurich North America and St. Paul Travelers, and insurance syndicates, such as those at Lloyd's of London, offer business insurance products through agents and commercial insurance brokers, such as Marsh. Because of the complexities involved in identifying business risks, assessing their impact, and choosing a plan of action, it is a good idea to secure the services of a risk management professional who is familiar with e-business.

TIP

A good source of information about risk management issues is the International Risk Management Institute (IRMI) Web site. Use a link on the student online companion to this text to check out this site. You can also search the Web using the keywords "risk management professional" to find businesses that focus on providing risk management services.

Most of the time, your e-business systems will run smoothly; however, there may be times when you must deal with operational problems, or when an unexpected and uncontrollable outside event disrupts normal business operations. Consider, for instance, what might happen to your e-business in the wake of a natural disaster, such as Hurricanes Katrina and Rita, which struck the Gulf Coast of the U.S. in 2005, or a human-made disaster, such as the 9/11 terrorist attacks. Such events could prevent you from operating your business for days, weeks, or even months. Preparing for the worst, for things such as a loss of data or a sudden inability to operate your e-business as a result of errors, accidents, or natural or human-made disasters, is another critical aspect of risk management.

BUSINESS CONTINUITY PLANNING

A **business continuity plan (BCP)** specifies how your e-business will resume partial or complete operations after a major disruption, such as a loss of a major vendor, serious failure of your products or services, management malfeasance, a natural disaster, or a terrorist attack.[9] Business continuity planning includes:[10]

- identifying events that might lead to a business interruption
- assessing how disruptive events might affect your e-business
- determining the resources you will need to maintain critical business functions after disruptive events
- developing a disaster recovery plan for critical business systems
- establishing procedures to communicate with stakeholders (employees, clients, vendors, bankers, emergency services, local authorities, and the public) after disruptive events

While the term "disaster recovery planning" is sometimes used interchangeably with business continuity planning, **disaster recovery planning** is generally associated with the *technological* aspects of a business continuity plan. Also known as contingency planning, disaster recovery planning deals with the specific resources and procedures needed to recover from hardware failures, loss of power, loss of communications lines, or other unexpected and catastrophic events that interrupt the operation of your business systems.

At a minimum, a business continuity plan for your startup e-business should designate the key employees who will make up your crisis management team; specify a plan of succession for management personnel (identify who is the next in line to assume responsibilities if a manager is not available); and describe procedures for notifying all employees where to report when a disruptive event occurs after work hours. Your business continuity plan should also provide for business-critical information to be maintained off-site and be accessible to members of your crisis management team. This information may include (but is not limited to) the following items:

- backup copies of system software, software programs, and critical data files
- instructions on how to access both paper and electronic records in off-site storage
- copies of electronic file backup and restore procedures
- computer system operations manuals and service agreements
- information on how computer network components are configured
- emergency contact information for all members of the crisis management team
- names, addresses, phone numbers, and e-mail addresses of employees, vendors, and customers
- emergency duty rosters for employees
- office space floor plans and lists of computer and office equipment
- copies of lease agreements and insurance policies
- copies of emergency service agreements with electrical, telephone, and Internet service providers

Your plan may contain other elements, such as arrangements for key employees to work from home after a disruptive event occurs at the workplace. As an added precaution, you may also choose to review your plan with your local emergency services personnel. If you outsource your Web site operations to an ISP or a Web hosting company, be certain to have a clear understanding of the ISP or hosting company's business continuity plan and disaster recovery procedures. The service agreement you sign should specify exactly how the contractor will resume your Web site operations and recover lost data after a disruptive event.

A number of online resources, such as the Disaster Recovery Journal or InfoSysSec, offer articles, tips, and templates for business continuity and disaster recovery planning. A risk management professional can also help you work out the details of an effective business continuity plan. Once you create your business continuity plan and disaster recovery procedures, don't wait for a disruptive event to see how well it all works. Instead, you should conduct a periodic assessment of your plan's elements so that you can reassess the risk of disruptive events and make any necessary adjustments.

TIP

Some large organizations employ a key member of the management team, sometimes called a contingency planner, to manage its business continuity planning efforts.

All businesses must keep both paper and electronic records of their day-to-day activities; additionally, various state and federal regulations and laws require that a number of transaction records, such as sales tax records, be maintained. Consequently, risk management also includes instituting policies and procedures to protect your vital business records.

BUSINESS RECORDS MANAGEMENT

Primary records consist of those records that document your key e-business activities, such as sales transactions, order fulfillment, and payment transactions. **Secondary records** include important information that is generated from e-business activities but that is not used in daily operations. Such information includes customer data, employee data, e-mail and instant message communications, Web server logs, and so forth. **Records management** involves the planning processes and actions your e-business must take to make certain your e-business's primary and secondary records (whether paper or electronic) are safely retained for an appropriate period of time, guarded against unauthorized access, and then destroyed when—and only when—they are no longer needed.[11] Typically, the records management process includes:[12]

- identifying records (paper or electronic documents, including e-mail and instant messages that support business activities and decisions or that are required for financial or legal purposes)
- storing the records in an online or offline environment where they are protected against unauthorized access or damage
- controlling access to the stored records
- searching for needed records

- maintaining records-retention schedules
- destroying records as scheduled

With both primary and secondary records, you must determine who has access to the records and how long the records should be retained. Making records available as needed is critical; but restricting access to only authorized individuals helps you keep your records from being misplaced, damaged, or stolen. The period of time business records should be retained depends on a number of factors, including financial reporting or auditing requirements, legal liability requirements, and local, state, and federal regulations and laws. Table 10-3 contains sample retention guidelines for the most common types of business records.[13, 14]

TABLE 10-3 Sample records-retention guidelines

Type of Record	Retention Guidelines
Bank statements	7 years
General correspondence	3 years
Insurance policies	Indefinitely
Sales records	7 years
Tax records	10 years
Payroll records	7 years
Trademark records	Indefinitely

Your attorney and accountant can provide you with the current records-retention guidelines that are appropriate for your e-business. You may also choose to seek the advice of a certified records management professional to help you develop your e-business's records management policies and procedures. The Web is a good source of information on hiring a certified records management professional.

TIP

The Sarbanes-Oxley Act of 2002 changed federal securities laws and put new reporting requirements on all publicly-traded companies. Born out of the financial scandals of the early 21st century involving major companies such as Enron, Arthur Andersen, WorldCom, and Global Crossing, Sarbanes-Oxley requires an annual report on the effectiveness of accounting controls, including the "electronic paper trails" that provide traceable data used in financial reports.[15] You can use links on the student online companion to this text to learn more about Sarbanes-Oxley and its effect on business records management.

Sound records management policies include procedures to secure your e-business from the loss of a business-critical operating system, software programs, and data files. Creating an electronic backup copy of your computer software and data files can protect you from losses resulting from processing errors, a hardware failure such as a hard disk crash, or a catastrophic event such as a building fire. Manual recovery of data lost as the result of these types of events can cost your e-business anywhere from a few hundred to several thousand dollars.

Backup procedures can vary, depending on the size of a business. For example, a small business may have its employees regularly back up the files stored on their own hard drives (these may be either files that are critical to the company or files that the employees use frequently, such as document templates) to a Zip disk, USB flash drive, CD-ROM, or magnetic tape. The company might assign its network administrator to be responsible for backing up system software and data files on shared network resources, such as file servers, to magnetic tape. The backup media would then be taken to a secure off-site location so that it is protected from fire or other catastrophes.

Very large businesses that have multiple offices connected by a network might create an off-site backup system by copying the backup data from servers at one location to servers at other locations. Depending on the individual business's needs, this backup could be performed once a day or weekly. On the other hand, an e-business that needs to be able to process online order transactions on a 24/7-basis might schedule its backup procedures for critical data to run continuously, often in real time. Today, it is easier than ever to back up system and data files by contracting with an e-business, such as USDataTrust or EVault, and then use the Internet to back up files on off-site servers.

Scheduling and conducting regular computer system and data file backups is only one step. You must also have procedures in place to regularly test the accuracy of the backup procedures and the validity of the backup media. For example, you—or one of your employees who is not involved with performing the actual backup process—should periodically attempt to retrieve and restore backed up files. This testing allows you to find out whether your files are being backed up correctly and on schedule, and to check that the backup media is error-free.

NETWORK AND WEB SITE SECURITY

Any business, whether it is a traditional brick-and-mortar business, a brick-and-click e-business, or a pure-play e-business, must be concerned about network security. The Internet is a public network consisting of thousands of private computer networks that are connected together in a myriad of ways. This means that your e-business's private computer network system is exposed to threats that may arise from anywhere on the public network. In addition to planning for network security issues, an e-business must be concerned about Web site security. You must always be prepared for known and unknown network and

Web site attacks—or else risk losing assets. New methods of attacking networks and Web sites and new network security holes are being discovered with disturbing frequency. By carefully planning your network and Web site security, you can reduce the risk of many known and as yet unknown threats.

Another very important reason to secure your network and Web site is to protect your e-business's relationships with its customers. Many Web users believe that there is a large risk to their privacy and security when they buy products and services or submit personal information online. Although this perception of risk may be greater than the actual risk, it is still a cause for concern. Your security measures must therefore address your customers' perceived risks just as much as they must mitigate any actual risks.

You cannot expect to achieve perfect security for your network and your Web site, but you must have adequate security to protect your e-business's reputation, assets, revenue stream, and customer privacy. Determining what level of security is adequate for an e-business depends on the e-business's particular situation. For example, an e-business that uses its Web site to provide information on flavors of dog food may not require the same level of security as an online banking Web site. You must determine your e-business's security needs according to potential risks, the value of the assets at risk, and the cost of implementing a network and Web site security system.

As you might expect, securing your e-business's network and Web site takes time, effort, and money. But keep in mind that the potential cost to your e-business for *failing* to provide adequate network and Web site security can be much higher than the cost of actually providing it. In July 2005, the Computer Security Institute (CSI) announced the results of its tenth annual Computer Crime and Security Survey, which was based on responses from 700 security managers from financial institutions, government agencies, medical institutions, universities, and businesses. The survey was conducted in association with the Computer Intrusion Squad of the San Francisco office of the Federal Bureau of Investigation (FBI). The survey results, some of which are summarized below, emphasize the ongoing vulnerability of organizations to computer security violations.[16]

- 56 percent of survey respondents experienced unauthorized uses of their computer systems within the past 12 months
- 95 percent of respondents reported Web site incidents
- respondents reported more than $130 million in estimated financial losses

According to the CSI 2005 report, more than 80 percent of reported estimated financial losses resulted from virus infections, unauthorized network access and theft of proprietary information, distributed denial of service attacks, and Web site defacement.[17]

Viruses, Worms, and Trojan Horses

Computer viruses and worms are the most common network security risk facing e-businesses today. A **virus** is a small program that inserts itself into other program files, just as a virus in nature embeds itself in normal cells that become "infected." The computer virus is then spread when an infected program executes and infects other programs. The effects of viruses are numerous and can range from being mildly annoying to debilitating. For example, viruses can prevent computers from booting, they can erase files or entire hard drives, and they can prevent users from being able to save or print files. Viruses can be introduced into a network through infected files shared by network users or through infected files attached to e-mail messages.[18]

An especially problematic type of virus often introduced into a network through data communication ports (the logical entry points into a network for incoming data) or e-mail is a **worm**. A worm is a type of virus that doesn't alter program files directly. Instead, a worm replicates itself and often goes unnoticed until its uncontrolled replication consumes network resources and thus slows down or crashes the network.[19] For example, in mid-August 2005, the Zotob worm, exploiting a vulnerability in some versions of the Windows operating system, infected networks belonging to government agencies, media companies (CNN, ABC, The New York Times, and others), and businesses (American Express and Caterpillar Inc.). This caused users at these organizations to experience repeated shutdowns and reboots (restarts) of their computers.[20]

Mass mailing worms, like the Sobig worm (and its variants such as Sobig-A and Sobig-F) that ran amok across the Internet in 2003 and 2004, can be especially damaging. A **mass mailing worm** hijacks the e-mail address book on an infected computer and then sends itself to all the e-mail addresses in the book; the process is repeated at each newly infected computer. For example, during the high point of the Sobig infections in the early fall 2003, America Online (AOL) found its e-mail traffic had almost quadrupled. In a single day, AOL identified more than 23 million e-mail messages infected with the Sobig-F worm. Some estimates put the damage (network downtime, lost productivity, lost revenues, and so forth) caused by the Sobig-F worm variant alone at more than $36 billion.[21]

Taking its name from a story in Homer's Iliad, a **Trojan horse** is a program that emulates a benign program, appearing to do something useful or entertaining. However, it performs harmful actions as well, such as destroying files, logging keystrokes, spreading viruses, creating a "back door" entry point to give an intruder access to a computer system, and so on.[22] A Trojan horse can be included in an e-mail attachment or it can be attached to downloaded software. Examples of Trojan horse programs include Trojan.Webus.I (which creates a backdoor entry point) and PWSteal.Flecsip.B (which steals passwords).

To secure your network against viruses, worms, and Trojan horses, you should:

- install updates or patches to your network operating system as soon as they are available, so that you prevent viruses and worms from exploiting known operating system vulnerabilities
- purchase antivirus software from a major vendor, such as Symantec or McAfee; install it on all the computers in your network, and keep the virus definition databases updated on a regular basis; set the software's preferences to scan all electronic files and e-mail messages for virus, worms, and Trojan horse threats
- close all unused network communication ports

Unauthorized Network Access and Theft of Proprietary Information

Your e-business's assets may be at risk from unauthorized access to network resources by individuals outside of your company and, possibly, by your own employees, contractors, business partners, and customers.

Hackers and Crackers

An entire glossary of words and phrases exists for identifying various network and Web security risks. In the previous section, you learned about the security risk terms "virus," "worm," and "Trojan horse." Just as there are terms for programs that invade networks, there are also terms for the people who access network resources without authorization. Originally, **hacker** was used to describe a gifted software programmer. Because of the wide press coverage of unauthorized network access resulting in security breaches, the term "hacker" is now generally used for those involved in malicious unauthorized computer system access.[23] Some software programmers prefer the terms **ethical hacker** or **white-hat hacker** to refer to hackers who use their skills to find weaknesses in computer systems and make the weaknesses known without regard for personal gain. In contrast, they use **black-hat hacker** or **cracker** to refer to a hacker who accesses network resources to attempt to disrupt service, to steal valuable information (such as credit card numbers), or to cause other damage.

The best way to recognize when a hacker is attempting unauthorized network access is to monitor network performance. Setting up, logging, and monitoring established network reference points or benchmarks on your networks can alert you to security problems. To do this, you will probably want the expertise of a skilled system administrator and other well-trained technicians who know how to use benchmarks to monitor and manage the network and servers. In addition to monitoring these benchmarks, the system administrator should regularly monitor the Web sites of software vendors whose products your company has purchased, as well as security-related Web sites (AntiOnline), blogs (Security Fix), and discussion groups (CastleCops), to stay abreast of network and Web site security issues.

Other tools to prevent intruders from hacking into a network include passwords, firewalls, and intrusion detection systems. A **password** is a code used to gain access to a computer network. Passwords are an effective means of preventing unauthorized network access, but only when they are chosen and used properly. Often an employee chooses a bad password, such as a short common word, a name or birthday of a person he or she knows, or pet's name, because the employee wants to be able to remember the password easily. Unfortunately, selecting a password based on this criterion—that it will be easy to remember—can be dangerous for several reasons. For one, outside intruders or hackers can penetrate a network by using software that "guesses" an authorized user's password by trying millions of common words until one of the words is accepted. Alternatively, someone inside an organization who is familiar with the employee may breach network security by "guessing" the employee's password by trying known names, birthdates, or pet's names. A password that is at least six characters in length and contains a mix of numbers as well as uppercase and lowercase letters is much more difficult to guess. Table 10-4 presents a sampling of good and poor passwords.

TABLE 10-4 Sample passwords

Good Passwords	Poor Passwords
W3jz73K	door
nr4tY32	Mary
zU80k3t	Fluffy

A **firewall** is software and/or hardware designed to isolate a private network from a public network. A firewall provides an easy-to-manage entry point to the network behind it. Firewalls can control the type of information that is allowed to pass from the public network to the private network, as well as the information that passes from the private network to the public network. Firewalls can also log activity to provide an audit trail in case the network is penetrated. Figure 10-2 illustrates a firewall.

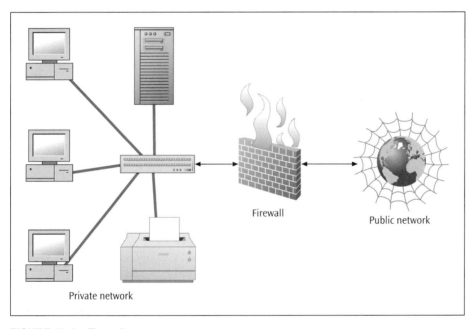

Firewall

Public network

Private network

FIGURE 10-2 Firewall

Intrusion detection is the ability to analyze real-time data to detect, log, and stop unauthorized network access as it happens. You can install an **intrusion detection system** that monitors your network for real-time intrusions and responds to intrusions in a variety of user-determined ways, such as sending e-mail to notify the network administrator, recording the event for future analysis, and stopping the intrusion. Cisco's Intrusion Detection System and Intrusion's SecureNet Intrusion Prevention are two examples of intrusion detection systems.

Unauthorized or Inappropriate Network Access by Employees and Other Insiders

You may think that your e-business's assets are at risk only from an attack on your network by an outside intruder. Unfortunately, however, your e-business's assets may also be at risk from the unauthorized and inappropriate use of network resources by your own employees or other e-business insiders. For a variety of reasons (for example, to establish a work environment built on mutual trust, professionalism, and so forth), businesses typically allow insiders (consisting both of employees and contractors) to have ready access to the business's network and data. Often the company does not monitor the use of these resources closely enough. Should business insiders become disgruntled, they would be able to perpetrate a variety of potentially debilitating computer-related crimes against the business, ranging from destroying electronic data and planting viruses to stealing the credit card numbers of the e-business's customers.[24] If you doubt that insider-perpetrated computer crime is a real threat, just visit the Department of Justice's Computer Crime and Intellectual Property Section (CCIPS) Web pages to review the wide range of real-world computer crimes, sometimes called **cybercrimes**, perpetrated by organization insiders.

In addition to the risk of insiders committing computer crimes through unauthorized network access, your e-business assets are at risk when employees and contractors access the Internet and Web for personal reasons while they are on the job. While it may seem innocuous, the inappropriate use of network resources can drain a firm's general resources, especially when the activity is widespread—which, by many indicators, it is. For example, a June 2005 survey of more than 10,000 workers by Salary.com and AOL indicated using the Internet and Web for personal use was the respondents' number one time-wasting activity at work.[25] This means that as an e-business entrepreneur you are likely to pay your employees thousands of wasted salary dollars. Examples of unauthorized or inappropriate use of network resources by employees include:

- surfing the Web to read news stories, shop, play games, and visit pornography sites
- sending and receiving personal e-mail messages during work hours
- circulating offensive jokes or other material using internal e-mail or instant messaging
- using the business's high-speed Internet connections to download personal music and video files

Not only does inappropriate use of network resources reduce employee productivity, it also increases the risk of your network becoming infected with viruses, worms, and Trojan horses. Inappropriate use of network resources can consume the high-speed bandwidth your e-business needs to conduct its business transactions, and, if offensive material is involved, may place your e-business at risk for costly and embarrassing workplace harassment lawsuits. Some methods you can use to protect your e-business against the threat of insider-related computer crime and inappropriate use are: [26]

- establishing and circulating clearly worded polices that define your users' security-based responsibilities and describe what activities on the Internet and Web constitute acceptable use
- enforcing your acceptable use policies when they have been violated
- restricting physical access to network facilities and data
- restricting employee access to the Internet
- installing software that monitors Internet use by your employees

At a minimum, you should publish security and acceptable use policies that detail how and when your employees may access your network, the Internet, and the Web while at work. Next, provide adequate explanation and training on the security and acceptable use policies so that your employees understand them. Finally, have your employees acknowledge their understanding of the security and acceptable use policies in writing. For added security, you should make certain to restrict physical access to your network and sensitive information (whether in electronic or paper form).

It may not be necessary or practical for your startup e-business to restrict Internet access to specific key employees, but as the number of employees grows, you may consider this measure. Also, at some point you may find it necessary to monitor your employees' network and Internet use by installing monitoring software from vendors such as SpectorSoft and Sepama Software. This type of software can block, filter, and report on your employees' network, Internet, and Web activities. While measures such as monitoring Internet use might seem insensitive or intrusive, their purpose is to protect your e-business from a real threat.

QUOTES ON SUCCESS

"After the first week [of using an Internet-monitoring software program] I did a spreadsheet report showing a minimum of 40 to 50 hours a week wasted, just on Internet surfing. The report easily projected savings of more than $50,000 a year [to be achieved by reducing personal Internet surfing]."

Keith Becker, Network Administrator for Illinois Wholesale Cash Register Corporation

When considering such measures, remember one important fact: typically, the weakest link in any security system is the people using it. This vulnerability can take a number of different forms and be exploited in various ways. For example, intruders or hackers often use a tried-and-true method called **social engineering**, which is based on the principle that people generally want to be helpful and tend to assume the best even from strangers. An intruder or hacker using social engineering tries to get helpful and trusting employees to divulge passwords, open locked doors, and so forth. Another vulnerability arises when a business's security policy is so confusing that the employees are not able to follow it, or they choose not to follow it because the policy makes it difficult for them to get their work done. For example, employees might resent a policy that requires them to make frequent changes to their passwords. It is imperative, therefore, that you make certain your employees understand your security policies and procedures and are aware of any penalties associated with failing to comply with them.

Distributed Denial of Service Attacks

A **distributed denial of service**, or **DDoS**, attack is an attack that is designed to disable a network by flooding it with useless traffic. To launch a DDoS, a hacker might first compromise multiple personal computers by installing Trojan horse programs that allow the hacker to control these computers remotely. Then the hacker would use the compromised or "zombie" computers to send a continual stream of traffic to a Web server. This stream not only disrupts the real traffic at the Web site, but it ultimately crashes the server, which tries to respond to the excess traffic.[27] Figure 10-3 illustrates a DDoS attack.

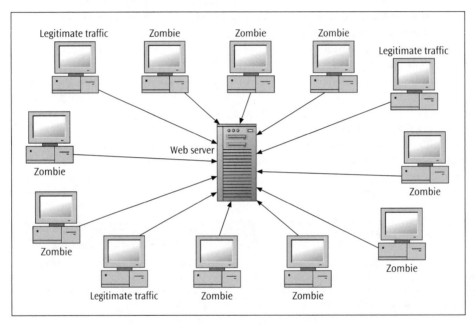

FIGURE 10-3   DDoS attack

While a DDoS attack does not do any technological damage, it can cause substantial financial damage, because every second your e-business's network or Web site is down can result in lost revenues. Meanwhile, the only reward to hackers for launching DDoS attacks seems to be the opportunity to show off their skills.

The typical measures taken to mitigate the effects of a DDoS attack include being prepared to block incoming traffic from the zombie sources (assuming the sources can be identified), having extra bandwidth and server processing capacity just in case of an attack, and redirecting legitimate traffic to backup servers. Unfortunately, these measures can be expensive and may not be viable options for a startup e-business operating its own servers. If this is the case for your startup e-business, your only option may be to wait out the attack. Given this situation, another advantage of a startup contracting its Web site operation to an ISP or Web hosting company instead of operating its own servers is that the ISP or hosting company should have the technological resources and expertise to mitigate a DDoS attack. If you contract with an ISP or hosting company, it is a good idea to have DDoS protection included in the service agreement.[28]

TIP

A "smurf" is a type of distributed denial of service attack. In a smurf attack, a network connected to the Internet is bombarded with Packet Internet Groper (PING) messages, which are normally sent to troubleshoot an Internet connection. The network attempts to reply to each PING message sent, as it's designed to do; in the process, it becomes overloaded. By sending hundreds or thousands of PINGs, a single smurf attacker can bring down a good-sized network.

Web Site Defacement

Web site defacement or vandalism occurs when a hacker maliciously alters a Web page.[33] Many businesses, colleges/universities, and government agencies have reported defacement of their Web sites. In fact, in May 2005, Zone-H.org, a Web site that tracks hacking events, reported it had recorded one million Web site defacement events in its archives.[34]

Web site defacement causes not only embarrassment and frustration for an e-business, but it can also have serious financial and legal repercussions. For example, in February 2000, Aastrom Biosciences, Inc., a Michigan-based medical products company, experienced a serious defacement designed to manipulate its stock price. A bogus news release announcing a merger with a California biopharmaceutical company, Geron Corporation, was posted on Aastrom's Web site. The result was an upward spike in both companies' stock prices. After discovering the defacement, Aastrom notified Geron, and representatives of both companies advised officials with the NASDAQ index, where both stocks are traded, that no merger was being planned.[35]

Hackers might deface Web sites for any number of personal or political reasons; taking steps to secure a network against hackers also protects it against Web site defacement.[36]

TIP

Web site **spoofing** or phishing is when criminals create a fake Web site and then redirect traffic from the legitimate Web site to the fake Web site by manipulating DNS records (IP addresses) or by luring visitors into clicking e-mail links. In Chapter 5, you learned about the risks to an e-business of not protecting its customers' data from theft, including personal and corporate identify theft from spoofing and phishing. Remember to remain vigilant against criminals who might steal your corporate identity and lure visitors to a fake Web site.

SECURITY AUDITS AND PENETRATION TESTING

A **security audit** performed by an outside auditing firm or security consulting firm can provide you with an overall assessment of your e-business systems' security by checking for vulnerabilities and providing recommendations for fixing them. A security audit can also make certain that your security efforts comply with applicable laws and regulations, such as FDIC regulations or HIPPA (Health Insurance Portability and Accountability Act) information-exchange requirements. A typical security audit involves a review of security policies, interviews with employees, analyses of network and operating system settings, and research into historical network operating data. Security auditors are looking for answers to questions such as the ones listed below.[37]

- Do published security policies exist? If so, how well do employees understand and comply with them?
- How easy or difficult is it to guess passwords?
- Are controls in place to restrict both physical and electronic access to data?
- Have the latest patches for operating system and software programs been installed?
- Have systems software programs and data been backed up?
- Where is backup media stored, who has access to it, and how current is the last backup?
- Does a business continuity plan exist, and who is responsible for its implementation?
- Have disaster recovery procedures been rehearsed?

Firms such as SecurityMetrics and TraceSecurity offer security auditing services. Accounting firms that offer security audit services often follow the American Institute of Certified Public Accountants (AICPA) SysTrust and WebTrust audit criteria. These audit criteria cover best business practices, site security, and customer information privacy. Displaying the AICPA SysTrust or WebTrust seals indicates to your customers that your computer systems and Web site have been audited and meet the specific security criteria.

E-PIONEERS

Getting "Poked and Prodded"

When Bank X, located in New England, rolled out its online banking Web sites, it realized that its e-business risks included more than just the various problems with network software that could be handled by its IT staff. Bank X, like many other e-businesses, understood that network and e-business risks were just as big a threat to its assets as any other business risk and thus deserved the attention of top management. Accordingly, Bank X involved its top management in the security process by forming a technology risk committee that included the bank's executive vice president, the senior vice president, the senior vice president of retail operations, the vice president of IT, and the chief financial officer.[38]

The technology risk committee realized that it was important for the bank to be able to withstand intruders attacking its network and Web sites, and so it wanted to know where the network and Web site security were vulnerable. To determine these vulnerabilities and how well its network and Web sites could stand up to an attack, in July 2000 the bank hired a security-consulting firm to perform a security audit that included penetration testing.

First, the security-consulting firm conducted a preliminary interview with the bank's internal financial auditor and the vice president of IT to establish the bank's objectives for the security audit. These objectives included testing the security of Bank X's internal network and its two Web sites, which were hosted by two different hosting companies—one that handled its customers' online banking and one that handled static information pages. The security firm did not ask the bank for any details about the Web sites or its internal network because it wanted to use only publicly available information—that is, information that would be available to potential hackers.[39]

The security-consulting firm's testing of the bank's internal network, including its e-mail server, indicated no problems. But tests of the security of the two hosted Web sites resulted in a very different outcome. Waiting until the early morning hours (when the system administrators were less likely to be monitoring their systems), the security-consulting company began attempting to penetrate the security at the two Web hosting companies. The company that hosted the bank's online transaction Web site responded quickly: within 30 minutes, it notified the bank by voice mail and by e-mail of the attempted intrusion. Also, within 12 hours, the hosting company had traced the attempted intrusion back to the security-consulting firm's IP address, found the security firm's phone number, and called it to find out what was going on. Bank X's vice president of IT and the internal financial auditor were thrilled with the hosting company's quick identification of the attack and its systematic response to it.[40]

continued

Unfortunately, the second hosting company, which hosted the bank's static information pages, did not detect the attack. In fact, the second hosting company was not even aware of the attempted intrusion until the vice president of IT called with the bad news. Although no customer transaction data was available at this part of the bank's site and therefore couldn't be stolen, the vulnerability of the site to intrusion made it susceptible to Web site defacement by hackers, which could embarrass the bank and damage its reputation.[41] As result of the security testing, the second hosting company upgraded its security in order to keep the bank's business. All in all, the bank's management considered the cost of a security audit on its network and Web sites to be money well spent.[42]

TIP

384

Many organizations recognize the importance of using security audits to protect their network resources and other assets. According to the 2005 CSI/FBI Computer Crime and Security Survey, 87 percent of respondents reported that their organizations conducted security audits.

As part of a security audit, your e-business's network and Web site may undergo a comprehensive penetration test. **Penetration testing** uses real-world hacking tools to test network and Web site security. Penetration testing provides an opportunity to see how your e-business's network and Web site security stands up to the most current hacking tools and techniques. Penetration testing can also measure the effectiveness of your e-business's intrusion detection systems and how well your employees respond to attempted intrusions. To better simulate a real-world penetration by a hacker, it is a good idea to arrange for the penetration testing to be conducted by outside security consultants instead of your own employees. When you evaluate a security consultant or consulting firm who may perform the penetration testing, consider the following actions.[43]

- Ask for evidence that the security-consulting firm has liability insurance to protect against accidental damage to your network, loss of proprietary data, and loss of revenues as a result of network downtime.
- Have everyone on the consultant's penetration team sign a nondisclosure agreement.
- Consider requiring a third-party background check on each member of the consultant's penetration team.
- Decide whether it makes sense to use a security-consulting firm that employs former hackers.
- Determine if the consultant's team is going to use packaged security scanning software that could be employed by the in-house staff, or if they are using custom tools.
- Develop a clear scope for the penetration test and a workable time frame.
- Determine whether to have the security-consulting firm perform a DDoS attack; if you decide to have one conducted, schedule the attack so that you minimize the disruption to customer access.

Make sure the final security audit report includes an accounting of all the attacks attempted and whether or not they were successful, a return of all the paper or electronic information gathered by the security-consulting firm, and recommendations on how to fix any problems discovered during the tests.

. . .OUT OF DISASTER: AN E-BUSINESS OPPORTUNITY

Because of his unfortunate experience, Phil Gilmour suspected that there was a market for products and services that could replace the old magnetic tape-based backup/recovery systems that were used by so many small- and medium-sized businesses. He began researching the backup/recovery options available to businesses in this size range. Gilmour's research led him to a company located in the Toronto, Canada area named VytalVault, Inc. VytalVault, Inc. provided business data backup, restore, and off-site storage services, and it had recently introduced a new network backup and restore service that could be accessed over the Internet.[44] Gilmour liked VytalVault's online backup/restore solution so much he formed a new company named EVault to license and sell the VytalVault service on the U.S. West Coast.

The relationship between EVault and VytalVault was successful, and in 2000 the two companies merged, retaining the EVault name. Today, EVault has six secure storage vaults located in the U.S. and Canada, with offices in California, Canada, and the United Kingdom. The company provides a variety of automated online data backup and restore services for more than 6,000 clients.[45]

QUOTES ON SUCCESS

"Right now, about 90 percent of small- and medium-sized companies don't have off-site data backup or they back up on tape on the premises, and then use trucks to ship those tapes to off-site storage. What that means is that we have a huge potential market for our products and services. SMBs [small- and medium-sized businesses] all need the basics but they just have not paid attention to security and recovery. There is lots and lots and lots of business out there for us."

Phil Gilmour, founder of EVault

Chapter Summary

- Risk management is a process that identifies a risk to your business assets, assesses the impact of the risk on your business, and determines how to manage the risk.

- Managing a risk can involve avoiding the risk completely, reducing the impact of the risk, retaining all or part of the risk, or transferring the risk to someone else through the purchase of insurance.

- A business continuity plan (BCP) specifies how your e-business will resume operations following a major disruptive event, such as a fire, flood, theft, or terrorist attack.

- The BCP for a startup e-business should include, at a minimum, crisis management team designations, a management succession plan, and contact procedures for notifying all employees where to report in case a disruptive event occurs after work hours.

- Records management involves identifying, storing, protecting, and searching for paper and electronic records, as well as destroying records according to a records-retention schedule.

- Tested software and electronic data backup and restore procedures are an essential part of records management.

- Planning for and implementing network and Web site security can protect your e-business's network resources and data from being damaged or lost due to computer viruses, worms, and Trojan horses. Network and Web site security can also protect against threats such as unauthorized access by outside intruders; unauthorized or inappropriate use of systems by employees, contractors, and other insiders; DDoS attacks; and Web site defacement attacks.

- A virus is a destructive program that inserts itself into another program; a worm is a type of virus that replicates itself and consumes vital network resources. A Trojan horse is a destructive program that enters a computer by masquerading as a useful or entertaining program.

- The term "hacker" is commonly used to identify individuals who breach network security to steal data, disrupt network services, or commit other malicious acts. To protect your e-business against hackers, you should implement tools such as well-designed passwords, software or hardware firewalls, and network intrusion detection systems.

- Insider computer crime is a serious threat that must be considered when creating security policies and procedures. Additionally, you must consider risks to your e-business assets from inappropriate use of your network, the Internet, and the Web by your employees.

- Distributed denial of service (DDoS) attacks involve using "zombie" computers to send continual streams of traffic to a network with the express purpose of crashing the network.

- Web site defacement is a type of vandalism that is not only embarrassing to the victim firm but also may have financial or legal repercussions.

- A security audit, which is usually conducted by an outside security-consulting firm or audit firm, provides you with an overall assessment of your e-business system's security vulnerabilities and offers recommendations for fixing them.

- Penetration testing is a security test in which security professionals attempt to invade a client's network in order to identify vulnerabilities that hackers could exploit.

Checklist

Securing your e-business:

- ❑ Have you used the risk management process to identify potential risks to your e-business, to assess the impact of each risk, and to take action(s) to manage each risk?

- ❑ Do you have a business continuity plan (BCP) with tested disaster recovery procedures in place? Is your BCP reviewed on a regular basis?

- ❑ Do you follow sound records management processes and procedures, including tested backup and restore procedures for electronic data?

- ❑ Do your managers and employees take a serious approach to network and Web site security?

- ❑ Have you published and circulated your e-business's security policies and procedures?

- ❑ Are all of your employees familiar with published security policies and procedures, and have they signed an acknowledgment that they understand and will comply with them?

- ❑ Have you installed antivirus software on all the computers on your e-business's network, and are the software's virus definitions updated on a regular basis?

- ❑ Have you installed a software or hardware firewall to block or filter incoming and outgoing network traffic?

- ❑ Have you restricted physical access to your network facilities?

- ❑ Do you have acceptable use policies in place for your network, the Internet, and the Web? Do your employees understand these policies, and are they aware of any penalties for failing to comply with the policies?

- ❑ Is it necessary to restrict employee online access to your network, the Internet, and the Web? If not, should you monitor the online activities of employees who do have access?

- ❑ Have you hired a security-consulting or auditing firm to perform a security audit with penetration testing? If so, have you complied with the audit's recommendations?

Key Terms

black-hat hacker	primary records
business continuity plan (BCP)	records management
cracker	risk management
cybercrimes	secondary records
disaster recovery planning	security audit
distributed denial of service (DDoS)	social engineering
ethical hacker	spoofing
firewall	Trojan horse
hacker	virus
intrusion detection system	Web site defacement
mass mailing worm	white-hat hacker
password	worm
penetration testing	

Review Questions

True/False Questions

1. Risk management is a three-step process for identifying, assessing, and managing risks to business assets. True or False?

2. A business continuity plan (BCP) is used to identify vulnerabilities in network security. True or False?

3. Penetration testing uses real-world hacking tools to test the security of a network or Web site. True or False?

4. A mass mailing worm is a program that appears to do something useful, but instead does something destructive or malicious. True or False?

5. Insurance can be an effective part of an e-business's risk management program. True or False?

Multiple Choice Questions

1. Which of the following uses "zombie" computers to attack a network?
   a. Trojan horse
   b. Web site defacement
   c. Sobig-F worm
   d. DDoS attack

2. Which of the following is part of managing risks identified during the risk management process?
   a. transferring all or part of the risk to someone else
   b. avoiding the risk
   c. reducing the potential loss from the risk
   d. all of the above

3. A firewall is:
   a. an established network performance reference point.
   b. software or hardware designed to isolate a private network from a public network.
   c. a virus that infects programs.
   d. off-site storage for electronic data backups.

4. Which of the following is a well-formed password?
   a. Rover
   b. 8/11/1942
   c. x24Gz80
   d. John

5. Web page defacement:

   a. enhances the appearance of a Web page.

   b. secures the Web page from intruders.

   c. embarrasses Web site owners.

   d. prevents identity theft.

Exercises

1. Using online search tools or other relevant sources, research the types of computer crimes that have been committed by outside intruders or by employees over the past three to five years. Then create a list of these computer crimes, including a description of each crime and its impact on the business victims. Use your list to discuss with your classmates the risks an e-business faces from computer crimes by outside intruders or by insiders, and describe how an e-business might protect itself against such risks.

2. Using links on the student online companion to this text or online search tools, identify at least three online storage service vendors. Create a table that compares the services and prices for each vendor. Suppose you are operating a startup e-business and have decided to use an online storage service vendor to support your system and data file backup plan. Use the table to discuss with two classmates the advantages and disadvantages of using each of the three vendors you researched.

3. Using links on the student online companion to this text or other relevant resources, review current news about Web site and network security issues. Then create an outline of topics you find in your research. Use the outline to direct a discussion with several classmates on the types of security issues an e-business currently faces and how it might protect itself against them.

4. Using online search tools or other relevant resources, locate and review case studies or articles that provide examples of real-world security audits and penetration testing and that give their results. Make a list of lessons to be learned from these real-world examples. Use your list to participate in a discussion on the importance of security audits and penetration testing.

5. Using online search tools or other relevant resources, research and identify at least two instances in which a real-world business suffered some type of loss because it lacked sound records management policies and procedures (including backup and restore procedures) or because its employees failed to follow the company's existing policies and procedures. List the business, the type of loss, and the actions that led to the loss. Then use the list to lead a discussion on the importance of records management.

Case Projects

1. You are the president of a startup e-business and are concerned that your employees are spending too much time using the Internet and the Web for personal use while they are at work. Using online search tools or other relevant resources, research and locate a few sample templates that businesses might use to create a policy that defines acceptable use of the Internet and Web. Then, use your research to draft an acceptable use policy for the Internet and Web.

2. Business is booming for your new B2B e-business, and you need to purchase insurance to minimize the risk of loss for both your physical and nonphysical assets. You are meeting with your insurance professional tomorrow. Before the meeting, you want to make certain that you can ask appropriate questions about the types of insurance available to protect your company against both traditional business risks and risks specific to e-businesses, such as problems with the network or Web site. Use online search tools or other relevant resources to learn more about potential risks to your physical and nonphysical assets and the types of insurance you might need. You may assume any information about your e-business that has not been specifically stated here. Then create a list of talking points you can use during the meeting.

3. You and your B2C e-business partner have decided to hire a security-consulting firm to help you understand network and Web site security issues and to perform a security audit including penetration testing of your e-business's internal network and its hosted Web site. You are interviewing representatives from three security-consulting firms this week and want to be prepared for each interview. Create an outline of the topics you need to discuss during the interviews.

Team Project

You and three classmates are part of the sales team for a B2B e-business. The purchasing manager for a potential major client has declined to order products through your Web site because of security concerns. You and your sales team are meeting tomorrow with the purchasing manager and the vice president of production to assure the client that there are adequate security measures in place.

Assume that your e-business has a risk management plan, has adequate network and Web site security, has survived a security audit, and schedules regular third-party penetration testing of your network and Web site.

Working with your team, use Microsoft PowerPoint or other presentation tool to prepare a 5–10 slide presentation you can use to reassure this major client that ordering at your Web site is safe. Then use the presentation to persuade two classmates, selected by your instructor to pose as the target client's purchasing manager and vice president, that ordering products online at your Web site is safe.

For Further Study

Here are some resources that might help you in further investigating the topics covered in this chapter.

Student Online Companion

Check out the *Creating a Winning E-Business, Second Edition* student online companion Web site for links to the sites discussed in this chapter and to other useful Web sites.

Articles and Books

Berinato, Scott and Scalet, Sarah. "The ABCs of Security." CIO.com. www.cio.com/research/security/edit/security_abc.html. March 20, 2002.

Davis, Rick. "Risk Management in the Digital Age." The CEO Refresher. www.refresher.com/!digitalage.html. 2001.

Duffy, Daintry. "Test Your Defenses." *Darwin Magazine*. www.darwinmag.com/read/120100/defenses.html. December 2000.

Emery, Priscilla. "Why Records Management." CMS Watch. www.cmswatch.com/Feature/127-RM-101. May 30, 2005.

Gordon, Lawrence et al. "2005 CSI/FBI Computer Crime And Security Survey." www.cpppe.umd.edu/Bookstore/Documents/2005CSISurvey.pdf. 2005.

Hayes, Bill. "Conducting a Security Audit: An Introductory Overview." SecurityFocus.com. www.securityfocus.com/infocus/1697. May 26, 2003.

Lowery, Jessica. "Penetration Testing: 'The Third Party Hacker.'" www.sans.org/rr/whitepapers/testing/264.php. February 2002.

Lucent Technologies. "Corporate Security: Business Continuity and Disaster Recovery Post 9/11." www.lucent.com/livelink/0900940380095c5b_White_paper.pdf. January 2005.

Murphy, Brian. "Sarbanes-Oxley Records Management Implications." *Sarbanes-Oxley Compliance Journal*. www.s-ox.com/feature/detail.cfm?articleID=924. July 21, 2005.

Posey, Brien. "Hiring Hackers as Security Consultants." www.windowsecurity.com/articles/Hackers-Security-Consultants.html. June 14, 2005.

The U.S. National Archives & Records Administration. "Typical Records Management [RM] Functions and Typical RM Program Activities." www.archives.gov/records-mgmt/policy/prod6a.html?. 2005.

Wikipedia. "Business Continuity Planning." en.wikipedia.org/wiki/Disaster_recovery_(business). August 17, 2005.

End Notes

[1] *National Post: Entrepreneur Section*. "EVault." www.canada.com/national/nationalpost/entrepreneur/evault.html. 2005.

[2] Ibid.

[3] Thomas, Susan L. "Gilmour Puts Momentum to Test Again." *East Bay Business Times*. eastbay.bizjournals.com/eastbay/stories/2003/06/02/smallb2.html. May 30, 2003.

[4] Montalbano, Elizabeth. "EVault: A Business Model Built From a Data Disaster." sanfrancisco.bizjournals.com/sanfrancisco/stories/2004/10/11/focus9.html. October 8, 2004.

[5] Ibid.

[6] Insweb Learning Center. "Glossary." www.insweb.com/learningcenter/glossary/general-r.htm. 2005.

[7] Wikipedia. "Risk Management." en.wikipedia.org/wiki/Risk_management. August 13, 2005.

[8] InsurTrust. "The Need for e-Business Insurance to Cover Technology, Digital Data, and Computer Hacker Risks." www.insuretrust.com/cyberRiskMgt-Need4eBusinessRiskMgt.cfm. 2005.

[9] Wikipedia. "Business Continuity Planning." en.wikipedia.org/wiki/Disaster_recovery_(business). August 17, 2005.

[10] MI5: The Security Service. "Business Continuity." www.mi5.gov.uk/output/Page267.html. 2004.

[11] Emery, Priscilla. "Why Records Management." www.cmswatch.com/Feature/127-RM-101?. May 30, 2005.

12 Ibid.

13 Magos, Alice. "Ask Alice About Record Retention." CCH Business Owner's Toolkit. www.toolkit. cch.com/advice/recordre.asp. 2005.

14 Yeo & Yeo: CPAs & Business Consultants. "Business Record Retention Schedule. www.yeoandyeo.com. 2005.

15 Sarbanes-Oxley 101.com. "Are You Sarbanes-Oxley Compliant Yet?" www.sarbanes-oxley-101. com/. 2005.

16 Gordon, Lawrence, et al. "2005 CSI/FBI Computer Crime and Security Survey." i.cmpnet.com/ gocsi/db_area/pdfs/fbi/FBI2005.pdf. 2005.

17 Ibid.

18 Beal, Vangie. "The Difference Between a Virus, Worm, and Trojan Horse." Webopedia. www.webopedia.com/DidYouKnow/Internet/2004/virus.asp 2005.

19 Ibid.

20 CNN.com. "Worm Strikes Down Windows 2000 Systems." www.cnn.com/2005/TECH/internet/ 08/16/computer.worm/. August 17, 2005.

21 Gaudin, Sharon. "Sobig's Birthday—Tracking the Most Damaging Virus Ever." IT Management. itmanagement.earthweb.com/secu/article.php/3297551. January 9, 2004.

22 Beal, Vangie. "The Difference Between a Virus, Worm, and Trojan Horse." Webopedia. www.webopedia.com/DidYouKnow/Internet/2004/virus.asp 2005.

23 Indiana University Knowledge Base. "What are Hackers and Crackers?" kb.iu.edu/data/agwt. html. July 1, 2005.

24 Bianco, David. "Enterprise Security: Who Do You Trust?" *Information Security*. infosecuritymag. techtarget.com/ss/0,295796,sid6_iss143_art311,00.html. October 2003.

25 Malachoski, Dan. "Wasted Time at Work Costing Companies Billions." Salary.com. www.salary. com/careers/layoutscripts/crel_display.asp?tab=cre&cat=nocat&ser=Ser374&part=Par555. June 2005.

26 Bianco, David. "Enterprise Security: Who Do You Trust?" *Information Security*. infosecuritymag. techtarget.com/ss/0,295796,sid6_iss143_art311,00.html. October 2003.

27 Wikipedia. "Denial-of-service Attack." en.wikipedia.org/wiki/DDoS. August 30, 2005.

28 Vijayan, Jaikumar. "Mydoom Lesson: Take Proactive Steps to Prevent DDos Attacks." Computerworld. www.computerworld.com/securitytopics/security/story/0,10801,89932,00. html. February 6, 2004.

29 Wired News Report. "Mafiaboy Sentenced to 8 Months." wired-vig.wired.com/news/business/ 0,1367,46791,00.html. September 13, 2001.

30 CNN.com. "Cyber-attacks Batter Web Heavyweights." www.cnn.com/2000/TECH/computing/_ 02/09/cyber.attacks.01/index.html. February 9, 2000.

31 CNN.com. "E*TRADE, ZDNet Latest Targets in Wave of Cyber-attacks." www.cnn.com/2000/ TECH/computing/02/09/_cyber.attacks.02/. February 9, 2000.

32 U.S. Department of Justice, Federal Bureau of Investigation Press Release. www.fbi.gov/ pressrel/pressrel01/mafiaboy.htm. January 18, 2001.

[33] Hollander, Yona. "Prevent Web Site Defacement." Internet Security Advisor. www.networkassociates.com/us/local_content/white_papers/wp_2000hollanderdefacement.pdf. December 2000.

[34] Zone-H.org. "1,000,000 of Defacements in Zone-H Archives." www.zone-h.org/en/news/read/id=205945/. May 22, 2005.

[35] Ramirez, Charles E. "Hackers Vandalize State Firm." *Detroit News*. www.detnews.com/2000/business/0002/18/02180092.htm. February 18, 2000.

[36] Newlin, Lew. "Site Defacements." PromotionWorld.com. www.promotionworld.com/internet/articles/sitedefacements.html. August 7, 2005.

[37] Hayes, Bill. "Conducting a Security Audit: An Introductory Overview." www.securityfocus.com/infocus/1697. May 26, 2003.

[38] Vaas, Lisa. "Security Checkup—One Bank's Experience at Having Its E-Biz Links Poked, Prodded, Scanned." *eWeek*. August 14, 2000.

[39] Ibid.

[40] Ibid.

[41] Ibid.

[42] Ibid.

[43] Lowery, Jessica. "Penetration Testing: 'The Third Party Hacker.'" SANS. www.sans.org/rr/whitepapers/testing/264.php. February 2002.

[44] Press Release. Altamonte Springs, Florida and Oakville, Ontario. "Columbia Data Products Partners With VytalNet and Offers Open Transaction Manager." www.cdp.com/articles/vytalpress.htm. January 10, 2000.

[45] EVault. "Company Management and Managed Services." www.evault.com. 2005.

CREATING A WEB SITE WITH MICROSOFT FRONTPAGE

LEARNING OBJECTIVES

In this appendix, you will learn to:

- Start the FrontPage program
- Describe the elements of the FrontPage window in Page view and Design sub view
- Use a FrontPage wizard to create and save a Web site
- View your Web site's pages in different FrontPage views
- Close the current Web site folder and the FrontPage program

STARTING FRONTPAGE

A complete overview of how Microsoft's Web authoring program FrontPage works and how its many features may be used to create Web sites and Web pages is beyond the scope of this book. However, this appendix is designed to help you gain some experience in the process of using FrontPage—in particular, its wizard tool—to create a small-business Web site and its related pages. A FrontPage wizard is a series of dialog boxes that lead you through the process step by step. In this appendix, you will start FrontPage and work through a wizard's steps to create a Web site. After you complete the wizard steps, you will use FrontPage's different views to review the new Web site and sample Web pages you've created.

As you have learned, a Web site consists of one or more related Web pages that are connected to each other via hyperlinks. FrontPage organizes all of a Web site's related pages in a single folder called a FrontPage Web site. When you install FrontPage, a folder called "My Webs" is created in the My Documents folder on your computer's hard drive. FrontPage then saves any new FrontPage Web site folders in the "My Webs" folder. While this is the default setting, you can set FrontPage to save your FrontPage Web site folders in another location. As you work through the exercises in this appendix, you will save the FrontPage Web site folder you create in a location specified by your instructor.

As you learned in Chapter 9, working with FrontPage is similar in many ways to working with Microsoft Word, the popular word processing program. In FrontPage, you work with menu commands, toolbar buttons, task panes, and document views that will be familiar to you if you use Word.

First, you will start the FrontPage program. When you start the program, a new, blank Web page opens in Page view. To start the FrontPage program and view the new, blank Web page:

Step 1.   Click the **Start** button on the Windows taskbar.

Step 2.   Point to **All Programs**.

Step 3.   Point to **Microsoft Office**, if necessary.

Step 4.   Click **Microsoft Office FrontPage 2003**.

Step 5.   Click **No** if asked to make FrontPage your default editor.

FrontPage features a number of different ways to view a Web page. In Chapter 9, you learned about the default Page view and Design sub view. The FrontPage program opens with a blank Web page in the Page view and Design sub view. The Getting Started task pane may also open; close it, if necessary, so that your screen looks similar to Figure A-1, which identifies the primary elements of the FrontPage window in Page view.

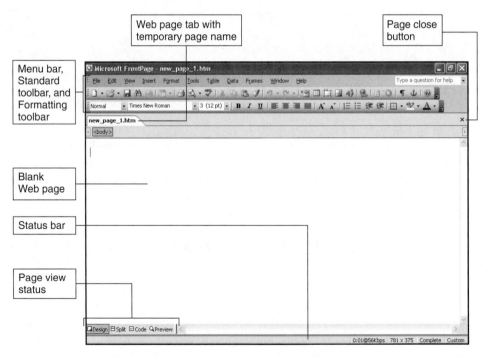

FIGURE A-1   FrontPage program window in Page view and Design sub view

Menu Bar, Standard Toolbar, and Formatting Toolbar

The menu bar, located below the title bar, has eleven drop-down menus that contain groups of related commands. For example, the File menu contains commands for opening, closing, previewing, and printing Web pages. You can use the mouse or the keyboard to select a command from the menu bar. In this appendix you will use the mouse to select menu bar commands.

The Standard toolbar is located under the menu bar and contains buttons that represent the most commonly used commands. For example, the Standard toolbar contains buttons for opening, saving, previewing, and printing a Web page. The Standard toolbar allows you to execute a command quickly through just one click of the corresponding button. You can customize the Standard toolbar (or any other toolbar) by adding or deleting buttons.

The Formatting toolbar is located under the Standard toolbar and contains buttons that represent commonly used formatting commands. For example, there are buttons that enable you to modify the appearance of a Web page by changing the font, font size, and color of its text; by aligning the Web page text horizontally; and by inserting bulleted or numbered lists.

Status Bar

The status bar appears at the bottom of the window, above the Windows taskbar, and it displays messages as you work on a Web page. For example, when you open a Web page, the status bar displays a message telling you how long it takes for the current page to download in a Web browser, the page size in pixels, and other option settings.

Page Sub Views

As you learned in Chapter 9, Page view has four different sub views: Design, Split, Code, and Preview. Buttons to switch between these views are located in the lower-left corner of the window above the status bar, as shown in Figure A-1. The Design sub view provides a WYSIWYG (What You See Is What You Get) environment. This allows you to create and edit your Web pages in the Design sub view much the way you would a word-processing document in Microsoft Word. As you create and edit a FrontPage Web page document, FrontPage inserts the appropriate HTML codes or tags that a browser will read to display the page. The Split sub view allows you to work in the Design sub view and, at the same time, view the automatically inserted HTML code or tags. You use the Code sub view to examine, insert, or edit the HTML codes or tags. The Preview sub view shows how the current Web page looks in a Web browser.

USING A WIZARD TO CREATE A WEB SITE

FrontPage contains a variety of wizards and templates you can use to create Web sites and Web pages. As you learned earlier, a wizard is a series of dialog boxes that takes you step-by-step through performing certain tasks. A template is a model document that contains page settings, formats, and other elements. Using a FrontPage wizard or template to create a new FrontPage Web site or Web pages saves you time because the wizard or template automatically provides the features you would most likely use to organize your pages' content.

Each FrontPage wizard or template is designed to create a Web site or page that serves a specific purpose, such as building a corporate presence or providing customer support. Because you will not be using the blank page that automatically opened when you started FrontPage, you will close it before you view FrontPage's various Web wizards and templates.

To close the blank Web page and view FrontPage's various Web wizards and templates:

Step 1. Click the **Page Close** button in the upper-right area immediately above the blank Web page to close the page.

Step 2. Click **File** on the menu bar.

Step 3. Click **New** to open the New task pane.

Step 4. Click the **More Web site templates** link in the New task pane to open the Web Site Templates dialog box.

The General tab in the Web Site Templates dialog box opens. In this dialog box, you first select the desired wizard or template and then specify where the new FrontPage Web site folder is to be saved. For this exercise, you will use the Corporate Presence Wizard and save the folder in the location specified by your instructor. To do this, you should:

Step 5. Click the **Corporate Presence Wizard** icon to select it.

Step 6. In the text box labeled Specify the location of the new Web site:, type the location specified by your instructor for storing your Web pages. Use the Web site folder name "Sample_Web." When typing the folder name, remember to include the underscore to represent a space, but do not type the quotation marks.

The dialog box on your screen should look similar to Figure A-2.

Step 7. Click **OK**.

At first, the Create New Web Site box opens to show you how the creation of the new FrontPage Web site folder will progress. In a few seconds, the first Corporate Presence Web Wizard dialog box opens. Your dialog box should look similar to Figure A-3.

You can use the buttons at the bottom of the wizard dialog boxes to cancel the wizard, go to the next step of the wizard process, return to the previous step, or finish the wizard process. To go to the next step:

Step 1. Read the dialog box text to prepare for continuing the wizard process.

Step 2. Click **Next**.

Corporate Presence
Wizard

FrontPage Web folder
name and location

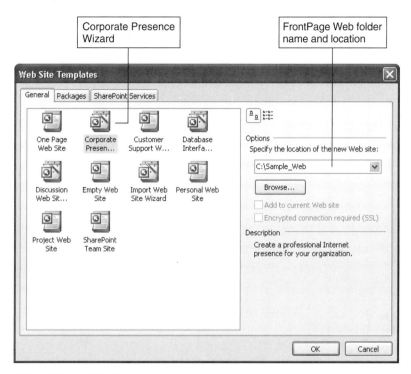

FIGURE A-2 Web Site Templates dialog box

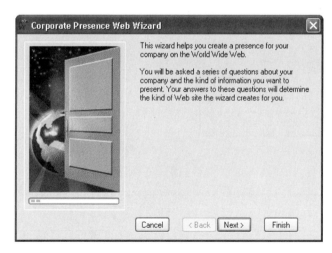

FIGURE A-3 First wizard dialog box

The second wizard dialog box opens. In this dialog box, you must select the type of pages you wish to include in your Web site folder. As you see in Figure A-4, a home page is required, but you can include or exclude the other pages listed in the dialog box by adding or removing a check mark from the check box to the left of each page name. To include only the Home page, Products/Services page, and the Feedback page:

Step 3.　Click the **What's New**, **Table of Contents**, and **Search Form** check boxes to remove the check marks, if necessary.

Step 4.　Click the **Products/Services** and **Feedback Form** check boxes to insert check marks, if necessary. Your dialog box should look similar to Figure A-4.

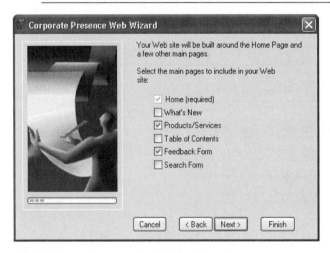

FIGURE A-4　Second wizard dialog box

Step 5.　Click **Next**.

The third wizard dialog box opens. In this dialog box, you must select the topics you wish to include on your home page. To select the Introduction, Mission Statement, Company Profile, and Contact Information topics:

Step 6.   Click the **Introduction, Mission Statement**, **Company Profile**, and **Contact Information** check boxes to add check marks, if necessary.

Your dialog box should look similar to Figure A-5.

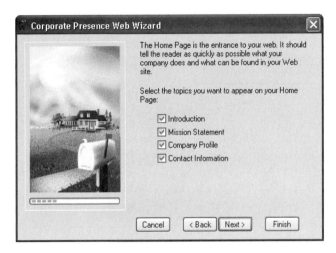

FIGURE A-5   Third wizard dialog box

Step 7.   Click **Next**.

The fourth wizard dialog box opens. In this dialog box you specify the number of Products or Services pages you would like your Web site to feature. To create one Products page and no Services pages:

Step 8.   Type the number **1** in the Products text box, if necessary.
Step 9.   Press the **Tab** key.
Step 10. Type **0** in the Services text box, if necessary.

Your dialog box should look similar to Figure A-6.

Step 11. Click **Next**.

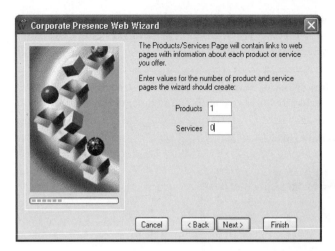

FIGURE A-6   Fourth wizard dialog box

The fifth wizard dialog box opens. This dialog box allows you to insert product images, pricing information, or an information request form. For this exercise, you will not be including any of this information on the Products page.

Step 12. Remove the check mark from each check box, if necessary.

Your dialog box should look similar to Figure A-7.

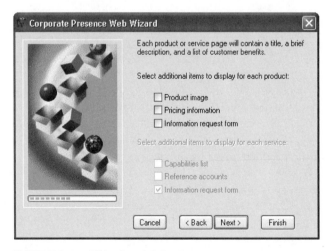

FIGURE A-7   Fifth wizard dialog box

Step 13. Click **Next**.

The sixth wizard dialog box opens. You must now select the elements that are to appear on the Feedback Form page. To capture as much information as possible about potential customers, you will include all the possible elements. To do this:

Step 14. Click each of the check boxes to insert a check mark, if necessary. Your screen should look similar to Figure A-8.

Step 15. Click **Next**.

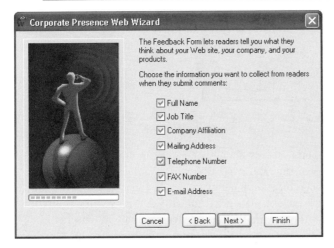

FIGURE A-8   Sixth wizard dialog box

The seventh wizard dialog box opens. In this dialog box, you select the format in which the Feedback Form stores the visitor data. You'll choose a format that separates the information elements by tabs. This makes it easier to export the data to other programs, such as Access and Excel. To select the tab-delimited data format:

Step 16. Click the **Yes, use tab-delimited format** option button, if necessary. Your dialog box should look similar to Figure A-9.

Step 17. Click **Next**.

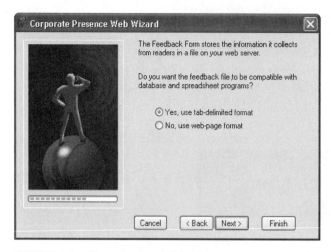

FIGURE A-9 Seventh wizard dialog box

The eighth wizard dialog box opens. In this dialog box, you can automatically add the elements that you would like to appear on each page at your Web site. One design objective could be to have the company logo on each page. Another design objective could be to have navigational hyperlinks to other pages at the site. Additionally, a well-designed Web page should include contact information for the Webmaster (the individual who manages a Web site), a copyright notice, and the page modification date. To indicate the desired elements:

Step 18. Click the lower **Links to your main web pages** check box to remove the check mark, if necessary.

Step 19. Click the remaining check boxes to insert a check mark, if necessary. Your dialog box should look similar to Figure A-10.

Step 20. Click **Next**.

The ninth wizard dialog box opens. A published Web site is one that is uploaded and stored on a Web server. Some Web authors add incomplete Web pages to a published Web site but mark them with some standardized text and an icon to indicate that the Web pages are "under construction." Other Web authors think that only completed Web pages should be added to a published Web site. For your sample Web site, you will not include incomplete Web pages. To turn off the "under construction" icon:

Step 21. Click the **No** option button. Your dialog box should look similar to Figure A-11.

Step 22. Click **Next**.

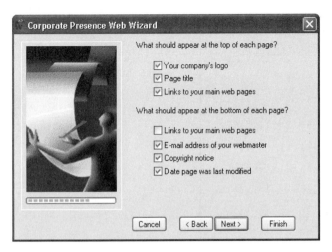

FIGURE A-10   Eighth wizard dialog box

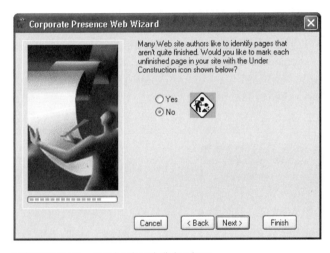

FIGURE A-11   Ninth wizard dialog box

The tenth wizard dialog box opens. In this dialog box, you add your company's name and address information. Choose a fictitious name and address for your small business. To add this company information:

Step 23. Type name and address information in the appropriate text boxes. Your dialog box should look similar to Figure A-12.

Step 24. Click **Next**.

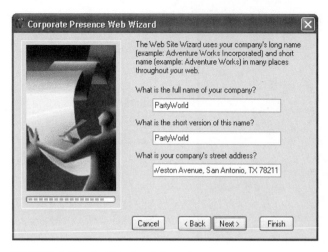

FIGURE A-12 Tenth wizard dialog box

The eleventh wizard dialog box opens. In this dialog box, you can add additional contact information for your company. For your sample Web site, create the necessary fictitious telephone and fax numbers and e-mail addresses. To add the information:

Step 25. Type phone numbers and e-mail addresses in the appropriate text boxes. Your dialog box should look similar to Figure A-13.

Step 26. Click **Next**.

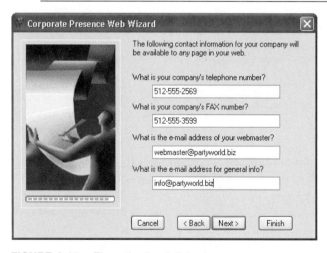

FIGURE A-13 Eleventh wizard dialog box

The twelfth wizard dialog box opens. At this point, you are almost finished with the wizard steps. FrontPage can automatically create a "to do" list of tasks that help you complete your Web pages. To allow FrontPage to create these task items and then to finish the wizard:

Step 27. Click the **Show Tasks View after Web site is uploaded** check box to add the check mark, if necessary. Your dialog box should look similar to Figure A-14.

Step 28. Click **Finish**.

You have just created a FrontPage Web site folder titled "Sample_Web" for a Web site that contains four pages. The four pages include the home page (index.htm), the feedback page (feedback.htm), and two products pages (prod01.htm and products.htm). You are now ready to view the site's structure, folders, and pages in different FrontPage views.

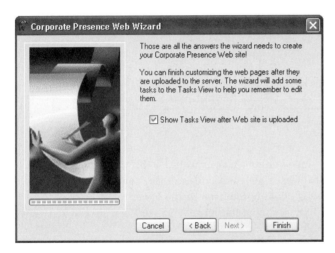

FIGURE A-14   Twelfth wizard dialog box

VIEWING A WEB SITE IN DIFFERENT FRONTPAGE VIEWS

Creating, editing, and maintaining a FrontPage Web site consists of many related tasks. To help manage a Web site effectively, you can create an electronic "to do" list of outstanding items to be completed. When completing some actions, FrontPage can automatically add a task to the task list for you. For example, when checking spelling in a FrontPage Web site, you can have FrontPage create a task for each page that contains misspelled words.

You have already learned about using FrontPage's Page view; FrontPage provides even more ways to view and manage a Web site. For example, when you finish the wizard steps described in the previous section, the Sample_Web site is first displayed in Tasks view. Tasks view allows you to see and edit or create reminders for the Sample_Web "to do" list. You can also manually add tasks to the list. Figure A-15 shows the tasks that were added during the wizard process to the "to do" list for the Sample_Web site.

Creating a Web Site with Microsoft FrontPage

Electronic task or "to do" list created for
Sample_Web during the wizard process

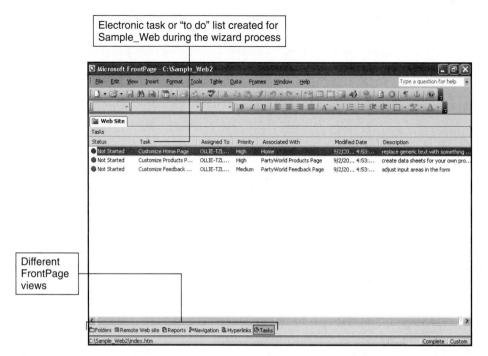

Different
FrontPage
views

FIGURE A-15 Tasks view

In addition to using Tasks view, you can manage your Web site through other views such as Folders view, Remote Web site view, Reports view, Navigation view, and Hyperlinks view. You can switch to these other views by clicking a button in the lower-left corner of the FrontPage window above the status bar, as shown in Figure A-15. You can also switch views by clicking a command on the View menu.

To manage the folders and files in a FrontPage Web site folder easily, you must be able to see them. The Sample_Web site folder contains additional subfolders created by FrontPage—for example, the _private subfolder and the images subfolder. The images subfolder can be used to store any picture, sound, or video files that will be featured at the Sample_Web site. Storing these files in the images subfolder reduces the clutter in the primary Web site folder. The _private subfolder and other subfolders contain files that FrontPage uses to manage your Web site and should be left alone. The Sample_Web in Folders view is shown in Figure A-16.

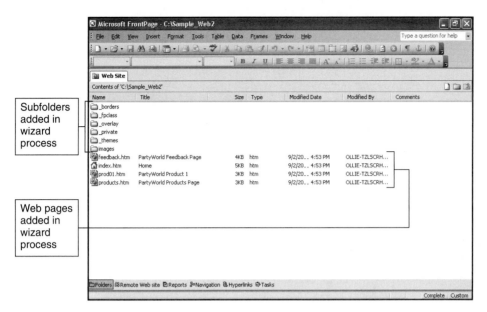

Subfolders added in wizard process

Web pages added in wizard process

FIGURE A-16 Folders view

You can use Remote Web site view to work with pages that are stored in a FrontPage Web site folder on a remote server. Reports view, as shown in Figure A-17, provides information about the status of a Web site. For example, it summarizes the number of files that are in the current Web site, and it indicates if the Web site contains broken hyperlinks. A broken hyperlink is one that no longer connects to a valid location.

Summary information about
current status of Sample_Web

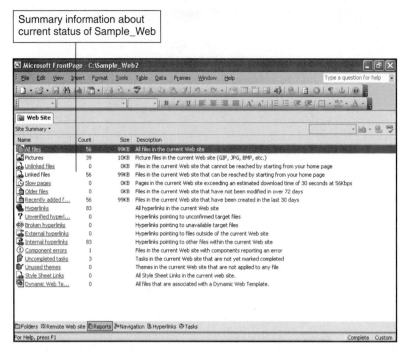

FIGURE A-17 Reports view

A FrontPage Web site consists of interrelated pages that are organized in a hierarchical structure. The home page appears at the top of the structure, and the related pages branch out at different levels below the home page. You use Navigation view, as illustrated in Figure A-18, to review the Web site's organizational structure. You can modify that structure in Navigation view by adding or deleting pages, or moving pages to a new position in the structure.

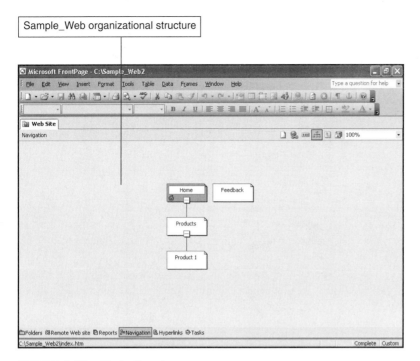

FIGURE A-18 Navigation view

Hyperlinks view, as shown in Figure A-19, depicts the linking relationships between the Web pages that were created during the wizard process.

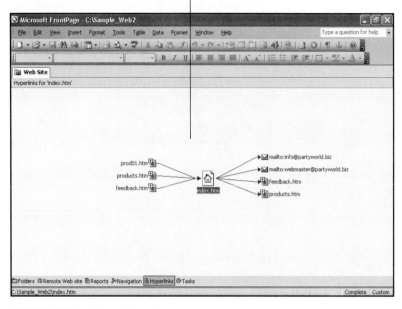

Linking relationships between home page (index.htm) and other pages in the Sample_Web site

FIGURE A-19   Hyperlinks view

To see the Sample_Web site in Reports, Navigation, Hyperlinks, Tasks, and Folders views, and then to view the Web site's home page (index.htm) in Page view (and the Design view sub view):

Step 1.   Click the **Reports** button.
Step 2.   Click the **Navigation** button.
Step 3.   Click the **Hyperlinks** button.
Step 4.   Click the **Tasks** button.
Step 5.   Click the **Folders** button.
Step 6.   Double-click the **index.htm** file name in Folders view to open the home page in Page view.

The incomplete Sample_Web home page is shown in Figure A-20.

The pages in the Sample_Web site contain partial navigation bars, sets of text or button hyperlinks to other pages in the Web site, placeholders for text and graphics, and comments that indicate where you should type certain text, such as the mission statement. A default color-coordinated visual theme, such as those you learned about in Chapter 9, was automatically applied during the wizard process.

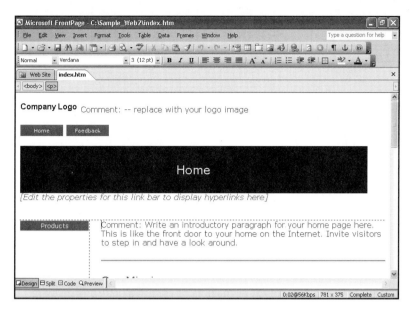

FIGURE A-20   Incomplete Sample_Web home page

At this point, only the basic framework for a Web site and its pages has been created. To create a complete Web site, you would edit the individual pages to include the appropriate text, graphics, and links. As you do this, you might find it necessary to add new Web pages. After you have edited existing pages and added new ones, you would then create the final hierarchical navigational structure in Navigation view. As you can see, using a FrontPage wizard to begin building a Web site is simply the first step in the process of creating a Web site.

CLOSING THE WEB SITE FOLDER AND THE FRONTPAGE PROGRAM

After you create and edit a FrontPage Web site, your next step is to publish the site's folder and pages to a Web server that others can access to view your Web pages. Your instructor may provide additional information on how to publish your sample FrontPage Web site folder and pages to a Web server.

When you are finished with the FrontPage program, you should close both the open Web site folder and the program. If you do not close the open Web site folder, the next time FrontPage starts, the same Web folder will reopen. To close the Sample_Web FrontPage Web site folder and the FrontPage program:

Step 1.   Click **File** on the menu bar.

Step 2.   Click **Close Site** to close the Sample_Web site folder.

Step 3.   Click **File** on the menu bar.

Step 4.   Click **Exit** to close the FrontPage program.

---

accessibility The process of designing Web pages so that Web resources are available to people with disabilities.

accredited investor An individual investor with a minimum net worth of $1 million, or an individual income of at least $200,000 per year, or a household income of $300,000 per year.

accredited registrars ICANN-approved private companies that register Internet domain names.

ACH check See *electronic check*.

Active Server Page (ASP) Technologies developed by Microsoft Corporation to embed server-side controls, such as JScript or ActiveX programs, in a Web page.

ActiveX controls Small programs, similar to Java applets, that are used to provide Web page interactivity primarily for pages viewed in the Microsoft Internet Explorer Web browser.

ActiveX technologies Object-oriented technologies developed by the Microsoft Corporation to compete with the Java programming language.

advertiser See *merchant*.

Advisory Board A group of outside advisors who add credibility to a startup e-business by supplying the experience and business background that might be missing from the startup business's management team.

affiliate An e-business that participates in an affiliate program.

affiliate agreement The formal agreement between a merchant and affiliate that specifies an affiliate program's terms, such as referral fees or commission rate, payment terms, requirements, and restrictions.

affiliate management network A third-party entity that recruits affiliates, manages the registration process, tracks and properly credits all of the referral fees and commissions, and arranges for payment.

affiliate marketing A revenue-sharing approach to marketing that involves paying other e-businesses (affiliates) to promote a merchant e-business's Web site and the products and services it offers.

affiliate program A marketing tool in which an e-business pays another business a referral fee or commission on sales, or other actions made by customers who click through from the other business's Web site. Also known as pay-for-performance program.

affiliate tracking system A process that allows a merchant to control how its affiliates are credited, whether per click, lead, or sale; monitors the window of time in which an affiliate gets credit when a visitor clicks through to the merchant site and takes an action; and records and stores affiliate information, and provides commission or fee reports.

angel investment club A group of individual angels who come together to identify new investment opportunities and sometimes combine their individual investments with other club members to spread the investment risk.

angel investor   Wealthy individual investors who enjoy investing in new business ideas.

applet   A small Java program embedded in a Web page that executes when the page is loaded in a visitor's Web browser.

application service providers (ASPs)   Companies that deliver and manage software applications and other computer services from remote data centers serving multiple customers.

associate   See *affiliate*.

associate program   See *affiliate program*.

authentication   The process of identifying an individual or e-retailer usually based on a combination of username and password.

B2B   An e-business model in which the e-business sells its products or services to other businesses or brings multiple buyers and sellers together.

B2B exchanges   B2B e-businesses that bring multiple business buyers and sellers together in a central marketspace.

B2C   An e-business model in which the e-business sells its products or services directly to the consumer.

B2G   An e-business model in which the e-business sells its products or services to local, state, and federal government agencies.

back-end systems   Those business processes such as accounting, inventory control, marketing, and order fulfillment that are not directly accessed by customers.

banner ads   Rectangular images, or banners, that appear fixed in place on a Web page and that usually link to an advertiser's Web site.

benchmarks   Established reference points used to monitor a computer network's operations; performance-based objectives.

bitmap graphics   Graphics created by defining the color of individual pixels in a bitmap.

black-hat hacker   A malicious hacker who accesses network resources in an attempt to steal valuable information (such as credit card numbers) or disrupt service. Also known as a cracker.

blogs   Online diaries; also called weblogs.

bootstrapping   A method for self-funding a startup business that avoids borrowing money and raising equity financing by using unique and inventive ways to acquire resources.

bots   See *spiders*.

brand   A combination of name, logo, and design that identifies a business's products or services and differentiates its products and services from those of its competitors.

breadcrumb trail   A horizontal list that appears on a Web page and shows all the levels of links between the page currently being viewed and the Web site's home page (or another major page); it can provide a site's visitors with feedback on where they are at a site and how they got there.

brick-and-click   Business that sells products or services from a physical location and on the Internet.

brick-and-mortar   Business that sells products or services from a physical building.

burn rate   The rate at which a startup business is using its cash reserves.

business accelerators   See *commercial business incubators*.

business continuity plan (BCP)   A plan that specifies how a business will resume partial or complete operations after a major disruption, such as a loss of a major vendor, serious failure of products or services, management malfeasance, a natural disaster, or a terrorist attack.

business description The section of a business plan that provides the reader with a summary of a business's background and business idea, the legal form of the business, when and where the business was formed, its history to date, key personnel, and future goals.

business incubators Traditionally, a non-profit organization, such as a government agency or university, that nurtures new businesses by providing access to investors, technology, and expertise.

business plan A formal business planning document that identifies the business and its mission, names the key players on the management team, describes the products or services to be offered, provides an analysis of the current marketplace in which the business operates, identifies customers and competitors, determines the resources necessary for profitable operations, and sets a timetable for profitability.

"C" corporation A common type of business organization in which the business's owners are its shareholders.

C2B An e-business model in which consumers name a price for a product or service and the targeted business accepts or declines that price.

C2C An e-business model in which consumers sell items or services directly to other consumers.

cardholder not present The risk of fraud associated with credit card payment transactions in which the card and the cardholder are not present at the point-of-sale, for example with online credit card payments.

cascading style sheets (CSS) A Web page development tool that contains the rules or codes that define style issues—fonts, color, and item positioning—for all the pages at a Web site.

cash cow A business that continues operations in order to generate cash.

catalog merchants Businesses that supplement a successful traditional mail-order business with an online shopping site, or move their catalog sales to an online store.

certificate authority A trusted third-party organization that guarantees the identity of the sender and issues digital certificates.

charge card A rectangular piece of plastic used instead of cash or checks to pay for goods and services; cardholders must pay the card issuer in full each month upon receipt of a statement of charges.

chargeback A consumer's refusal to pay a charge on his or her credit card account resulting from returned products, billing errors, non-delivery of product, or an unauthorized charge.

click fraud The act of creating fraudulent pay-per-click data; phony clicks may be generated by repeatedly clicking competitors' ads to run up advertising expenses or by billing advertisers for click-throughs that were not really made.

clickable table of contents A section near the top of a Web page that provides text links to individual sub-topics throughout a Web page.

click-through rate The percentage of viewers who click on a banner ad to view the advertised Web site.

co-locate To rent space and Internet connectivity for an e-business's servers.

commercial business incubators A for-profit business incubator.

Common Gateway Interface (CGI) A standard or set of rules that determines two-way communication between a Web server and a Web browser.

content management system (CMS) A type of business management software that controls all the processes involved in Web content development, including authoring, reviewing, editing, and then publishing the content in a timely way.

content repositories Content management system (CMS) databases that contain Web page templates and style sheets, commonly used graphics, text documents, and syndicated content used to build Web pages "on the fly."

cookie stuffing An unethical act committed by an affiliate intending to defraud merchants into paying unearned affiliate fees or commissions; it involves placing multiple cookies containing the affiliate's commission codes on an unsuspecting visitor's computer during a single visit to an affiliate's Web site.

cookies Small text files stored on the hard drives of Web visitors and used to gather information about visitors, store passwords and Web site customization options, and so forth.

copyright A form of legal protection for the author of a published or unpublished original work.

corporation A taxable entity owned by shareholders who are protected from liability.

cost-per-click affiliate program See *pay-per-click affiliate program*.

cost-per-lead affiliate program See *pay-per-lead affiliate program*.

cost-per-sale affiliate program See *pay-per-sale affiliate program*.

cover sheet The top page of a business plan; it usually includes the title of the document, the preparer's name, the plan copy number, and a "Confidential" notation.

cracker See *black-hat hacker*.

crawlers See *spiders*.

credit card A rectangular piece of plastic used instead of cash or checks to pay for goods and services; cardholders pay the card issuer in full or in part upon receipt of a monthly statement.

customer relationship management (CRM) A type of e-business management software that uses the Internet to help a business manage its customer base.

debit card A rectangular piece of plastic used instead of cash or checks to pay for goods and services; cards are issued in connection to an existing bank account and charges are immediately deducted from that bank account.

digital cash See *electronic cash*.

digital certificate An electronic message attachment that verifies the sender's identity.

digital wallet See *electronic wallet*.

directory A Web search tool whose index is created using manual submissions of company information and Web site URLs.

disaster recovery planning The portion of a business continuity plan concerned with the technical aspects of recovering computer systems and/or other systems after a catastrophic event.

discount rate A merchant account fee charged to a business for each credit, debit, or charge transaction processed.

distributed denial of service (DDoS) An attack on a computer network that involves multiple computers and designed to disable the network by flooding it with useless traffic.

domain name An address that identifies a Web site on the Internet, and also recognizes the server that handles Web browser requests, the specific organization associated with the domain name, and the general category in which the organization operates.

double opt-in A type of opt-in e-mail marketing in which a confirmation e-mail message is sent to the consumer who signs up; the consumer's e-mail address is not added to the e-business's database until he or she responds to this confirmation e-mail message.

Dreamweaver A professional-grade WYSIWYG Web authoring tool developed by Macromedia.

Dynamic HTML (DHTML)   A combination of technologies including HTML tags, style sheets, and a scripting language (for example, JavaScript) used together to add animations and interactive features that appear when a Web page loads in a Web browser.

e-business   A broad spectrum of business activities conducted on the Internet, such as buying and selling products or services, providing customer service and support, collaborating with business partners, and enhancing internal productivity.

e-cash   See *electronic cash*.

e-check   See *electronic check*.

e-commerce   See *electronic commerce*.

electronic cash   A method of transmitting a unique electronic number or other identifier that carries a specific value to pay for goods or services purchased online.

electronic check   An electronic version of a paper check.

electronic commerce   The process of buying and selling products or services across a telecommunications network; see *e-business*.

electronic data exchange (EDI)   The process of exchanging electronic data between trading partners over a private telecommunications network.

electronic funds transfer (EFT)   The process of exchanging funds electronically between banks instead of exchanging paper money.

electronic wallet   Encryption software that stores a user's payment information much like a physical wallet and may reside on the user's PC, the card issuer's server, or an e-business's server.

elevator pitch   A short (two- or three-minute) pitch for a new business idea that can be used to raise the interest of a potential investor when time is a premium.

embedded text link   A link positioned inside a text paragraph.

encryption   The translation of data into a secret code.

enterprise resource planning (ERP)   A type of e-business management software that integrates all aspects of a business, including planning, manufacturing, human resources, accounting, finance, sales, and marketing.

entrepreneur   Someone who assumes the risks associated with starting and running his or her own business.

entrepreneurial abilities   Specific abilities needed by a person who wishes to start and operate a business, such as leadership, high-energy personality, self-confidence, organizational skills, and the ability to act quickly and decisively.

entrepreneurial process   A series of steps by which a person determines whether he or she has the abilities to be an entrepreneur and then decides whether it's best to start a new business or purchase an existing one. The steps are as follows: (1) decide if someone is an entrepreneur; (2) if yes, decide to purchase or start a business; (3) if starting a business, then define the business idea, create a business plan, and secure financing; (4) if purchasing or starting a business, operate that business; and (5) harvesting the business.

entry page   See *splash page*.

e-retail   E-business based on the B2C e-business model.

e-retailers   Retailers who participate in B2C retail e-business by maintaining online stores.

ethical hacker   See *white-hat hacker*.

executive summary   The third or fourth page of a business plan; it provides a brief overview of the entire business plan.

exit strategy   A section of a business plan that describes the entrepreneur's long-term plans for a new business and how potential investors will recover their investment.

Extensible Hypertext Markup Language (XHTML) A markup language created through the reformulation of HTML 4.01 into XML and designed to offer more strict coding rules to help eliminate coding errors and omissions, better structured documents that display in a Web browser more quickly, the flexibility of creating custom tags, and more control over how Web pages are viewed by wireless devices and screen readers.

Extensible Markup Language (XML) A markup language that is a streamlined subset of SGML that was developed in 1996 by the XML Working Group of the World Wide Web Consortium (W3C) and is used to describe what data is rather than how data looks.

external link A link that directs a visitor outside the Web site the visitor is currently viewing.

extranet A private network consisting of two or more intranets that are connected via the Internet to allow participating business partners to view each other's data and to complete business transactions.

financial plan The section of a business plan that explains how a business idea, as well as goals and strategies associated with this idea, translate into profits.

firewall Software or hardware used to isolate a private computer or computer network from the public network.

Fireworks Multimedia software developed by Macromedia to draw vector graphics, edit bitmap graphics, and create animated GIF files.

first-mover advantage An advantage inherent in being the first business of one's kind in the marketplace.

Flash Multimedia software that was originally developed by Macromedia to provide animation for vector graphics but now is often used to create Web page ads or online product tutorials.

floating ads See *Shoshkele ads*.

forms A common element of a Web page that is used to gather information from visitors; an electronic version of a paper form, with text boxes, check boxes, option buttons, and drop-down lists.

forward auctions Online marketspace in which many buyers bid on products or services offered by a single seller.

frames Sections of a Web browser's display area that can contain different Web pages.

FreeHand A design tool developed by Macromedia to plan the layout for a printable or online document that contains graphics, such as a brochure, product catalog, or a Web page.

front-end systems Those business processes, such as viewing information and ordering products, that customers can interface with and control.

FrontPage An easy-to-use WYSIWYG Web authoring tool developed by Microsoft Corporation.

fulfillment house An independent company that warehouses, picks, packages, and ships products for other businesses.

general partnership A legal business entity that consists of multiple co-owners of a for-profit business operating under a partnership agreement.

GIF images A compressed image file format used for simpler Web images and animated images.

going public The act of issuing a public stock offering or IPO.

hacker Originally a slang term for a gifted software programmer; now a slang term for someone who deliberately gains unauthorized access to individual computers or computer networks.

hierarchical structure   A type of Web site structure that presents carefully organized information in different levels, beginning with a top level (general information) followed by multiple levels of increasingly more detailed information.

home page   The primary page at a Web site.

hosted storefront software   Software that can provide everything a small e-retailer needs to build an online store: Web site hosting services, easy-to-use templates to build online store Web pages (such as product catalogs), and shopping cart software.

hyperlinks   Text or pictures appearing on a Web page that are associated with the location (path and filename) of another Web page.

hypertext   An organization scheme in which text on one page links or connects to text on another page.

Hypertext Markup Language (HTML)   An easy-to-use markup language that was originally developed by Tim Berners-Lee, the inventor of the World Wide Web, to allow documents with headings, text, tables, bulleted or numbered lists, images, links, and other elements to be published on the Web and displayed properly in a Web browser.

impressions   The number of times an online ad is viewed.

industry   Businesses that make or sell similar, complementary, or supplementary products or services.

informal investors   The network of family members, friends, and angel investors who invest in a new business.

internal link   A hyperlink between two Web pages at the same Web site.

International Organization for Standardization (ISO)   An international body of worldwide organizations that set standards for numerous industry and government products and services.

Internet   A worldwide public network that connects many smaller private networks.

Internet accelerators   Commercial business incubators that focus on startup e-businesses.

Internet service provider (ISP)   An e-business that provides access to the Internet for a fee.

intranet   An internal company network that uses the Internet and Web technologies to allow employees to view and use internal Web sites not accessible by the outside world.

intrusion detection system   A system that analyzes real-time data to detect, log, and stop unauthorized computer network access.

issues analysis   A portion of a business plan that identifies the threats or opportunities a new business faces.

Java   An object-oriented programming language that was developed by Sun Microsystems and is used to create applets and servlets.

JavaScript   A scripting language that was developed originally by Netscape Corporation and is used to add interactivity and dynamic content to Web pages.

JPEG images   A compressed image file format used for complex images such as photographs.

JScript   A scripting language that was developed by Microsoft Corporation and functions like JavaScript; it is used to add interactivity and dynamic content to Web pages.

legacy systems   Existing computer systems that must be integrated with new e-business systems.

limited liability company (LLC)   A business organization that combines the limited liability feature of a corporation with the tax status of a partnership or sole proprietorship.

limited partners   Participants in a limited partnership.

limited partnership A partnership in which the general partners assume management responsibilities and unlimited liability for the partnership and other partners have no management responsibility and are legally liable for their capital contribution.

linear structure A type of Web site structure in which visitors typically view a series of pages in sequential order.

link exchanges A marketing tool in which e-businesses exchange links with complementary Web sites.

link farm A network of linked Web sites whose sole purpose is to boost link popularity for the linked sites.

link popularity A measure of the number of quality incoming links to a Web site.

link stuffing See *link farm*.

liquid design A Web page development technique using tables or cascading style sheets that allows pages to automatically resize to fit the size of the browser window in which they are being viewed.

market research The process of collecting and analyzing the data needed to make informed decisions about how to go about selling a business's products or services in a specific marketplace.

market segment A subgroup of an overall market that can be distinguished according to a number of factors, including geographic region, demographic characteristics of consumers (such as age, gender, education, family size, and income), psychographic characteristics of consumers/regions (such as buying behavior and price sensitivity), and so forth.

market size A measure of the number of potential consumers, consumer purchasing power, or projected sales volume of a specific market or market segment.

marketing The process of developing the mutually satisfying relationships between a business and its customers that result in sales and profits.

marketing budget The portion of a business plan that estimates the costs for all marketing strategies.

marketing mix A concept that identifies those elements of marketing strategies and tactics over which a business has control: product, place, promotion, and price (also known as the Four Ps).

marketing plan A component of a business plan that establishes, directs, and coordinates marketing efforts for a specific period of time.

marketing strategies The section of a marketing plan that describes how the products and services will be priced, promoted, and distributed.

marketplace analysis The section of a business plan that includes information about the specific industry that the business is a part of plus a description of the business's targeted customers and competitors.

marketspace An electronic marketplace where buyers and sellers come together to conduct e-business.

markup language A set of rules and instructions embedded in an electronic document that can describe a document's data and specify how document elements, such as its text and images, are handled by computers.

mass mailing worm A type of replicating virus that hijacks the e-mail address book on an infected computer and then sends itself to all the e-mail addresses in the book; the process is repeated on each newly infected computer.

Mathematical Markup Language (MathML) A markup language that is used to structure mathematical expressions on a Web page.

merchant An e-business that operates an affiliate program.

merchant account An account at a financial institution set up by a business to process credit, debit, and charge card payments from its customers.

merchant account provider A business that partners with financial institutions and payment gateway services in order to provide the entire payment card processing package: a merchant account, payment processing software and equipment, and a connection to a payment gateway.

meta search engine A Web search tool that allows users to search the indexes of multiple search tools at one time.

meta tags Small segments of HTML code that add information to a Web page that only a Web browser can see; the information is not visible on a Web page but can be read by search engines.

micropayments Electronic payments for online purchases ranging from a few cents to $5.

middleware Software designed to enable a business to connect two different computer systems, for example between the business's existing (legacy) systems and its new e-business systems.

mission statement A statement of challenging but achievable actions a business takes to realize its vision.

mixed hierarchical structure The most commonly used Web site structure; it allows for cross-linked pages within a hierarchy.

name identification The advantage a business enjoys when its name is not only readily recognized by potential customers, but identified with the business's specific products or services.

navigation bar A series of icon- or text-based internal hyperlinks to major pages at a Web site.

navigation menus A drop-down or expandable list of internal links, similar to a list of commands from a software application menu.

navigation tabs Navigational elements on a Web that are similar to file folder tabs and represent links to additional resources on a Web page.

network A group of two or more computers linked by cable, telephone lines, or other wired or wireless media.

network effect The increasing value of a network to each participant as the number of total participants increases.

newsgroup A topic-specific electronic "bulletin board" accessed over the Internet. Newsgroup participants submit messages and respond to others' messages using a Web browser or e-mail software plug-in called a newsreader.

non-profit business incubators See *business incubators*.

objectives Goals that are specific, attainable, measurable, realistic, and time-specific.

operational plan The section of a business plan that describes the business's location as well as the equipment, labor, and other processes (such as Web site operation) required to sell the business's products or services.

opt-in e-mail advertising The process of getting consumers to voluntarily learn more about a company and its products and services that is based on sending e-mail advertising messages or e-mail newsletters to potential customers who have agreed in advance to receive them.

P2P payment systems Internet payment systems that allow a sender to transmit a given quantity of funds from a bank account or credit card to a receiver via e-mail; the receiver can then add these funds into his or her bank account or credit card.

page tagging   A technique for recording Web site visitor behaviors by adding to Web pages hidden JavaScript images or tags that store data about visitor behaviors in a database.

paid-inclusion program   A search tool program in which an e-business is guaranteed accelerated inclusion of its Web pages in a search tool's index for a fee.

parasiteware   A type of spyware that redirects affiliate links on a user's computer or replaces the content of the user's existing affiliate-tracking cookies.

partnership   A form of business organization in which two or more owners of a business operate under a partnership agreement.

password   A code used to gain authorized access to a computer network.

pay-for-performance program   See *affiliate program*.

payment gateway   A secure online service that submits payment card transactions to the merchant bank's card processing network and notifies the e-business's customer if the payment transaction is approved or declined.

pay-per-click   An online advertising program in which an e-business pays for online ads only when a viewer clicks through to its Web site via the ad.

pay-per-click affiliate program   An affiliate marketing program in which the merchant pays the affiliate a set fee each time a visitor clicks through to the merchant's site regardless of whether or not the visitor takes any action at the site.

pay-per-lead affiliate program   An affiliate marketing program in which the merchant pays the affiliate a set fee for each visitor who clicks through and takes an action at the merchant's site, such as completing an online survey, registering at the site, or opting-in to receive e-mail.

pay-per-sale affiliate program   An affiliate marketing program in which the merchant pays the affiliate a percentage of the sale when a visitor clicks through to the merchant's site and makes a purchase.

penetration testing   Using real-world hacking tools to test computer security.

performance measures   The element of a marketing plan that describes the ways in which a business will measure the success of its marketing strategies and tactics.

permission marketing   The process of getting consumers to voluntarily learn more about a company and its products and services.

Photoshop   Image-editing software that was developed by Adobe Systems and is used to create animated GIFs, draw and edit vector graphics, and edit bitmap graphics.

PHP: Hypertext Preprocessor   A scripting language that can be embedded in HTML and used to build Web pages "on the fly" with elements (text, images, audio, and video) stored in databases.

pitch document   A short marketing document based on a business plan's executive summary and used to market a startup business to potential investors.

planned cash flow statement   A statement of expected cash flows, describing both the cash coming in to the business and how that cash is spent.

podcasting   The use of RSS to access audio files such as music or radio programs.

pop-under ads   An online ad that opens in its own window on the desktop underneath a Web page.

pop-up ads   An online ad that opens in its own window on top of a Web page.

press release   A short announcement of a newsworthy item that is sent to members of the press.

primary records   Records that provide supporting documentation for key e-business activities such as sales, order fulfillment, and payment.

primary research   The process of physically collecting marketplace, consumer, and competitor data; organizing and manipulating it; and then analyzing and publishing the results.

private placement memorandum   Discloses the benefits and risks of an investment to potential private investors.

pro forma balance sheet   A financial statement that provides a "snapshot" of a business's assets, liabilities, and capital at a specific point in time.

products section   The portion of a business plan that describes the products or services the business will offer, including details about what the products or services are, what they do, and how they benefit the customer.

projected income statement   A financial statement that shows estimates of revenues, expenses, taxes, and net profits or losses over a specific period of time.

protocol   A standard or agreed-upon format for electronically transmitting data.

public relations (PR)   The process of establishing and maintaining a business's public image.

publisher   See *affiliate*.

pure-play e-retailers   E-businesses that offer traditional or Web-specific products or services only over the Internet. Also known as virtual merchants.

qualitative research   The process of collecting data that requires informed interpretation and cannot be analyzed using statistical methods.

quantitative research   The process of collecting data that can be analyzed using statistical methods.

raster graphics   See *bitmap graphics*.

records management   The planning, processes, and actions a business must implement to make certain its primary and secondary records (whether paper or electronic) are safely retained for an appropriate period of time, guarded against unauthorized access, and then destroyed when—and only when—they are no longer needed.

return on investment (ROI)   The benefit an e-business gets in return for the investment it makes in its Web site.

reverse auctions   Online marketspace where a single buyer offers to purchase products and services from multiple competing sellers.

rich media   Online ads that contain interactive elements, audio, and video; for example, streaming media or Shoshkele ads.

risk management   A process through which a business identifies a risk of a business loss; assesses the risk's potential impact on the business; and then determines how to handle the risk—that is, whether to avoid it, minimize it, retain all or part of it, and/or transfer all or part of it to someone else.

rollover links   Animated graphics with links that appear or disappear as a visitor moves the mouse pointer over the graphic.

RSS (Really Simple Syndication)   An XML technology used to publish blogs and to syndicate other types of content (such as headline news stories) from Web sites.

"S" corporation   A common type of business organization that gives the business's owners partnership tax status and corporate liability protection.

scalability   The ability of a business to continue to function well regardless of how large the business becomes; also the ability of a server to handle increased traffic loads without crashing.

search engine optimization (SEO) The art of building Web pages and Web site links to maximize the chances of the pages getting positioned at or near the top of as many search results lists as possible.

search engines Web search tools whose index is created automatically by software programs, called spiders, that browse the Web looking for new pages.

search function A tool that allows site visitors to perform a keyword search for products, services, or other information at a Web site.

search tools Search engines, meta search engines, and directories used to locate resources on the Internet and Web.

secondary records Records containing important information not used in daily operations but generated from an e-business's activities, usually consisting of logs of customer data and Web site activity.

secondary research The process of collecting data through secondary sources, such as market research companies, who perform the primary research.

Secure Sockets Layer (SSL) A security protocol that provides server-side encrypted transactions for electronic payments or other secure Internet communications.

security audit A professional assessment of a business and its systems' security that involves checking the systems for vulnerabilities and providing recommendations for fixing them.

self-incubation Startup businesses that share office space, ideas, and a network of advisors during the early stages of their business development.

server log files A record of all the events (server requests) taking place on a server.

servers Special computers that provide users access to shared resources such as files, programs, printers, and communications lines.

service marks Distinctive symbols, words, or phrases used to identify a business's services and distinguish them from other businesses' services.

services section See *products section*.

servlet A small Java program that is similar to an applet but is executed at the server, before the Web page is loaded in a browser.

shopping cart software Software that manages an online customer's purchases by temporarily storing information about the items selected for purchase and then handling the complete checkout process, including shipping and tax calculations, credit card authorization, and payment processing.

Shoshkele ads Online ads that appear to float across a Web page for a few seconds then settle in a specific area of the page or automatically close.

sidebar ad A vertical online ad that is typically 600 pixels high by 120 pixels wide; also called a skyscraper ad.

site index See *site map*.

site map A Web page that shows a summary of all the linked pages at a Web site and depicts how those pages are categorized in the site's organizational structure.

smart card A small electronic device approximately the size of a credit card that contains electronic memory and is used for a variety of purposes, including storing medical information, network identification, and electronic cash.

sole proprietorship A form of business organization in which the business and the owner are one and the same for tax and liability purposes.

spiders Software programs that browse Web pages to automatically update search engine indexes with new URLs.

splash page A Web page that usually contains big, flashy graphics and perhaps sounds used to create a showy entrance to a Web site.

spoofing The act of creating a fake Web site and then redirecting traffic from the legitimate Web site to the fake Web site for criminal purposes.

spyware A general term used to describe software that has been installed on a personal computer without the owner's permission.

staging server A temporary server used to host Web pages that need to be tested before they can be published to a final destination server.

Standard Generalized Markup Language (SGML) The ISO standard (ISO 8879:1986) that defines a set of rules for encoding documents so that the documents are transportable between computer systems.

stored value card A smart card that stores cash; examples include prepaid telephone cards, government benefit cards used to purchase food, and gift cards.

storefront software Software that enables an e-business to create an online store through which it can accept order and payment information and then process that information.

storyboard A blueprint commonly used in movie and television production to design (or depict) the important points of a narrative; it is also used in Web site development as an organizational tool in designing the structure of a Web site.

strategy A plan of action to accomplish a goal.

streaming media ads Online ads that use streaming audio and video.

subscription model A business model followed by e-businesses that provide high-value content for a subscription fee.

sweat equity The time, effort, and personal money an entrepreneur spends to get a new business started.

Synchronized Multimedia Integration Language (SMIL) A markup language used to synchronize multimedia elements on a Web page.

table A Web page layout element consisting of columns and rows within which text or data or images can be positioned.

table of contents The third or fourth page of an business plan; it provides page number references for the major sections and subsections of the plan.

tactics Specific actions taken to follow strategies and accomplish goals.

tags Markup language codes or instructions enclosed in angle brackets, for example, <title> and </title>.

target market(s) A group of potential customers that share a common set of traits that set the group apart from other groups.

term sheet A list of the major points of proposed financing being offered by an investor.

text editor Software, such as Notepad, that is used to create text documents but lacks the special document-creating and formatting features found in a word processing program.

Three-Click Rule A rule of thumb that suggests that Web site should be designed such that site visitors are able to find useful information or make a purchase in no more than three clicks from the home page.

thumbnail A small version of an image that downloads in a Web page in place of the larger image to reduce download time; clicking a thumbnail image allows a visitor to view the full-sized image.

title page The second page of an business plan; it repeats information from the cover sheet and adds the preparers' contact numbers and the name of the person receiving this particular copy of the plan.

top-of-page link Text or graphic image that is linked to a position at the top of the currently viewed Web page and thus allows viewers to quickly return to the top of the page.

trademark A distinctive symbol, word, or phrase used to identify a business's products and distinguish them from other businesses' products.

trading partners Business partners that exchange data electronically over a private telecommunications network.

Trojan horse A special type of virus that emulates a benign program by appearing to do something useful or entertaining.

Uniform Resource Locator (URL) See *domain name*.

usability Web design guidelines created to help all Web site visitors accomplish their goals quickly and easily.

value activities The primary and support activities that make up a company's value chain.

value chain All the primary and support activities performed to create and distribute a company's goods and services.

value-added networks (VANs) Private telecommunications networks over which business or trading partners exchange data electronically.

vector graphics Graphics created by drawing lines, curves, and polygons.

venture capitalist (VC) Investors that raise hundreds of millions of dollars from other organizations such as endowments, insurance companies, and pension funds to fund new businesses.

vertical market A specific industry in which similar products or services are developed and sold using similar methods.

viral marketing Marketing a business's products or services by electronic word of mouth; a marketing approach that exploits the network effect to spread information about a business's products or services rapidly.

virtual credit card numbers Disposable credit card numbers that can be used only for a specific period of time or only at the original point-of-sale.

virtual malls B2C e-businesses that host many different online e-retailers.

virtual merchants See *pure play e-retailers*.

virus A small program capable of harmful actions that inserts itself into other program files; these files then become "infected" in the same way a virus in nature embeds itself in normal cells and spreads when an infected program executes and infects other programs.

vision statement A business's statement of long-term dreams and goals.

Web analytics Various measurements of Web site visitor behaviors, such as page views or conversion rate.

Web authoring software Software that provides Web developers a "What You See Is What You Get," or WYSIWYG, environment in which Web pages are created much the same way documents are created in word processing software.

Web browser Software program used to access and view Web pages.

Web hosting companies Companies that host commercial Web sites from Internet data centers.

Web metrics See *Web analytics*.

Web pages Linked documents that can contain text, graphics, video, audio, and hyperlinks and are stored on Web servers and viewed in a Web browser.

Web ring An online promotional tool in which an e-business links its Web site to various related Web sites, which in turn link to other Web sites, thereby creating a circular chain of complementary hyperlinks.

Web servers Computers that store hyperlinked documents, also known as Web pages.

Web site A collection of related Web pages.

Web site defacement Vandalism that changes the appearance of Web pages.

Web-based forums Discussion groups that allow participants to read and post messages through a Web site and Web browser.

webbed structure A type of Web site structure that allows pages to be linked to each other without regard to how the content of these pages fits together logically.

white-hat hacker A hacker who uses his or her skills to find weaknesses in computer systems and make the weaknesses known, without regard for personal gain.

word of mouth The process of one customer telling others of his or her experience with a business's products or services.

World Wide Web (Web) A subset of the Internet consisting of computers called Web servers that store documents called Web pages linked together by hyperlinks.

worm A type of virus that doesn't alter program files directly but replaces a document or program with its own code and then uses this code to replicate itself.

WYSIWYG An acronym for What You See Is What You Get.

XML schema A document type definition or specification for the XML tags used in a specific document.

INDEX

Index

TABLE OF PERMISSIONS

FIGURE	CAPTION	CREDIT LINE
1.3	EarthLink	Reprinted by permission of EarthLink.
1.4	Mozilla	Reprinted by permission of Mozilla Firefox.
1.5	The Tattered Cover Book Store	Reprinted by permission of The Tattered Cover Book Store.
1.6	Shopzilla	Reprinted by permission of Shopzilla.
1.7	Edmunds.com	© Edmunds.com, Inc. Reprinted by permission of Edmunds.com, Inc.
1.8	CARFAX	Source: www.carfax.com
1.9	Travelocity	Source: www.travelocity.com
1.10	The Internet Truckstop	Reprinted by permission of The Internet Truckstop.
1.13	eDiets.com	Reprinted by permission of eDiets.
1.14	Yahoo! Shopping	Reproduced with permission of Yahoo! Inc. © 2005 by Yahoo! Inc. YAHOO! and the YAHOO! logo are trademarks of Yahoo! Inc.
1.15	eBags	Reprinted by permission of eBags.com.
1.16	Harry and David	Reprinted by permission of Harry and David. © 2005 Harry and David.
1.17	Elance	Reprinted by permission of Elance®Marketplace.
1.18	Dairy.com	Copyright Dairy.com. All rights reserved. Dairy.com is a trademark of Dairy, LLC.
1.19	atla.org	Reprinted by permission of Association of Trial Lawyers of America.
1.20	Business.com	Reprinted by permission of Business.com.
1.21	HedgeHog	Reprinted by permission of HedgeHog, Inc.
1.22	Bidmain	Reprinted by permission of Bidmain.
1.23	American Boat Listing	Reprinted by permission of American Boat Listing.
1.24	eBay	These materials have been reproduced with the permission of eBay Inc. COPYRIGHT © EBAY INC. ALL RIGHTS RESERVED.

FIGURE	CAPTION	CREDIT LINE
1.25	TraderOnline.com	Reprinted by permission of Trader Publishing Company.
1.26	AllExperts	Reprinted by permission of AllExperts.com.
1.27	priceline.com	Reprinted by permission of Priceline.com.
1.28	Egghead.com	Source: www.amazon.com
2.2	uBid	Reprinted by permission of uBid.
2.3	Bluetooth	Source: www.bluetooth.org
2.4	Groove Networks	Source: www.groove.net
2.5	MSN Hotmail	Source: www.hotmail.com
2.6	eBay	These materials have been reproduced with the permission of eBay Inc. COPYRIGHT © EBAY INC. ALL RIGHTS RESERVED.
2.7	Amazon.com	Source: www.amazon.com
2.8	Ticketmaster	Reprinted by permission of Ticketmaster.
2.9	General Motors	Reprinted by permission of General Motors Corporation.
2.10	Ford Motor Company	Reprinted by permission of Ford Motor Company.
2.11	Chrysler	Reprinted by permission of Chrysler.
2.12	Southwest.com	Reprinted by permission of Southwest Airlines.
2.13	Grainger	Source: www.grainger.com
2.14	Cybersettle	Reprinted by permission of Cybersettle.com.
2.15	Half.com by eBay	These materials have been reproduced with the permission of eBay Inc. COPYRIGHT © EBAY INC. ALL RIGHTS RESERVED.
2.16	Cisco Systems	Source: www.cisco.com
2.16	Cisco Systems/2 photos	Reprinted by permission of Achille Bigliardi Photography.
2.17	Microsoft	Source: www.microsoft.com
2.18	Rackspace Managed Hosting	Reprinted by permission of Rackspace Managed Hosting.
2.19	Digital Witness	Reprinted by permission of Digital Witness.
3.1	SCORE	Reprinted by permission of SCORE Association.
3.2	SBA	Source: www.sba.gov
3.3	E-Future Centre	Reprinted by permission of The E-Future Centre.
3.4	BizPlanIt	Reprinted by permission of BizPlanIt.
3.5	Bplans.com	Reprinted by permission of Bplans.

FIGURE	CAPTION	CREDIT LINE
3.9	Round Table Group	Reprinted by permission of Round Table Group.
3.22	WeddingChannel.com	Reprinted by permission of WeddingChannel.com, Inc.
4.1	Band of Angels	Reprinted by permission of Band of Angels Fund, L.P.
4.2	Sendmail	Source: www.sendmail.com
4.3	Active Capital	Reprinted by permission of Active Capital.
4.4	Draper Fisher Jurvetson (DFJ)	Reprinted by permission of Draper Fisher Jurvetson.
4.5	Small Business Investment Companies (SBIC) Program	Source: www.sba.gov
4.6	ZipRealty	Reprinted by permission of ZipRealty, Inc.
4.7	National Venture Capital Association (NVCA)	Reprinted by permission of National Venture Capital Association.
4.8	Austin Technology Incubator (ATI)	Reprinted by permission of ATI/CEI/IC2 Institute, The University of Texas at Austin.
4.9	Advanced Technology Development Center (ATDC)	Reprinted by permission of Advanced Technology Development Center, Georgia Institute of Technology.
4.10	Houston Technology Center	Reprinted by permission of Houston Technology Center.
4.11	Illinois Technology Enterprise Center (ITEC)	Reprinted by permission of Illinois Technology Enterprise Center.
4.12	Women's Technology Cluster (WTC)	Reprinted by permission of Women's Technology Cluster.
4.13	Batavia Industrial Center (BIC)	Reprinted by permission of Mancuso Business Development Group.
4.14	Veritas Medicine	Reprinted by permission of Veritas Medicine.
4.15	Guru.com	Reprinted by permission of Guru.
5.1	U.S. Copyright Office	Source: www.copyright.gov
5.2	U.S. Patent and Trademark Office	Source: www.uspto.gov
5.3	TRUSTe	Reprinted by permission of TRUSTe.
5.4	Heidrick & Struggles	Reprinted by permission of Heidrick & Struggles.
5.5	Korn/Ferry International	Reprinted by permission of Korn/Ferry International.
5.6	SpencerStuart	Reprinted by permission of Spencer Stuart.

FIGURE	CAPTION	CREDIT LINE
5.7	U.S. Department of Labor, Bureau of Labor Statistics	Source: www.dol.gov
5.8	Salary.com	Reprinted by permission of Salary.com.
5.13	Visa Cash cards	Reprinted by permission of Visa U.S.A. Inc.
5.14	MasterCard prepaid cards	Source: www.mastercard.com/aboutourcards
5.15	American Express Travelers Cheque cards	Reprinted by permission of American Express.
5.17	Peppercoin	Reprinted by permission of Peppercoin.
5.18	Yaga	Source: www.yaga.com
5.19	BitPass	Screen Shot reprinted by permission of BitPass, Inc.
5.20	BidPay	Source: www.bidpay.com
5.21	PayPal	These materials have been reproduced with the permission of eBay Inc. COPYRIGHT © EBAY INC. ALL RIGHTS RESERVED.
5.24	Yahoo! Small Business	Reproduced with permission of Yahoo! Inc. © 2005 by Yahoo! Inc. YAHOO! and the YAHOO! logo are trademarks of Yahoo! Inc.
5.25	GoECart	Reprinted by permission of GoECart.
5.26	StoreFront.net	Reprinted by permission of LaGarde, Makers of StoreFront E-Commerce Solutions.
5.27	OpenText Corporation	Source: www.opentext.com
5.28	Interwoven	Reprinted by permission of Interwoven.
5.29	Vignette	Reprinted by permission of Vignette.
5.30	Webgistix	Reprinted by permission of Webgistix Corporation.
5.31	Turnaround	Reprinted by permission of WeShip4You.com.
6.3	Dell	Reprinted by permission of Dell Inc.
6.4	Original Ask Jeeves search tool	Reprinted by permission of Ask Jeeves, Inc.
6.5	The Namestormers	Reprinted by permission of The Namestormers.
6.6	NameLab	Reprinted by permission of NameLab.
6.7	Network Solutions	Reprinted by permission of Network Solutions.
6.8	GoDaddy.com	Reprinted by permission of GoDaddy.com.
6.9	Register.com	Reprinted by permission of Register.com.
6.10	Forrester	Reprinted by permission of Forrester.

Table of Permissions

FIGURE	CAPTION	CREDIT LINE
6.11	eMarketer	Reprinted by permission of eMarketer.
6.12	NPD Group	Reprinted by permission of The NPD Group.
6.13	STAT-USA Internet	Source: www.stat-usa.gov
6.14	Bitpipe	Bitpipe.com is part of the TechTarget network of IT Websites (www.techtarget.com). Reprinted by permission.
6.15	The Direct Marketing Association	Reprinted by permission of Direct Marketing Association.
6.16	Google	Reprinted by permission of Google Inc.
6.17	Teoma	Reprinted by permission of Ask Jeeves, Inc.
6.18	MSN Search	Source: www.search.msn.com
6.19	metacrawler	© 2005 InfoSpace, Inc. All rights reserved. Reprinted with permission of InfoSpace, Inc.
6.20	Dogpile	© 2005 InfoSpace, Inc. All rights reserved. Reprinted with permission of InfoSpace, Inc.
6.21	KartOO	Reprinted by permission of KartOO.
6.22	Mamma	Reprinted by permission of Mamma.com.
6.23	Yahoo! Directory	Reproduced with permission of Yahoo! Inc. © 2005 by Yahoo! Inc. YAHOO! and the YAHOO! logo are trademarks of Yahoo! Inc.
6.24	LookSmart Directory	Source: www.looksmart.com
6.25	Open Directory Project	The dmoz.org/Open Project Directory page is © 2005 Netscape Communications Corporation. Used with permission.
6.29	Marketleap Link Popularity Check	Reprinted by permission of Digital Impact, Inc.
6.30	Webmaster Toolkit Link Popularity Checker	Source: www.webmaster-toolkit.com/link-popularity-checker.shtml
6.31	LinkPopularity.com	Reprinted by permission of The PC Edge, Inc.
6.32	United Virtualities	Reprinted by permission of United Virtualities.
6.33	Yahoo! Search Marketing	Reproduced with permission of Yahoo! Inc. © 2005 by Yahoo! Inc. YAHOO! and the YAHOO! logo are trademarks of Yahoo! Inc.
6.34	goClick.com	Reprinted by permission of goGlick.com. © 2005 goClick.com, Inc. All Rights Reserved.
6.35	Google Adsense	Reprinted by permission of Google Inc.

FIGURE	CAPTION	CREDIT LINE
6.36	LinkPartners.com	Reprinted by permission of LinkPartners.com.
6.37	RSS in Government	Source: www.rssgov.gov
7.2	Amazon.com Associates Program	© 2005 Amazon.com, Inc. All Rights Reserved.
7.3	Dell Home Affiliates Program	Reprinted by permission of Dell Inc.
7.4	Barnes&Noble AffiliateNetwork	Source: www.bn.com
7.5	1-800-FLOWERS.COM	Reprinted by permission of 1-800-FLOWERS.COM.
7.8	Commission Junction	Reprinted by permission of Commission Junction.
7.9	LinkShare	Source: www.linkshare.com
7.12	BuyCostumes.com	Reprinted by permission of BuySeasons, Inc.
8.1	Rackspace Managed Hosting home page	Reprinted by permission of Rackspace Managed Hosting.
8.8	Managed Hosting page	Reprinted by permission of Rackspace Managed Hosting.
8.9	Network page	Reprinted by permission of Rackspace Managed Hosting.
8.10	Leadership page	Reprinted by permission of Rackspace Managed Hosting.
8.11	The World Factbook – Tanzania (top of page)	Source: www.cia.gov/cia/publications/factbook
8.12	The World Factbook – Tanzania (bottom of page)	Source: www.cia.gov/cia/publications/factbook
8.13	Top of the USAJOBS page	Source: www.usajobs.opm.gov
8.14	Bottom of the USAJOBS page	Source: www.usajobs.opm.gov
8.15	Rackspace Managed Hosting	Reprinted by permission of Rackspace Managed Hosting.
8.16	Medicare	Source: www.medicare.gov
8.17	FirstGov.gov	Source: www.firstgov.gov
8.18	Federal Communications Commission (FCC) Site Map	Source: www.fcc.gov
8.19	Internal Revenue Service	Source: www.irs.gov
8.20	eLuxury.com	Source: www.eluxury.com